Drought Policies: Case Studies on Mega-droughts for the High Level Experts and Leaders Panel on Water and Disasters (HELP)

In Focus – Special Book Series

Drought Policies: Case Studies on Mega-droughts for the High Level Experts and Leaders Panel on Water and Disasters (HELP)

Editors
Jerome Delli Priscoli and Kenzo Hiroki

PUBLISHING

Published by **IWA Publishing**
Alliance House
12 Caxton Street
London SW1H 0QS, UK
Telephone: +44 (0)20 7654 5500
Fax: +44 (0)20 7654 5555
Email: publications@iwap.co.uk
Web: www.iwapublishing.com

First published 2019

British Library Cataloguing in Publication Data
A CIP catalogue record for this book is available from the British Library

ISBN: 9781789061178 (paperback)
ISBN: 9781789061185 (eBook)
ISBN: 9781789061192 (ePUB)

Contents

Introduction

Jerome Delli Priscoli[a] and Kenzo Hiroki[b]

[a] *TEC Chair Global Water Partnership (GWP), Senior Advisor USACE IWR (ret.), Editor in Chief Water Policy*

[b] *Professor, National Graduate Institute for Policy Studies (GRIPS), Coordinator, High Level Experts and Leaders Panel on Water and Disasters (HELP)*

This is the second special issue of *Water Policy* produced for the UN Secretary General's High Level Panel (HELP) on water-related disasters. It focuses on drought policies. In 2015 the first special issue of *Water Policy* (volume 17, issue s1) focused primarily on flood policies. During 2016 HELP members solicited case studies on drought disaster policies. Cases were also solicited from the Global Water Partnership (GWP) and the World Water Council members. After reviews, these 12 contributions comprise this special issue of *Water Policy*.

The special issue begins with 'Drought aspects – fostering resilience through insurance' by Kron, Schlüter-Mayr, and Steuer of the Geo Risks Research at Munich Reinsurance Company (Kron *et al.*, 2016). They review definitions of droughts and describe their views on the appropriate role of insurance. Like other articles they note the difference between reactive and proactive approaches to drought. Kron *et al.* say that the problem of drought as opposed to permanent aridity is that neither nature nor humans have adapted to such temporary dryness, and consequently have difficulty coping with it. The authors also say that droughts must not be viewed as a merely physical phenomenon or natural events; that drought impacts on society result from the interplay between natural conditions and the demands people place on the water supply, and on the vulnerability of the region concerned. The authors note that reducing risk means acting in advance, before a crisis starts, while efforts (actions, but often only plans) to mitigate consequences are typically undertaken when the crisis is already manifest – which is therefore less efficient and carried out under time pressure.

They suggest that insurance is the last but not least important component of drought management. As a remedy, it covers the loss, but as a prophylactic measure it also provides planning security. A farmer will not be ruined by a disastrous drought, but will be given a financial cushion with which to restart his life. Sovereign risk insurance schemes can be a first step towards implementing a market system.

Kron *et al.* conclude that if societies, from governments to individuals, are prepared for frequent and normal extremes, and the residual risk from rare events is transferred to the finance industry, we can achieve enough resilience to guarantee sustainable living.

Specific cases follow this global overview and begin with 'Water policy responses to drought in the MDB, Australia' by James Horne (Horne, 2016). The article reviews water policy responses to drought

doi: 10.2166/wp.2016.325

in Australia, focusing on the Murray-Darling Basin (MDB) during the two decades from 1997, a period that includes the decade-long Millennium Drought. The drought brought a much sharper focus to discussions of scarcity and value of water, and became a crisis in the MDB in 2006.

The article relates how public and private actors came to understand core notions as preconditions for serious reform. These were that an understanding of the value of water was required if agricultural output was to continue to expand, and if ecological outcomes were not to be further compromised.

Over time a reform package was constructed that strengthened the MDB water market infrastructure, including enhanced regulatory oversight, enhanced transparency and upgraded information, new institutions, enhanced water resource planning, enhanced transparency and greater community involvement. It allowed irrigators to manage their water assets more flexibly. It included upgraded water resource planning, new sustainable diversion limits in the MDB and resourcing to ensure the environment would have access to a share of basin resources.

As Horne tells the story, the completeness of the package can be attributed not only to the severity of drought, but also to political leadership, a disrupting strategy in the form of national legislation and a strong national budget that provided financial resources. In short, the crisis was also seen as an opportunity.

An article on the drought in São Paulo, Brazil follows. Brazil is one of the most richly water endowed countries in the world yet in 2014-2015 it has suffered severe drought. Braga and Kelman in 'Facing the challenge of extreme climate: the case of Metropolitan São Paulo' describe the hydrologic conditions that resulted in this severe but highly unlikely drought (Braga & Kelman, 2016). They talk of how structural and non-structural initiatives were integrated across a variety of sectors and undertaken by water authorities to avoid social chaos in the Metropolitan Region of São Paulo; a region of 21 million inhabitants. Furthermore, they discuss how the drought affected the existing water resources development plan, especially the anticipation of investments on water security. As in other articles, they raise the issue of risk assessment and ask whether such a drought should be seen as a statistical outlier or something that can happen again in the future. They report that their experience suggests more infrastructure and good governance through strong public institutions, efficient regulatory agencies, adequate legal frameworks and raising the consciousness of the population to use water efficiently.

Tomoyuki Okada in 'Integrated water resources management and drought risk management in Japan' describes Japan's three types of drought countermeasures: water resources planning and development as supply side measures; promotion of efficient water use as demand side measures; and coordinated drought risk management as risk control measures (Okada, 2016). But even with its preparations system in place, he notes that Japan cannot rely solely on developing new water resources facilities because social and economic costs are increasing. He recommends that existing water resources facilities need to be utilized efficiently. The society should more focus on risk management for secured water supply.

In 'National drought management framework for drought preparedness in Korea (lessons from the 2014–2015 drought), Hong, Lee and Cho analyze the Korean government's response to the 2014–2015 drought under the National Drought Management Framework, which is composed of four stages: prevention, preparedness, response, and recovery (Hong *et al.*, 2016). The findings indicate that the Korean government effectively addressed the drought, yet there are no measures currently in place to cope with mega droughts that last for more than 5 consecutive years. Korea's drought from 2014–2015 was by far the worst incidence of drought in the past 50 years. The government's early response, appropriate drought management policy, and well-defined risk management manual helped minimize adverse drought impacts. Nevertheless, the authors note that Korea must establish a more practical link between a real-time drought monitoring and prediction scheme and a drought impact assessment system. Also, central government must tailor drought management measures for administration by the local governments.

Although historically Mexico has suffered recurrent droughts with serious social and economic consequences, measures to deal with vulnerability to drought have received limited attention. Aguilar-Barajas *et al.* (2016) next show that drought has gained a slow but steady importance in water resources in Mexico since the 1989 water commission. They note however that the main focus remains on compensating economic losses after the event rather with limited achievement on reducing vulnerability to drought. In 'Drought policy in Mexico: a long, slow march toward an integrated and preventive management model', they note that there is a need to increase public awareness and boost the scientific knowledge necessary for the design of appropriate preventive measures. Like other articles in the special issue they call for more focus on preventative and less on compensation and reactive measures. Also like other articles they call for more risk management policies to deal with drought vulnerability in Mexico and for raising public awareness of droughts. They conclude that in Mexico, drought policy has to be understood as a matter of water security; in other words to ensure a steady, predictable and sufficient supply of water to cities, industry and agriculture. As such, drought prevention policies are central to national development and welfare.

In 'Evolution of drought management policies and practices in the United States', Stakhiv and Werick explain that the United States is a federalist country and the States have first sovereignty over water (Stakhiv *et al.*, 2016). They note there is no 'national' drought management policy. However, there is a strong federal policy that attempts to coordinate the management responses of numerous federal agencies and that drought and flood management comprise the key components of assuring national water security. In the USA, Federal drought management policies and programs focus mainly on demand reduction strategies, supported by strong environmental regulatory programs that strive to attain sustainable resource use.

They observe that as the uncertainties of climate change begin to manifest themselves, especially in the semi-arid western USA California's drought conditions could become worse. The major discord between the federal agencies and states lies in the relative inflexibility of existing federal laws governing the operation of federal water projects and federal regulatory standards, to rapidly and effectively assist the states in meeting unforeseen circumstances during a drought.

The USA is a large and geographically diverse country. Thus the authors note that the humid areas of the eastern USA have a different type of drought management problem than the Midwest or south-western USA. Concentrated population centres require water deliveries from distant watersheds. The problem of eastern municipal areas is one of adequate water storage and delivery reliability. The 17 western states, mostly semi-arid, are closer in drought management practices to those of developing nations, and can be used as models for effective responses to water security, and inherently to drought management.

Stakhiv and Werick, like others in the special issue, suggest that climate uncertainty will again require greater focus on supply-side options to increase water supply robustness and resiliency, especially in growing urban areas of the semi-arid west. The authors conclude with 10 major suggestions.

Estrela and Sancho in 'Drought management policies in Spain and the European Union: from traditional emergency actions to Drought Management Plans' provide a general overview of drought management policies in the EU and Spain (Estrela & Sancho, 2016). In Spain, policies have evolved from traditional emergency actions to drought management plans and to initiating a new national indicator system for droughts. Drought management plans have helped alleviate negative effects of recent droughts, improved the aquatic ecosystems and avoided public water supply restrictions. The management plans must also be updated periodically. They note that the high impacts of droughts have also raised awareness that drought is a normal and recurring fact of climate variability leading to acceptance of the need for management plans.

The Seyhan River Basin is the second largest basin after the Nile in Eastern Mediterranean, and is agriculturally one of the most productive regions of Turkey and Europe. In 'Drought planning and management: experience in the Seyhan River Basin, Turkey', Tuncok reviews the use of drought indicators, historical data on drought, and speculates on future droughts and responses (Tuncok, 2016). He finds that historical droughts are characterized by less frequent but more severe droughts, and future droughts are characterized by more frequent but less severe droughts. Hydropower and agriculture will have to adjust and a structured drought management plan needs to be developed. To do this he describes a drought cycle management plan.

Southern Africa's worst drought in decades has left 40 million people, 14% of the region's total population, food insecure this year. In 'South Africa and the drought that exposed a young democracy', Turton describes the politics and decision-making processes in South Africa that have been exposed by the 2015/16 El Niño event, which is the worst in recorded history (Turton, 2016). He does this by using three independent drivers; overarching political dynamics; climatic reality; and transition from an extractive economy.

The article describes the interplay between political structures and technical assessments of water resources. Turton shows how political decay in the system has inhibited the ability to confront drought and assured only reactive possibilities to deal with it.

Despite the importance of water to South Africa's transition, the high worldwide visibility of the innovative approaches the new democracy was taking regarding water, Turton cites a growing body of evidence suggesting that the state response has been inadequate. This has been driven by the broader political milieu in which water resources drought policy and decision-making is embedded. Science has become politicized, with an unfortunate linkage having been made to colonial rule. This has undermined the scientific community and has meant that evidence-based policy reform has failed.

Turton says the major lesson to be learned is that the El Niño severity is likely to increase, so the next event can be better planned for. But this will only be possible once the institutions of state have been reformed.

These case studies are followed by descriptions of two international programs dealing with drought polices. Authors Pischke and Stefanski note that droughts have generally been addressed in a reactive manner, only responding after drought impacts have occurred (Pischke & Stefanski, 2016). The Integrated Drought Management Program (IDMP) of the World Health Organization and the GWP provides a framework and commensurate technical support to countries on drought management. The actual development and implementation of national drought plans and policies still need to be done by the governmental ministries and national stakeholders. In 'Drought Management Policies – from global collaboration to national action', they emphasize how information from different sources is used to support countries to shift from only reacting to droughts when they occur to adopt proactive national drought policies that focus on improved collaboration and the mitigation of drought impacts.

The IDMP has many partners and they have been used to create several early outputs in assisting countries to develop more proactive drought policies and plans with practical guidance and applications tailored to regional and national circumstances. The article uses examples of the programs from Central and Eastern Europe and the Horn of Africa, including Kenya.

The UNESCO International Hydrological Programme (UNESCO-IHP) has been providing support to enhance human capacity, policy guidance and tools to the countries to address drought-related challenges through the International Drought Initiative and the Water and Development Information for Arid

Lands – A Global Network (G-WADI). In 'Strengthening drought risk management and policy: UNESCO-IHP's case studies from Africa and Latin America and the Caribbean', Verbis *et al.* (2016) report on the programs, approaches and results in selected countries of Latin America, Africa and the Caribbean.

Mirroring a common theme throughout this special issue, the focus has been on helping Africa, Latin America and the Caribbean to build proactive capacities to deal with drought. This article by notes the main gaps in both regions include a lack of access to locally relevant drought monitoring and early warning information for decision making, the identification of vulnerable communities within the countries and the integration of these two components into drought management policies. Regional drought monitoring, real time satellite precipitation estimations, drought atlas, national drought observatories, and case studies have been produced.

Several recurrent themes emerge from these diverse case studies and descriptions of programs. For example, most articles discuss the necessity to move from reactive (compensatory) to preventive policies. This theme has implications for use of insurance in developing countries, e.g. is insurance encouraging investments to help countries avoid disasters or is it acting mostly in a humanitarian way to compensate for loses to help people? Several authors point to the importance of risk assessment and to developing risk based policies for drought. This raises statistical issues of how such assessments of uncertainty and risks are done and how they relate to actual occurrence of events. Most articles call for more inter-sectoral policies, policies which integrate water resources management approaches and to the necessity of raising public awareness of droughts in times of no drought. The issue of structural versus nonstructural is clear in most cases. While often cast as 'either/or' the message that emerges is more one of how do you integrate these approaches. Finally a few articles bring to light how prevention is needed for national security as well as water security.

References

Aguilar-Barajas, I., Sisto, N. P., Magaña-Rueda, V., Ramírez, A. I. & Mahlknecht, J. (2016) Drought policy in Mexico: a long, slow march toward an integrated and preventive management model. *Water Policy 18*(S2), 107–121.

Braga, B. & Kelman, J. (2016) Facing the challenge of extreme climate: the case of Metropolitan São Paulo. *Water Policy 18*(S2), 52–69.

Estrela, T. & Sancho, T. A. (2016) Drought management policies in Spain and the European Union: from traditional emergency actions to Drought Management Plans. *Water Policy 18*(S2), 153–176.

Hong, I., Lee, J.-H. & Cho, H.-S. (2016) National drought management framework for drought preparedness in Korea (lessons from the 2014–2015 drought). *Water Policy 18*(S2), 89–106.

Horne, J. (2016) Water policy responses to drought in the MDB, Australia. *Water Policy 18*(S2), 28–51.

Kron, W., Schlüter-Mayr, S. & Steuer, M. (2016) Drought aspects – fostering resilience through insurance. *Water Policy 18*(S2), 9–27.

Okada, T. (2016) Integrated water resources management and drought risk management in Japan. *Water Policy 18*(S2), 70–88.

Pischke, F. & Stefanski, R. (2016) Drought management policies – from global collaboration to national action. *Water Policy 18*(S2), 228–244.

Stakhiv, E. Z., Werick, W. & Brumbaugh, R. W. (2016) Evolution of drought management policies and practices in the United States. *Water Policy 18*(S2), 122–152.

Tuncok, I. K. (2016) Drought planning and management: experience in the Seyhan River Basin, Turkey. *Water Policy 18*(S2), 177–209.

Turton, A. (2016) South Africa and the drought that exposed a young democracy. *Water Policy 18*(S2), 210–227.

Verbis, K., Amani, A., Mishra, A. & Cisneros, B. J. (2016) Strengthening drought risk management and policy: UNESCO International Hydrological Programme's case studies from Africa and Latin America and the Caribbean. *Water Policy 18*(S2), 245–261.

Foreword by Chairman Han Seung-soo for the 2nd Flagship Document of HELP Special Edition of *Water Policy* To be announced at COP22, Marrakech, Morocco

Water is life. But water is also a threat to life. During the past decade, water-related disasters have not only struck more frequently but have also been more severe, hampering sustainable development by causing political, social, and economic upheaval in many countries. Water-related disasters, such as floods, droughts, storm surges and tsunamis, account for 90% of all disasters in terms of the number of people affected.

The issue of water and disasters must be addressed if we hope to achieve sustainable development. We must share our experiences and lessons learned, strengthen regional coordination and collaboration, and set common goals and targets in order to lay a foundation for weathering the water-related disasters to come, and make progress towards creating a better-prepared and resilient society. All these elements should be translated into clear-cut messages and practical advice for decision makers to create effective policies and mechanisms that address water and disaster issues appropriately at all levels.

Droughts are slow onset disasters but with severe disasters. They affect far larger areas and regions than most other disasters. Millions of people have been affected by a single drought on countless occasions. Even the rise and fall of dynasties and governments has been brought on by mega-droughts throughout history. Cascade effects of mega-droughts can hamper global economy through food shortages, commodity price hikes, mass exodus, and political instability.

GDP growth is predicted to decrease by more than 6% due to acute water scarcity not only in the regions of Sub-Saharan Africa, North Africa and Middle East, but also the majority of Asia, including Central Asia, China, India, Korea and Japan, according to the World Bank Report 'High and Dry'. This area covers more than two thirds of the world population.

Droughts occur as a consequence of interaction between extreme hydrological events and the people and society. They are deeply rooted in social, economic, environmental and even political conditions of the affected areas. Compartmentalized water use and management often intensifies negative impacts of the disasters. Intricate causes of droughts, however, should not be an excuse for non-action. We should analyze and find clues for solutions by learning from the past.

This document focuses on mega-droughts of the past 20 years. Twelve cases from both developed and developing countries are elaborated in the document. Its intention is to draw lessons from the cases of

doi: 10.2166/wp.2016.024

extremely severe water shortages so that countries and stakeholders can be better prepared for extreme drought events in the future.

The volume is the second edition of Flagship Document by High-level Experts and Leaders Panel on Water and Disasters (HELP). It was published as a sequel to the compendium of cases on mega-floods, which was published in March 2015 (*Water Policy 17*(S1)). The panel is grateful for the support provided by USACE to publish these special issues.

HELP was originally established in 2007 as HLEP/UNSGAB under the recommendation of the UN Secretary General's Advisory Board on Water and Sanitation (UNSGAB) in order to raise top political awareness and galvanize global actions on issues of water and disasters. The group of 23 Members, 10 Advisors and two Coordinators, comprising eminent leaders and experts from politics, administration, academia, civil society, and private sectors as listed in the table below, has worked together to make a tangible contribution to the global community in this field. Our actions are spearheaded by what we call Flagship Initiatives (FI), specifically focused activities to achieve the group's objective. FI includes this volume as well as the UN Special Thematic Sessions on Water and Disasters which HELP and UN Secretary General co-organized in 2013 and 2015.

This action was recognized in September 2016 as a Priority Action in Action Plan of the High Level Panel on Water (HLPW) co-convened by the UN Secretary-General and the President of the World Bank Group earlier this year. HLPW is co-chaired by the Presidents of Mauritius and Mexico with the Presidents of Hungary, Peru, Senegal, Tajikistan, South Africa and the Prime Ministers of Australia, Bangladesh, Jordan, and the Netherlands as Members and myself as Special Adviser.

I sincerely hope that this document will help many countries, organizations, and stakeholders to make progress in understanding and addressing droughts, particularly mega-droughts, that may come in the future, even tomorrow.

Members of High-level Experts and Leaders Panel on Water and Disasters as of October 11, 2016

Chair

Dr Han Seung-soo, Chair of the High-level Experts and Leaders Panel on Water and Disasters

Vice Chairs

Mr Kittiratt Na-Ranong, Former Deputy Prime Minister and Minister of Finance, Kingdom of Thailand

Ms Melanie Schultz van Haegen, Minister of Infrastructure and the Environment, Kingdom of the Netherlands

Members

Dr Shamshad Akhtar, Executive Secretary, The United Nations Economic and Social Commission for Asia and the Pacific

Mr Rolf Alter, Director for Public Governance and Territorial Development, Organisation for Economic Co-operation and Development

Ms Irina Bokova, Director General, United Nations Educational, Scientific and Cultural Organization

Dr Benedito Braga, President, World Water Council

Ms Kristalina Georgieva, Vice-President of the European Commission

Dr Robert Glasser, Special Representative of the Secretary-General for Disaster Risk Reduction, United Nations Office for Disaster Risk Reduction

Dr Basuki Hadimuljono, Minister of Public Works and Housing, Republic of Indonesia
Mr Kang Hoin, Minister of Land, Infrastructure and Transport, Republic of Korea
Dr Keizrul bin Abdullah, Chairperson, Network of Asian River Basin Organization
Mr Shinichi Kitaoka, President, Japan International Cooperation Agency, Japan
Dr David Korenfeld Federman, Mexico
Mr Gerson Lwenge, President of the African Ministers' Council on Water and Minister of Water and Irrigation, Tanzania
Dr Kuntoro Mangkusubroto, Former Head of President's Delivery Unit for Development Monitoring and Oversight, Republic of Indonesia
Mr Masafumi Mori, Vice-Minister for Engineering Affairs, Ministry of Land, Infrastructure, Transport and Tourism, Japan
Ms Rosa Pavanelli, General Secretary, Public Services International
Ms Ségolène Royal, Minister of Ecology, Sustainable Development and Energy, French Republic
Dr Oyun Sanjaasuren, Chair, Global Water Partnership
Lt Gen. Todd T. Semonite, Commanding General, US Army Corps of Engineers, USA
Mr Bambang Susantono, Vice President for Knowledge Management and Sustainable Development, Asian Development Bank

Advisors

Mr Rae-kwon Chung, Former Principal Advisor on Climate Change to the Secretary-General of the UN, Climate Change Ambassador of the Republic of Korea
Dr Jerome Delli Priscoli, Chair of GWP Technical Committee
Mr Loïc Fauchon, Honorary President, World Water Council
Dr Gretchen Kalonji, Dean, Institute for Disaster Management and Reconstruction, Sichuan University
Dr Khin Ni Ni Thein, Secretary, Water Advisory Group & Member, National Water Resources Committee, Myanmar
Prof. Toshio Koike, Director, International Centre for Water Hazard and Risk Management
Mr Kenyu Komura, President, Japan Water Agency
Dr Wolfgang Kron, Head of Research for Hydrological Hazards, MunichRe
Mr Hideaki Oda, Former Co-chair of Working Group on Water and Disaster, United Nations Secretary-Generals' Advisory Board on Water & Sanitation
Dr Ursula Schaefer-Preuss, Former Chair, Global Water Partnership
Mr Koos Wieriks, Councilor to Ministry of Infrastructure and the Environment, Kingdom of the Netherlands

Coordinators

Dr Kenzo Hiroki, Coordinator of the High-level Experts and Leaders Panel on Water and Disasters
Dr Ilpyo Hong, Research Fellow, Korea Institute of Civil Engineering and Building Technology, Korea

Drought aspects – fostering resilience through insurance

Wolfgang Kron*, Sabine Schlüter-Mayr and Markus Steuer

Geo Risks Research, Munich Reinsurance Company, Königinstrasse 107, Munich 80791, Germany
**Corresponding author. E-mail: wkron@munichre.com*

Abstract

Droughts are silent killers, with the potential to cause enormous losses to society as a whole and to the insurance industry. Many loss-bringers are unseen, and the result of secondary events. This makes it difficult to assess the losses involved. Insurance against drought is particularly important in the agricultural sector, but in practice only feasible with governmental involvement. Some crop insurance schemes have proven quite successful and have gained importance in recent years, both in developed and developing countries. As drought is not only a consequence of unusual weather conditions, but also subject to the way in which water resources are managed, overall awareness is a key factor in being prepared to cope with the resulting risk, and in avoiding catastrophic consequences.

Keywords: Droughts; Drought disasters; Drought insurance; Drought losses; Loss statistics; Resilience; Water resources' management

1. Introduction: the significance of droughts

In the past half-decade, water-related natural disasters have dominated the headlines much more frequently than earthquakes and tropical cyclones. However, while floods provide spectacular images and volleys of loss estimates for short periods of time, droughts tend to trundle along in the background of the world's media coverage, and only pop up briefly in the news every now and then before fading into oblivion again. Nevertheless, drought is globally one of the most significant natural disasters in terms of spatial extent, duration and long-term socioeconomic and environmental impact (United Nations Development Programme (UNDP), 2013) and 'the unremitting stress of drought, famine and deepening poverty threatens to create social strains, in turn creating the potential for involuntary migration, the breakdown of communities, political instability and armed conflict' (Ban, 2010). It seems that despite modern technologies, from satellite observation and communication to cooperation partnerships and computerized optimization of water use, drought events are still occurring frequently all over the world and in almost every region.

doi: 10.2166/wp.2016.111

The Global Risks Report 2016 (World Economic Forum (WEF), 2016) states that water crises will constitute the biggest risk to the world in the coming decade. This topic has already been recognized as the front-runner with respect to impacts in 2015, and it ranks third in 2016. 'Water crises' risk is followed by 'failure of climate change mitigation and adaptation', 'extreme weather events', 'food crises', and 'social instability'. These five items look like they have been copied from a 'wanted poster' describing droughts. They are closely interconnected.

2. What constitutes a drought?

2.1. The nature of drought

Drought cannot be defined in absolute terms. It is an exceptional deviation from the regional rainfall norm, and signifies a prolonged and abnormal moisture deficiency. Annual rainfall of just 500 mm in one region instead of the normal 1,000 mm can lead to a serious drought there, while 500 mm in another region would mean twice the normal rainfall. Yet deviation from the mean is not the only relevant factor, with variability in water yield and deviation from normal seasonal distribution being equally important. Therefore, drought must always be seen in the context of normally prevailing local and regional conditions.

Droughts are caused not only by the current weather situation. They are often the result of previous conditions, and have a very strong tendency to persist. The probability of a drought occurring depends on many factors – among other things, on the supply of water stored in aquifers, the soil, and in reservoirs and lakes at the beginning of a dry period. Deficits accumulate in dry years, which consequently cannot be compensated for in subsequent years with just average precipitation levels; above-average rainfall is needed to achieve recovery. This means that an extreme shortage could easily arise in the event of a moderate – albeit non-critical – shortage of water in the previous year. Droughts are usually exacerbated by higher rates of evaporation due to excessive temperatures, intense sunshine, low humidity and sometimes high winds.

Droughts are events that unfold slowly and usually affect a relatively large area. Problems arise gradually, and people tend to become aware of a drought only when it has fully developed. Hence, no steps are taken to adapt to the situation unless a critical stage has either already been reached or is imminent. To a large extent, however, whether a dry period becomes a drought depends on how water resources are managed in the area concerned. Typically, suitable measures are taken in those areas in which droughts are relatively common. Areas that are rarely affected by drought, but in which water is permanently in short supply, pose a greater problem. In such areas, a crisis may arise within a short span of time as there is no buffer. Given the large-scale impact, losses in many, if not all, sectors of society can be very severe.

There are also cases when drought affects only a limited area. Several years ago the Spanish city of Barcelona had to be supplied with water by tankers as the local water sources had fallen dry (Keeley, 2008), and most recently Atlanta in the USA was brought to its knees by drought conditions in a relatively small watershed from which the city drew its water supply (Goodman, 2007). Another example is the drought in Brazil's metropolitan São Paulo (Braga & Kelman, 2016). In these and many other cases there was a lack of water management systems flexible enough to overcome unexpected periods of shortage – but sometimes also simply mismanagement.

The problem of drought as opposed to permanent aridity is that both nature and humans, in particular with respect to agriculture, have not adapted to temporary dryness, and consequently have difficulties coping with it. Drought therefore must not be viewed as a merely physical phenomenon or natural event. Its impacts on society result from the interplay between natural conditions and the demands people place on the water supply.

More than any other climate feature, climate change will alter the occurrence of extremely dry and hot weather conditions, in many (most) parts of the world (Intergovernmental Panel on Climate Change (IPCC), 2013). Climate change is responsible for a shift to warmer days, including a change in the proportion of extremely high temperatures. This makes the topic even more worthwhile to concentrate on when we prepare for a livable and sustainable future for billions of people.

2.2. Drought definitions

Due to the relative nature of drought, there are several definitions, each of them relating to a specific point of view.

Meteorological drought is the temporary imbalance between actual precipitation (or soil moisture) and the average over a given time span. It tends to be localized and short-term, but its influences may be severe on plants (crops). For example, a country such as Bangladesh, which is known for its severe floods, actually fears the absence or delay of (some) rainfall even more.

A hydrological drought affects the entire water resources' system in a region, consisting of surface water, snowpack and artificial storages (reservoirs), as well as soil moisture and groundwater. It represents the cumulative effect of prolonged deficits in precipitation. Meteorological droughts usually precede hydrological droughts, but are not a prerequisite. They are often 'co-sponsored' by humans overexploiting the water resources, which may take a long time to recover, as aquifers, in particular, are often replenished very slowly. There are also sub-types of hydrological droughts, such as soil moisture drought and groundwater drought.

Agricultural drought relates to a specific kind of crop, depending on various factors including the type of crop and its ability to store water, the stage of the growth cycle, soil type, temperature, etc. It describes the water stress on a crop determined by water supply and evapotranspiration, which may arise from both unexpected lack of rainfall or bad planning of the water demand during the growth cycle, e.g. by planting the crop too early or too late. Agricultural drought may lead to poor yield, but also to total crop loss.

Socioeconomic drought links human activity with the above drought definitions, which are described by purely physical/biological characteristics. In very general terms, it may be described as a failure of the water supply to meet the water demand of a – local, communal, regional or national – society. Both sides may contribute to this type of drought: the supply side through lack of rainfall, and the demand side through overuse of existing resources. This is the type we are talking about in the context of drought disasters.

Further characterizations of droughts refer to duration, geographical extent and severity. While seasonal droughts usually cause only material losses and temporary, but potentially severe, inconvenience, multi-year droughts ('megadroughts') have led to mass migrations of people throughout history and an irreversible loss of species. The larger drought-affected regions become, the more difficult it is to assist them with external supplies, in particular in areas with poor transport infrastructure. Recurrent droughts in a region may lead to desertification, i.e. the degradation of land that transforms originally productive land into wasteland.

Drought severity is commonly described by way of indices, which are mostly based on meteorological parameters. The indices consider relative rather than absolute conditions and enable comparisons of spatially different regions and their 'normal' climate.

The Palmer Drought Severity Index (PDSI) (Palmer, 1965) is one of the most commonly used indices. Precipitation and temperature data for the current and preceding months, as well as the locally available water content of the soil, are included in the calculations. The PDSI can be used to define long-term droughts (months) and evaluate their impact on agriculture. It is a standardized parameter that is calibrated to the local climate. Its advantage is that different parts of a country can be compared with each other. Integrating the PDSI over a year and over the entire area of a country, and then presenting it as a time series of annual values, allows us to identify drought periods extending to several years.

3. Droughts do not come alone

The principal consequence of drought – lack of water – is not the only impact. And it is not only lack of precipitation that causes it. As a rule, droughts are linked to heatwaves. Hence, excessive temperatures, intense sunshine, low humidity and sometimes high winds often exacerbate evaporation rates and thus 'consume' water. The simultaneous occurrence of heat and lack of water also intensifies the physiological stress for humans, animals and plants. One immediate consequence is the drastically increased wildfire hazard. The loss of fertile soil – if not expressed in the form of a dust storm – is a less spectacular, yet still severe, feature. In certain regions, subsidence occurs when soils shrink, potentially causing great damage.

Lack of drinking water for people and animals is the most critical potential consequence of a drought. In developed countries, this problem can practically always be avoided with the help of technical means (e.g. water trucks). In sparsely populated, remote areas, there may be no solution, thus making drought a deadly threat to people – and animals (pets and livestock).

Households and businesses are the first to feel the impact of restrictions placed on the use of water, e.g. for non-essential cleaning and watering purposes, such as washing the car or watering lawns and gardens. These restrictions gradually become more and more drastic until they finally affect hygiene.

Some branches of industry require enormous amounts of water for the production process, especially for cleaning and cooling. Food products cannot be produced without water.

As regards infrastructure, it is mainly heat that can cause physical damage. The asphalt pavement on roads and highways becomes soft and buckles. Metal items, including railway tracks and pipes, expand and buckle, with the consequence that trains may derail and water lines burst. Electric power transformers can fail or sustain damage, causing interruptions to power supplies. Levees may dry out and crack, with the damage inflicted potentially remaining unknown until the next flood.

High temperatures are also responsible for many other impacts. They can impair human health and even cause death. Dehydration, strokes, cardiovascular and breathing problems, lack of sleep, fatigue, nervousness and general stress are among the most common complaints (Steuer & Kron, 2012).

In agriculture, water is vital for plant growth and survival, as well as for a good harvest. This sector is probably the most affected of all.

Insofar as river transportation can be operated at all, only smaller cargoes can be carried when water levels are low. The vessels are also obliged to travel at lower speeds, and there is often not enough width for two vessels to pass one another owing to narrowing of the navigable channels.

Hydroelectric power production is impaired, as rivers carry less water, and water levels in reservoirs are low. Thermal power plants require large amounts of cooling water, and its unavailability consequently leads to a reduction in power generation. High temperatures in summer increase demand for power.

In addition to the shortages created by falling water levels in aquifers, rivers and lakes, the process of providing a safe drinking water supply also becomes more complex at higher water temperatures. Biological purification of sewage can be impaired if dilution is inadequate. Discharge rules may be violated. This results in greater pollution levels in rivers and lakes.

Heat and drought are a source of stress for wildlife and plants. In extreme cases, plants wither, animals die, and ecosystems are affected.

The water quality in rivers and lakes suffers in general. If the discharge drops while the load of pollutants remains the same, their concentration in the water increases, potentially harming aquatic life and plants. Water temperature rises. Pollutants and toxic material increasingly settle, as flows decrease. In future floods, these constituents are reactivated from the sediments, and may cause poisonous surges.

Access to rivers and lakes can be impeded by low water levels, making some recreational activities impossible. Pathogen counts in the water may also rise at high temperatures.

Dry soil is very easily eroded by wind, and the resulting dried-out vegetation no longer acts as a buffer against erosion. When it subsequently rains, the water causes erosion damage as it runs off. Prolonged droughts may lead to irreversible loss of fertile soil and eventually to desertification.

Apart from the direct adverse effect of dust on human and animal health, dust storms may contain pesticides, pollen, fungi and other substances that irritate human lungs and eyes. Dust particles can cause a variety of health problems, including asthma, especially in children, the elderly and those already suffering from respiratory or cardiovascular disease. While most particles from windblown soils are large and deposited quickly, very fine dust remains in the air for long periods.

The danger of wildfires rises with each hot and dry day.

Clayey soil shrinks when it dries. This can lead to subsidence damage.

The degradation of groundwater near coasts allows salt water to intrude and to lastingly spoil the aquifer. Coastal wetlands may be affected, damaged, or even disappear, which can have dramatic consequences for natural coastal protection.

Finally, water rights' disputes arise. These may occur between farmers, businesses, communities and entire nations, potentially leading to warlike conflicts.

While many of the above are temporary consequences and no longer relevant when the period of heat and drought ends, they may be irreversible in some cases.

4. Drought loss statistics and related problems

Hardly any other type of natural disaster is as difficult to comprehend as drought. For a hazard event such as landslide, the beginning and the end of the event, the area affected and the loss generated are completely clear, and the event can be described individually.

A drought is just the contrary. The problem starts with the following question: when does a drought begin, and when does it end? It is definitely not the last rainfall that marks the beginning, nor the first rainfall that marks the end. Droughts – in the sense of damaging events – develop slowly from dryness. It is only when looking back in time that we can determine when the one gradually transformed into the other. To specify the exact date is never possible, although this is usually quite an important factor in the context of insurance. The area concerned cannot – in most cases – be defined with accurate boundaries. And even the overall consequences are subject to uncertainty and subjective assessment.

With no time window for accumulation defined, and given that indirect losses (long-term business influences, loss of jobs, etc.) are sometimes much greater than direct losses (failure of crops, direct consequences of water shortages, etc.), estimating drought event losses becomes very difficult. Therefore, all loss figures relating to droughts have to be treated with great caution.

This considerable uncertainty must be taken into account when drought events are compared and represented in statistical terms. However, the order of magnitude of severity can nevertheless be gleaned from rough estimates.

4.1. Statistical data on droughts

As a globally operating reinsurance group, Munich Re reports droughts (i.e. those that cause losses) in its NatCatSERVICE database, together with the losses from all other types of natural disasters (Kron *et al.*, 2012; Wirtz *et al.*, 2012). Currently more than 30,000 individual loss events are stored in this global database. The information on each event comprises direct losses (damage), represented by overall losses and insured losses. In addition, it describes in a semi-quantitative way the characteristics of an event, such as the area, people, and infrastructure affected, special features such as noteworthy business interruption cases, and others. Some examples of entries are shown in Figure 1.

The number of fatalities, which is reported for other types of disasters, is not included for drought because associating the death of a person directly to drought is rarely possible; it is usually heat or famine that kills people.

Drought events listed in the database are very often linked to heatwave entries, and losses comprise many of the features displayed in Section 3, except wildfire and subsidence. In particular, it is not possible to distinguish and separately record the material losses. Tables 1 and 2 show drought events since

Date	Country	Region	Direct overall losses (US$ m)	Insured losses (US$ m)	Remarks
Jun – Sep 2012	United States	AR, CO, GA, IA, IL, IN, KS, KY, MO, MS, MT, NE, OH, OK, SD, TN, TX, WI, WY	25,000	12,000	Moderate to extreme drought conditions, >60% of the United States abnormally dry. Lack of rain, high temperatures. Low water levels in rivers, reservoirs, wells. Severe losses to agriculture (esp. soybeans, corn).
Dec 2013 – Mar 2015	Brazil	São Paulo, Minas Gerais, Rio de Janeiro; Espirito Santo	5,000		Severe drought (worst drought in 80 years), lack of rain, high temperatures. Reservoirs, rivers drying up. Hydroelectric power facilities deactivated, blackouts. Severe losses to agriculture (sugar cane, coffee). Water rationing for 4 million people. Affected: >27.6 million.
Jun 2010 – Feb 2012	Somalia	Entire country, esp. Bay; Bakool; Galguduud; Lower Shabelle; Middle Shabelle	>150		Severe drought. Lack of rain, two poor rainy seasons since October 2010. Worst drought in 30 years. Thousands of km² of cropland destroyed, pasture land affected, thousands of head of livestock killed, dairy affected. Severe food and water shortages. Famine, malnutrition, starvation, diseases. Displaced: 250,000, Affected: 4 million.

Fig. 1. Entries of drought events (selected examples) in the NatCatSERVICE database. *Source:* Munich Re NatCatSERVICE 2016.

1980 that were significant in terms of material losses, and the famine and heatwave events with the highest death tolls.

The sequence of inflation-adjusted (2015 values) annual losses from droughts/heatwaves over the past 36 years displays an upward trend (Figure 2). As most of the losses can be attributed to agriculture, the causes of the trend should mainly be found there. One of the main reasons for this is that yields and cultivated areas are significantly greater today than in the past; therefore, the loss potential has also increased. On the other hand, modern agricultural management practices can have a risk mitigating effect.

The two high loss peaks in the time series in Figure 2 are attributable to the severe droughts in the United States in 1988 and in 2012, which alone produced original, i.e. not inflation-adjusted, losses of US$15 bn and US$25 bn respectively. In the period shown, more than a third of all losses (36%) were generated in the USA (Figure 3). This is the same amount as for the remainder of America, Europe and Africa combined. With an 11% share, China is the country with the second-highest losses.

Table 1. Significant droughts in the period 1980–2015.

Period	Affected area	Direct overall losses in US$ m (original values)	Insured losses in US$ m (original values)
Jun–Sep 2012	United States (esp. Midwest)	25,000	12,000
Jan–Dec 1988	United States (esp. Midwest)	15,000	650
Jul–Aug 2003	Europe	14,000*	1,100*
Apr–Oct 2002	United States (esp. Great Plains)	10,000	2,000
Jan–Dec 2011	United States (esp. TX)	8,000	2,400
Jan–Jul 1996	United States, Mexico	6,200	500
May 1999 – Nov 2001	Iran	5,100	
Nov 2011 – Feb 2012	Argentina, Brazil, Paraguay	5,000	185
Dec 2013 – Mar 2015	Brazil (esp. South)	5,000	
Sep 1992 – Jan 1995	Spain	4,500	

*Including heatwave and subsidence losses.
Note: The table should not be interpreted as a ranking of the costliest droughts, because 1-year and multi-year events are listed together. *Source:* Munich Re NatCatSERVICE 2016.

Table 2. Fatalities from selected heatwaves and famines in the period 1980–2015.

Period	Event	Affected area	Fatalities
Apr 1983 – Oct 1984	Famine	Ethiopia	300,000
Jan 1980 – Dec 1985	Famine	Sudan, South Sudan, Chad	153,000
Feb 1981 – Nov 1985	Famine	Mozambique	100,000
Jul–Aug 2003	Heatwave	Europe	70,000
Jul–Sep 2010	Heatwave	Russian Federation (esp. Moscow)	55,000
Jun 2010 – Feb 2012	Famine	Somalia	50,000
Jan 1989 – Apr 1992	Famine	Angola	10,000
Jan–Dec 1988	Heatwave	United States	5,000
Jun–Aug 2015	Heatwave	Europe (esp. France, Belgium, Italy)	3,850
May–Jun 2015	Heatwave	India, Pakistan	3,670

Note: The figures listed are only rough estimates; famine-related fatalities may be influenced by political instability, civil war, etc. *Source:* Munich Re NatCatSERVICE 2016.

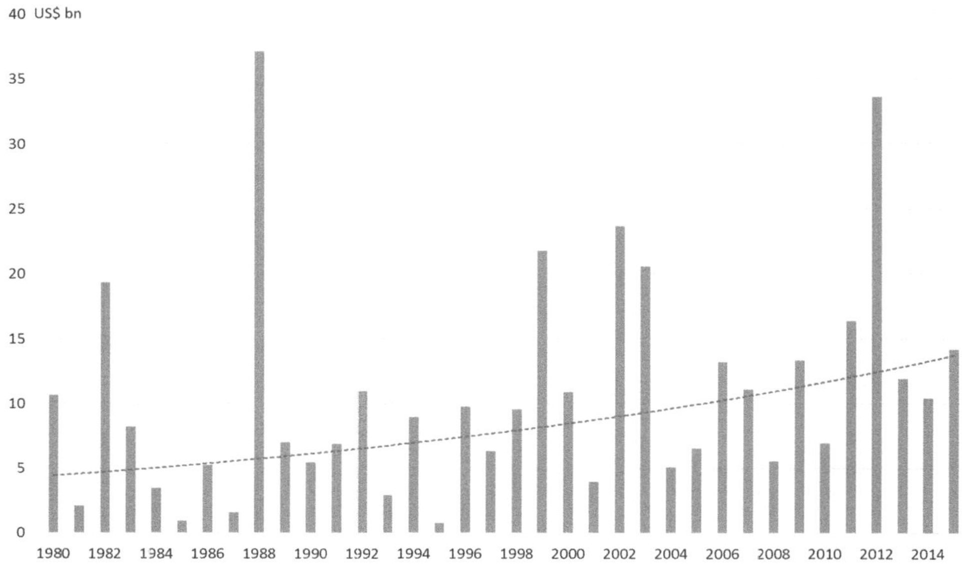

Fig. 2. Direct overall annual losses from droughts and heatwaves in the period 1980–2015 and exponential trend[1]. Contributing values are inflation-adjusted using the local consumer price indices. Wildfire and subsidence losses are not included. *Source:* Munich Re NatCatSERVICE 2016.

Figure 3 is highly biased with respect to the severity of drought in different regions. A drought event in the USA primarily has an economic impact (and therefore a large percentage in the pie chart), whereas a drought of the same magnitude in an African country may constitute a humanitarian disaster involving famine and migration, but produce relatively small losses in monetary terms. In Africa, hundreds of thousands of people died due to drought-induced famines in the 1980s. But even in the current century – in 2010–2012 – a famine claimed the lives of some 50,000 people in Somalia (see Table 1).

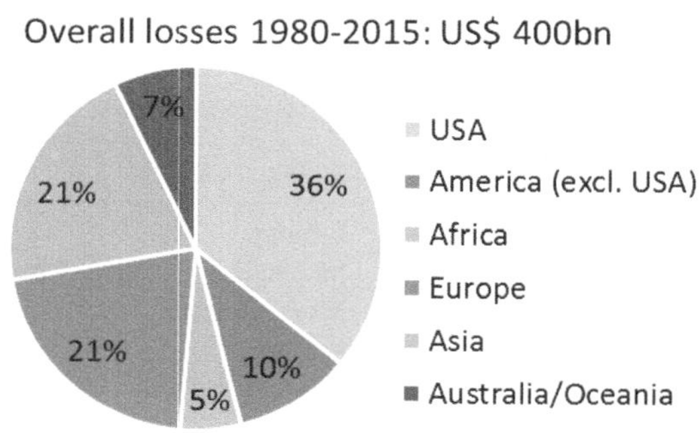

Fig. 3. Contribution of the USA and the five continents to the accumulated direct overall losses from droughts and heatwaves (excluding wildfire/subsidence) in the period 1980–2015. The figures are based on inflation-adjusted values using the local consumer price indices. *Source:* Munich Re NatCatSERVICE 2016.

---

[1] Throughout the paper, the term 'billion' or 'bn' indicates 10^9.

Fatalities in developed countries are caused by heatwaves. This is one of the natural hazards which – though not recognized by most people – is responsible for the highest death tolls, topped only by massive earthquakes and coastal floods. The heat and drought summer of 2003 in Europe claimed about 70,000 lives in excess of the normal mortality figure (Robine *et al.*, 2007), and in Russia the excess mortality figure amounted to over 55,000 in 2010 (Katsafados *et al.*, 2014). No other single events have claimed as many lives in these regions in a century.

4.2. Loss estimation methods for droughts

The values that were used to create Figures 2 and 3 are based on loss estimates compiled in the NatCatSERVICE database. The estimation is performed using different approaches, depending on the information available. Three information situations can be distinguished (when dealing with agricultural losses):

4.2.1. Information on insured losses is available. The loss estimation procedure involving the least uncertainty is scaling up a – known – insured loss (L_i [\$]). If the percentage of the insured part of the agricultural area (a_i [-]) and the typical portion of the loss retained by the farmers (D [-], estimated value) are known, we can calculate the overall loss (L_o [\$]) of agricultural crops and thus the non-insured loss as the difference between L_o and L_i:

$$L_o = \frac{L_i}{(1 - D) \cdot a_i} \tag{1}$$

This procedure implies that average insurance penetration in the country regarded is representative of the drought-affected region. In countries with high insurance penetration (e.g. USA, Canada), this method involves only small uncertainties, because the insured portion assumes a great part of the overall loss.

4.2.2. Quantitative information on affected values is available (lost/damaged crops, affected livestock). In countries with low insurance penetration, or where drought losses are not covered by insurance, overall losses (L_o [\$]) can be estimated using figures published on the lost crop percentage (l_c [-]) in the agricultural area (A [ha]) affected, the average expected yield (Y [tons/ha]), and the instantaneous prices (P [\$/ton]) for the respective agricultural products:

$$L_o = A \cdot l_c \cdot Y \cdot P \tag{2}$$

This bottom-up method can also be applied to livestock. The number of perished animals multiplied by the market price for one animal yields the figure for the overall loss in this category. Another approach uses macroeconomic figures, such as the destroyed portion of the countrywide harvest yield, in a top-down method.

4.2.3. Qualitative information on the extent of drought impacts is available. If quantitative information on losses and agricultural data and livestock is not available, the uncertainties may become very large. This applies to most African countries. In such cases, we use qualitative information on the event, such as the type of affected crops including special crops, dairy cattle and livestock, timber, aquaculture, etc. In addition, we consider indicatory parameters such as affected industry, commerce, private property, lifelines, environment, society, people, as well as duration and spatial extent of the drought, death toll, plus other soft indicators.

The number and degree of these indicators allow us to associate a severity level with a drought event, and thus to compare it with other events whose overall loss is known. From this, we can derive the loss for the current event. Moreover, we might need to modify the resulting value by a country-specific vulnerability factor. As the uncertainty of this estimation method is quite high, we regard the loss value obtained in this way with caution and interpret it as an estimate reflecting an order of magnitude, not a reliable figure.

In some cases, detailed scientific studies are conducted in the aftermath of a drought disaster, or civil/governmental entities investigate impacts and loss figures. In the instances where an external loss estimation has been made, we check its plausibility and completeness by applying one of the above estimation procedures. In practice, it may well be that political motivation leads to biased loss estimates: they are either too high in order to obtain more disaster assistance or too low in order to play down the severity and its impact on the respective region.

5. Insurance of drought

5.1. Affected lines of business

Insurance is the best tool for overcoming a situation in which an extreme event exceeds the normal resistance and protection capacity. It allows people and enterprises to get back on their feet financially and to start to resume normal lives again, the latter being the core factor of resilience. It also protects a state's budget from extraordinary expenses on top of its own losses in a catastrophic situation and – with respect to globally operating reinsurance companies – spreads the national loss globally, thus mitigating the regional situation.

Agriculture is the sector that uses and benefits most from drought insurance. Firstly, the sector is more directly affected than most others, and therefore more aware of the risk. Secondly, in selected countries crop insurance has a comparably long history with respect to this natural hazard. Agriculture (and related sectors) assumes by far the highest share of insured losses during droughts (note: wildfire losses to property are not considered drought losses). Crop insurance has been established successfully in many countries, and therefore efforts to introduce such schemes are ongoing in many others. Some examples are presented in the case studies described in Section 6.

Property insurance, the line of business that is normally most affected by natural disasters, hardly suffers from droughts – with two exceptions: losses caused by wildfires and subsidence, the latter only applying to the United Kingdom and France. Subsidence resulting from shrinking soils during drought periods can cause billions of dollars of damage to buildings that settle unevenly in these two countries, where the vast majority of houses are covered against this type of loss (see Section 6.6).

Wildfires are conditioned only by drought (Steuer, 2012, 2013) and are a secondary phenomenon, hence they are not discussed here in detail. Other property damage – typically caused by heat rather than by dryness, such as damage to roads and railway tracks – may occur, but does not normally accumulate to significant loss amounts.

More recently, the energy and transportation sectors have become increasingly important, and have shown more and more interest in insurance cover. They seek compensation when they have to stop or reduce business and thus lose profits. However, insurance solutions are still rare and not yet a major factor. Nevertheless, this field of business is most likely to develop in future as extreme summer weather situations are expected to increase in frequency and intensity with global warming.

Recent droughts in Eastern Asia have shown that drought can disrupt the production of electricity and the power supply. Particularly prolonged area-wide power failures have a huge loss potential. A power outage affecting the infrastructure of large cities can have serious consequences for the population as only a few days may suffice to severely disrupt the area-wide supply of essential (or even vital) goods.

Industrial and manufacturing production can grind to a standstill if industrial plants and workshops cannot bridge the power failure with electricity from their own generators. Production stoppages in a supplier's plant can disrupt entire supply chains and impair production by other firms dependent on the supply of precisely these products. Supply shortfalls can also occur when low river levels have a negative impact on barge traffic. Transportation firms and those that depend on the goods transported may suffer losses. The lost earnings may be covered under business interruption and contingent business interruption policies.

In addition to causing economic losses, functional disruptions and disrupted supply chains pose a major threat to public order and security. Consequences such as looting can result in considerable insured losses.

Of course, health (and to a smaller extent, life) insurance is also affected by extreme drought–heat-wave events, but it is difficult to come up with quantitative data on how much they cost the health insurance sector. Similarly, casualty insurance may be affected (e.g. if legal requirements such as emission limits of pollutants are exceeded during extreme weather conditions), but one can assume that the contribution of this line of insurance is relatively small.

For the aforementioned two high losses in the USA in 1988 and 2012, the insured portions were 5% (US$650 m) and 50% (US$12 bn) respectively. This large difference reveals the dramatic change in agricultural insurance penetration in the USA in just 25 years. In other markets, penetration has increased significantly too, as the relatively high portions of insured losses from drought disasters in North-east China (US$440 m representing 18% of the overall loss of US$2.5 bn) and in India (US$880 m, or 24% of US$3.7 bn) in 2015 show. The fact that insurance penetration for drought-related events has been increasing rather steeply in recent years is a sign that its positive effects are recognized. In some countries, governments have even implemented subsidy programs to incentivize farmers to buy insurance cover.

5.2. *Seasonal and annual climate variability*

Contracts between reinsurance and insurance companies are drafted on an annual basis. It is therefore of great interest to know whether the probability of drought in the area to be insured in the upcoming year/season is higher or lower. While other natural hazards may also become more or less likely in a certain climate phase, drought as a slow onset event – and thus one that tends to 'integrate the weather' over a longer period – promises better results in long-term forecasting (i.e. with a time horizon greater than 6 months). In the procedure of negotiating contracts, a knowledge edge may be very helpful. One of these climate variations, the El Niño-Southern Oscillation, has been investigated thoroughly, and we are able to produce elevated or reduced drought probabilities for certain regions.

6. Existing insurance schemes related to drought

6.1. Drought insurance for agricultural production

Drought can cause high losses in agricultural production, and may ruin a farmer or even have a significant influence on food security. Thus, drought risk mitigation is of major importance both on farm level and on national level; it is a devastating risk for the farmer and a systemic risk both for the government and for insurers. The term 'agriculture' in this context comprises not only crop-growing, but also other types of farming such as forestry (timber), plantations (palm oil, rubber, etc.), animal husbandry, aquaculture and similar branches.

If farmers are aware of their risk, they can implement on-farm risk management measures, such as investment in technology. Crop and livestock production are mainly exposed to drought risks. Crop yield depends on the combination of individual weather-related phenomena encountered during the growing period and applied farming practices. But farmers also have to cope with catastrophic events. In such cases, insurance can contribute to income stabilization, avert ruin and foster resilience.

Depending on the susceptibility of a given crop plant to drought, persistent water shortage causes varying degrees of drought stress and weakens growth. In addition to missing precipitation, high temperatures may adversely affect harvests in several ways. They may impair pollination, increase oxygen uptake at night resulting in increased carbon release and loss of biomass, impede the absorption of fertilizers, etc. Strong sunlight can burn the leaves, reduce carbon assimilation and stunt growth. The full extent of the losses can be established only at the end of the growing season, which is in stark contrast to the situation in conventional property insurance where a loss in most cases is apparent immediately after the damaging event. In extreme cases, plants may suffer irreversible damage or even die.

There are two crop insurance products: individual indemnity-based programs and index-based insurance programs. In multi-peril crop insurance on an indemnity basis, drought is usually covered together with other natural perils. Indemnity-based means that the actual loss incurred is the basis for compensation, usually reduced by a deduction. Other approaches are index-based insurance systems where the indices used are based on meteorological parameters or on regional yield amounts. Weather-index policies for drought pay out if a specific meteorological value, e.g. precipitation, is not achieved in a specific period – irrespective of the actual yield.

The cost of insurance is a major drawback of drought coverage. The insured generally requests a fairly frequently recurring period of payouts, which corresponds to a considerable coverage level and hence to costly premium rates of 5% to 10% of the sum insured. In comparison, earthquake premium rates are typically below 1%.

6.2. Drought insurance for national resilience

Assessing agricultural risks and the risk transfer through insurance is complex. Besides flood and excessive rain events, agricultural drought exposure is one of the most demanding risks to insure due to its systemic risk nature. Both crop and livestock production are key to national resilience. Unfortunately, they are largely exposed to drought and have to be the focus of a country's resilience strategy.

Public–private partnerships have proved to be successful, as they apply risk-sharing and risk-financing mechanisms between agricultural producers, insurance and reinsurance companies, and governmental public entities. These agricultural insurance schemes are available in many parts of the world, with a very wide range of development stages regarding structures and penetration rates. The largest agricultural insurance markets are in North America (USA, Canada, Mexico), followed by Asia. Europe, the rest of Latin America, Australia and, finally, Africa only add up to a very small portion of the world's premium volume.

Figure 4 shows the development of total premium volume from agricultural insurance in the United States and China, the two countries with the largest agricultural insurance markets, and India, ranked fourth. In the USA, annual volume has increased almost exponentially during the past 25 years; in China a steady increase has been observed in the past decade, while in India, after several years of increase, take-up currently appears to be stagnating. These increases point to success stories in the three countries, where the agricultural areas amount to roughly four (USA), five (China) and two (India) million square kilometers.

In most public–private agricultural insurance systems, catastrophe risks are financed by governments. With the African Risk Capacity (ARC) (see Section 6.5) for example, sovereign drought risks are transferred to the insurance market. This contributes to the public ability to cope with losses from droughts, especially in developing markets without robust access to international financing systems after disasters.

6.3. United States – national crop insurance system

As the world's biggest producer of staple foods, the US agricultural sector plays a major part in safeguarding supplies, not only in the United States but throughout the world. Drought losses in agriculture amount to billions of dollars in severe drought years. The peril is the underlying cause for more than a

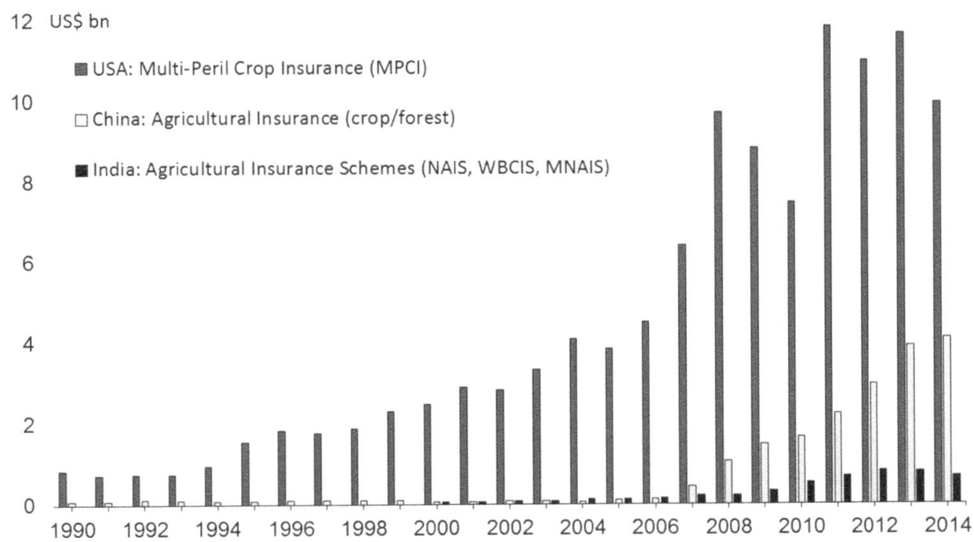

Fig. 4. Development of crop insurance premiums in the United States, China and India. Data for the insurance schemes of India are not available for all years (National Agricultural Insurance Scheme [NAIS] from 1999, Weather Based Crop Insurance Scheme [WBCIS] from 2007 and Modified National Agricultural Insurance Scheme [MNAIS] from 2010). *Source:* Munich Re based on United States Department of Agriculture, Chinese Insurance Regulatory Commission, Department of Agriculture and Cooperation (India) and Guy Carpenter.

third of indemnity payments to agriculture in the United States. Consequently, obtaining adequate financial security against such losses is a key support factor for the US farming sector, global food security and agricultural commodity price stability.

The United States has a well-working agricultural insurance scheme that provides multi-peril coverage. This scheme, the Federal Crop Insurance Program, involves the federal government and the private insurance industry. Used in a standardized and centralized way as a policy tool by the government, which subsidizes it, it is advantageous for all parties – the farmer, the government, and the (re)insurance industry.

The fact that crops are so heavily exposed to large-scale, systemic risks, such as drought, that cannot be borne entirely by the private sector has prevented the establishment of a purely private crop insurance market in the United States. The crop insurance scheme offers US farmers a number of key risk management advantages:

- High level of creditworthiness: many farmers are faced with a negative cash flow and have to take out a loan every year to fund the upcoming season's production costs.
- Excellent marketing opportunities, because it covers the risk that farmers will default on their sales contracts, which normally involve forward sales. In other words, they can at least forward-sell that portion of the yield guaranteed by the crop insurance policy knowing that they will be able to fulfill the contract. If the actual yield is then lower than expected, the indemnity paid under the policy can be used to offset the shortfall.
- In the event of a loss, the indemnification is guaranteed and paid out quickly.

The farmer assumes a part of the costs in the form of the premium and a deduction. The burden on the taxpayer, who entirely funds relief programs, is reduced by this contribution.

The Federal Crop Insurance Corporation (FCIC) administers the system and is directly accountable to the US Department of Agriculture. The FCIC issues guidelines dealing with claims' regulation, selects the products, and determines standard conditions of insurance and rates. The system is significantly subsidized by the federal government. Crop insurance providers calculate premiums netted by administration expenses. The expenses are borne by FCIC as a fixed rate. All providers apply the same actuarially sound premium rates, varying by crop type and region. This prevents harmful rate competition among the providers and adverse selection by the farmers. Along with the fact that crop insurers are not permitted to decline cover, this system secures an adequate level of market penetration. Without subsidization, farmers would be unable to afford technically adequate rates, and market penetration would be insufficient. The risks assumed are subject to a profit-and-loss-sharing system between the insurer and FCIC. Apart from underwriting and assuming risk, one of the primary functions of the (specialized) crop insurers is to give farmers competent advice on the various issues involved. The insurers also market and service the policies, estimate losses, and settle claims.

Currently, there are various crop insurance products covering around 100 different crops. Around 300 million acres (1.2 million km^2) are insured under the Federal Crop Insurance Program. This corresponds to a market penetration of more than 80%.

This system of providing cover against natural hazards in the US agricultural sector is far more efficient than governmental aid following disasters in other cases. It ensures that cover and indemnity are tailored to the farmer's individual risk situation, and provides fast claims' settlements. However, it also benefits the state because farmers pay much of the premium themselves, whereas aid payments are funded entirely by taxpayers. Public–private partnerships of this type also eliminate the need for state-run disaster relief infrastructure, because this part is provided by the private-sector insurance industry. Despite the contribution of governmental reinsurance, a substantial share of the risk is borne by the individual crop insurers. In a nutshell, the well-functioning – but expensive – US system relies heavily on the private insurance and reinsurance market.

6.4. India – public–private area-yield and weather-index insurance scheme

Index-based insurance is the most common form of agricultural insurance for crops in India. The programs offered are either an area-yield index or a weather-based index insurance. On 1 April 2016, the government launched a new system called Pradhan Mantri Fasal Bima Yojana (PMFBY). PMFBY is set to improve the National Agriculture Insurance Scheme (NAIS), which has operated for over 20 years and is currently the world's largest single crop insurance program insuring about 30 million of the 120 million Indian farmers in 2015.

NAIS relies on an index-based area approach. Indemnity is based on the realized (harvested) average yield of a defined area, such as a county or district. The insured yield is established as a percentage of the average yield for the area, and typically ranges from 50% to a maximum of 90%. An indemnity is paid if the realized average yield for the area is less than the insured yield, regardless of the actual yield on a policyholder's farm. This type of index insurance requires historical area-yield data from which the normal average yield and insured yield can be derived.

PMFBY will be implemented by the national government in cooperation with the Indian states. The government has set a steep target of doubling the number of farmers covered under crop insurance to 50% under the new scheme. The scheme provides insurance coverage and financial support to farmers in the event of failure of any of the specified crop plants as a result of natural calamities, pests and diseases. The insurance plan will be handled by the Agriculture Insurance Company of India, the public insurer for agriculture insurance, and ten other private insurance companies.

With PMFBY, the government has started to move from a claims-subsidy regime to a premium-subsidy regime. The premium rates for farmers have been fixed at 2% for summer crops, 1.5% for winter crops, and 5% for annual commercial and horticultural crops. The program is highly dependent on government subsidies, as the fixed premium rates are only about 30% of the actuarially required rates. Like its predecessor, premiums for the PMFBY program will be heavily subsidized, with 50/50 contributions by the respective state and the central government.

With respect to yield loss assessment under the new scheme, there is a high emphasis on the use of new technologies. Smartphones, remote sensing and global positioning system technologies will be used to optimize claims' payments. The scheme itself is future-proofed but implementation is challenging given the large size of the country and small landholdings.

The second pillar of Indian index insurance is based on weather-station data. Operated by private insurance companies and publicly subsidized, weather-based index insurance policies amount to approximately one-third of the entire premium volume.

6.5. African Risk Capacity (ARC)

Motivated by recurrent catastrophic drought events in the Sahel, ARC Ltd was established in 2014 as a specialized agency of the African Union (AU). The aim of this new sovereign insurance pool is to help AU Member States, by means of modern finance mechanisms, to improve their capacities to better plan, prepare and respond to extreme weather events and natural disasters, and thus prepare for the food shortage events that affect their vulnerable populations. ARC was established to use risk pooling and risk transfer to create pan-African climate-related response systems for African countries. The scheme currently (2016) covers drought risks in seven countries within the ARC Risk Pool: Gambia, Kenya, Malawi, Mali, Mauritania, Niger, and Senegal.

ARC Ltd was established as a mutual insurance company in the Bermudas, capitalized by KfW and DFID, the German and British Development Banks. ARC enables parametric insurance policies to be acquired by African countries, providing them with post-disaster event financing, which is predictable and has a rapid payout mechanism due to the index-based nature of the policies. The trigger is based on a Water Requirement Satisfaction Index within the Africa Risk View (ARV) model. The model was developed by the UN World Food Programme to calculate the estimated crop losses and drought-response costs from satellite-based rainfall information. As the ARV also serves as an early-warning mechanism, it becomes clear before the end of the season whether a payout will occur. If the early-warning mechanism exceeds a certain level, a 'Final Implementation Plan' detailing the use of the payout has to be submitted by the insured country. The ARC pays out on a nationwide level, and the payout is provided only if the government ensures allocation of indemnity to the affected people through this implementation plan. The recipe for success of ARC is the combination of risk assessment and risk modeling, contingency planning (implementation), risk transfer, and its rapid availability of funds. There is considerable potential for growth as other African countries are likely to join the program. Furthermore, there are plans to extend the cover to other hazards such as excess rain and wind.

6.6. United Kingdom and France – subsidence risk

In Western Europe, there is another feature of drought that can become very costly: subsidence.

Some clay minerals, in particular montmorillonite, shrink when they dry and swell when they become moist. A sample of pure montmorillonite – also known as bentonite or smectite – can alter in volume by a factor of 15. Natural soils do not contain a high percentage of expansive minerals, but can still change volume by more than 20%. This is potentially damaging for buildings, and can even be destructive.

Vast areas with soils containing a considerable fraction of expansive clays are found in the United Kingdom and in France. The shrinking of soil in a very dry year or in a sequence of dry years often happens unevenly underneath a building due to the irregular distribution of the clay percentage. The consequence is that the building settles unevenly, cracks and/or tilts, and windows and doors jam.

In both countries, damage by subsidence is insured. In France, it is included in the compulsory governmental CatNat insurance scheme, which compensates losses after a government-declared natural disaster – to which a drought with high subsidence damage belongs.

In the UK, subsidence losses are covered by most private property insurance policies. The history of subsidence insurance started there in 1971 when the peril was included in buildings' insurance, at no extra premium charge. In 1976, an exceptionally dry summer brought a first wake-up call. Insurance companies responded by introducing premiums and deductibles dependent on geographical location; highly exposed properties were barely insurable any longer. Nevertheless, a new peak in losses occurred in 1990–1991 after some dry years, with threefold losses compared with 1976.

Figure 5 shows that there is a clear correlation between summer rainfall and subsequent subsidence losses – typically with a time lag of 1 year. This lag results from the fact that the consequences of settlement often become visible – and are reported to the insurance company – only several months later.

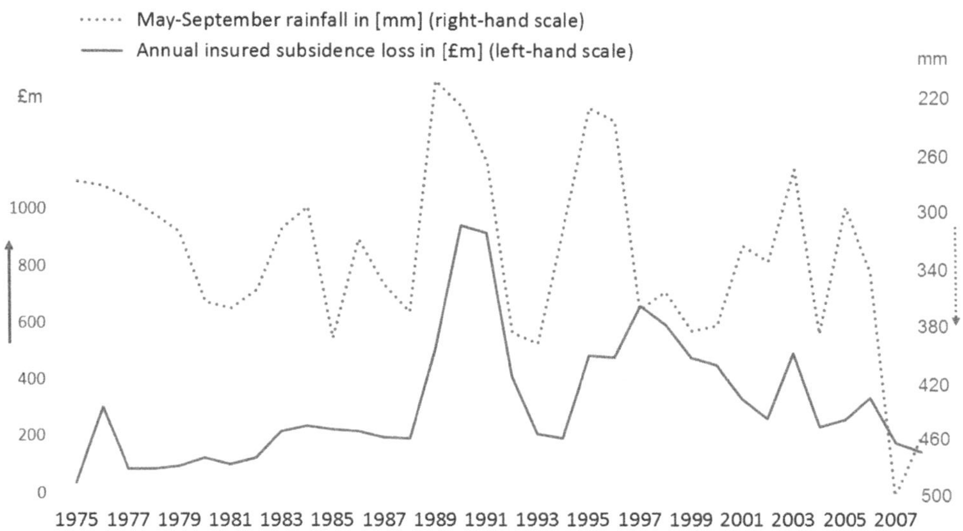

Fig. 5. Summer rainfall in southern England and Wales and insured subsidence losses 1975–2008 (in 2009 values, adjusted for inflation).

Drought-related subsidence was considered the most expensive natural hazard for the UK insurance industry until the 2007 floods.

7. Drought is not only a natural hazard

> *'Droughts are hard to avert, but their effects can be mitigated.* [...] *The price of preparedness is minimal compared to the cost of disaster relief. Let us therefore shift from managing crises to preparing for droughts and building resilience.'*

These words of Ban Ki-moon on the World Day to Combat Desertification 2013 (Ban, 2013) clearly state that dry weather conditions are not necessarily the path to drought disaster.

The problem of drought as opposed to permanent aridity is that neither nature nor humans have adapted to such temporary dryness, and consequently have difficulty coping with it. Drought therefore must not be viewed as a merely physical phenomenon or natural event. Its impacts on society result from the interplay between natural conditions and the demands people place on the water supply, and on the vulnerability of the region concerned.

In most cases, drought crises and drought disasters are the consequence of improper management of existing resources – and often even mismanagement. Reasons for these may be poorly developed societal and governmental structures and responsibilities, but also overuse and overexploitation of existing water resources. In this regard, drought is not a hazard that affects societies at different stages of development with different degrees of severity; it can seriously affect any society. The only difference is that in the one case (poor countries), people may die or get sick, in the other (rich countries), they may have to pay.

Many states, regions and cities must take the blame for water-related problems themselves, not shift it to nature. It is a lack of awareness, and sometimes even ignorance, that precedes a crisis. If the people in charge are sufficiently aware and practice good governance, the risk of drought can be reduced considerably. The term *risk* should be stressed here, in contrast to *consequence*. Reducing risk means acting in advance, before a crisis starts, while efforts (actions, but often only plans) to mitigate consequences are typically undertaken when the crisis is already manifest – which is therefore less efficient and carried out under time pressure.

Everywhere, except in naturally dry countries such as those in the Sahel in Africa where every drop of water is appreciated, the most important feature in this context is permanent awareness – from the top level of government to every single citizen – that the availability of sufficient water is not a law of nature. Water is a natural resource that – like most other resources – is limited. If societies include this perception in their everyday actions, the biggest step towards the sustainable use of water resources has already been taken. This behavior includes avoiding the careless, wasteful use and pollution of water (even if it is available in plentiful quantity), water-consuming production lines (crops, industries), excessive settling in areas with water scarcity (for example Las Vegas), overexploiting groundwater and surface water bodies, etc. To stop or slow such developments, let alone reverse them, is always more difficult than not being too lavish in the first place.

Contingency plans, monitoring the weather situation, and early-warning systems with strictly enforced rules and restrictions are also needed. The capacity to better forecast upcoming dry conditions, for instance, enables farmers to prepare for possible droughts in advance. Natural climate variabilities such as El Niño or La Niña are associated with distinct regions where reduced precipitation is more likely, and thus allow the regions concerned to prepare for and react to possible consequences at an early stage. New technologies such as remote-sensing are developing rapidly and can assist in agricultural management regarding plot identification, yield estimations, assessment of the vegetation status and loss estimations, to name but a few.

In many developing countries, farmers retain the risk of crop losses. Their risk management often only consists of diversifying their income sources by planting a variety of crops and by additionally breeding livestock. Implementing further risk management systems and tools is key for drafting sustainable development strategies, but also for adapting to the changing climate.

Insurance is the last but not least important component of drought management. As a remedy, it covers the loss, but as a prophylactic measure it also provides planning security. A farmer will not be ruined by a disastrous drought, but will be given a financial cushion with which to restart his or her life. Sovereign risk insurance schemes can be a first step towards implementing a market system.

8. Conclusion

Droughts are silent killers, with the potential to cause enormous losses to society as a whole and to the insurance industry in particular (Höppe, 2013). Outside the agricultural sector, losses may initially remain insignificant. However, they can suddenly spiral if developments triggered by a drought, especially in combination with extreme heat, cause events such as power blackouts. Climate change will certainly intensify extreme drought situations in many parts of the world.

In agricultural insurance, a system approach in the form of public–private partnerships is needed. This structure can provide the adequate legal, institutional and organizational framework in which insurance products and other risk management tools can work efficiently, and in favor of all parties involved. A national agricultural insurance system must involve the different production sectors, and address the interests of all stakeholders (producers, government, lending institutions and the insurance industry). Its main objective is to make insurance cover available to the majority of production sectors and farmers.

Overall awareness is a key factor in being prepared to cope with drought and avoid catastrophic consequences. While extremes of dry weather are not avoidable, disasters are. They are the net result of the effects of extreme weather events and the response to them. Effective prevention measures are both achievable and indispensable, but they will never provide complete protection. If societies, from governments to individuals, are prepared for frequent and normal extremes, and the residual risk from rare events is transferred to the finance industry, we can achieve enough resilience to guarantee sustainable living.

References

Ban, K. (2010). *Secretary-General's Message for 2010. World Water Day to Combat Desertification*, 11 June 2010. Available at: http://www.un.org/press/en/2010/sgsm12955.doc.htm (accessed 5 October 2016).

Ban, K. (2013). *Secretary-General's Message for 2013. World Water Day to Combat Desertification*, 17 June 2013. Available at: www.un.org/en/events/desertificationday/2013/sgmessage.shtml (accessed 5 October 2016).

Braga, B. & Kelman, J. (2016). Facing the challenge of extreme climate: the case of Metropolitan São Paulo. *Water Policy*.

Goodman, B. (2007). Drought-stricken South facing tough choices. *The New York Times*, 16 October 2007. Available at: http://www.nytimes.com/2007/10/16/us/16drought.html?_r=2 (accessed 5 October 2016).

Höppe, P. (2013). Drought – an underestimated natural hazard. In: *Topics Geo Natural catastrophes 2012 – Analyses, assessments, positions*, Munich Reinsurance Company, Munich, p. 25.

IPCC (2013). Climate change 2013: the physical science basis. In: *Contribution of Working Group I to the Fifth Assessment Report of the Intergovernmental Panel on Climate Change*. Stocker, T. F., Qin, D., Plattner, G.-K., Tignor, M., Allen, S. K., Boschung, J., Nauels, A., Xia, Y., Bex, V. & Midgley, P. M. (eds). Cambridge University Press, Cambridge, UK and New York, NY, USA.

Katsafados, P., Papadopoulos, A., Varlas, G., Papadopoulou, E. & Mavromatidis, E. (2014). Seasonal predictability of the 2010 Russian heat wave. *Natural Hazards and Earth System Sciences 14*, 1531–1542.

Keeley, G. (2008). Barcelona forced to import emergency water. *The Guardian*, 14 May 2008. Available at: https://www.theguardian.com/world/2008/may/14/spain.water (accessed 5 October 2016).

Kron, W., Steuer, M., Löw, P. & Wirtz, A. (2012). How to deal properly with a natural catastrophe database – analysis of flood losses. *Natural Hazards and Earth Systems Sciences 12*, 535–550. doi: 10.5194/nhess-12-535-2012.

Palmer, W. C. (1965). *Meteorological Drought. US Weather Bureau Research No. 45*, Washington, DC.

Robine, J. M., Cheung, S. L., Le Roy, S., Van Oyen, H. & Herrmann, F. R. (2007). *Report on excess mortality in Europe during summer 2003*. EU Community Action Programme for Public Health, Grant Agreement 2005114.

Steuer, M. (2012). Wildfires. In: *Severe Weather in North America*. Munich Reinsurance Company, Munich, pp. 134–145.

Steuer, M. (2013). Wildfires. In: *Severe Weather in Eastern Asia*. Munich Reinsurance Company, Munich, pp. 172–184.

Steuer, M. & Kron, W. (2012). Heatwaves and droughts. In: *Severe Weather in North America*, Munich Reinsurance Company, Munich, pp. 120–133.

UNDP (2013). *UNDP support to countries for implementing the Ten-Year-Strategy (2008-2018) of the UN Convention to Combat Desertification (UNCCD)*. Available at: http://www.undp.org/content/dam/undp/library/Environment%20and%20Energy/sustainable%20land%20management/UNDP%20support%20to%20countries%20for%20implementing%20the%20Ten-Year-Strategy%20(2008-2018)%20of%20the%20UN%20Convention%20to%20Combat%20Desertification%20(UNCCD).pdf (accessed 5 October 2016).

WEF (2016). *The Global Risks Report 2016. World Economic Forum (2016)*. Available at: http://weforum.org/reports/the-global-risks-report-2016 (accessed 5 October 2016).

Wirtz, A., Kron, W., Löw, P. & Steuer, M. (2012). The need for data: natural disasters and the challenges of database management. *Natural Hazards 40*, 135–157. doi: 10.1007/s11069-012-0312-4.

Water policy responses to drought in the MDB, Australia

James Horne

Australian National University, Canberra, Australia. E-mail: jameshorne@iinet.net.au

Abstract

This paper reviews water policy responses to drought in Australia, focusing on the Murray-Darling Basin (MDB) during the two decades from 1997. This period, which includes the decade long Millennium drought, brought a much sharper focus to discussions of scarcity and value of water. The drought initially focused attention on rising salinity and environmental water availability, as action on both was supported by strong science, and resonated politically. The drought became a crisis in 2006. Short-term planning focused on ensuring communities did not run out of water. For the longer term, the national government responded by announcing a major package of reform measures addressing sustainability and underlying scarcity, and recognising climate change. The package strengthened MDB water market infrastructure, upgraded water resource planning and the ability of irrigators to manage their water assets more flexibly, established new sustainable diversion limits and provided funding to ensure the environment received a larger share of basin water resources. But its completeness as a package can be attributed not only to the severity of drought, but also to political leadership, a disrupting strategy in the form of national legislation and a strong national budget that provided financial resources. The drought provided a crisis, but other ingredients were necessary to ensure effective action.

Keywords: Drought; Markets; MDB; Politics; Water governance; Water policy

1. Introduction

There is no universal definition of drought but there is also no debate that drought is an integral part of the Australian landscape (Botterill, 2003; Bond *et al.*, 2008; Khan, 2008; Connor & Kaczan, 2013; Bureau of Meteorology, 2016a; National Academies of Sciences Engineering and Medicine, 2016). In this paper we adopt the approach that drought constitutes a prolonged dry period where the available water is insufficient to meet the region's usual use. Droughts are thus much more than a rainfall insufficiency, but go to how a society manages the overall demand and supply of water in a particular region.

In Australia governments have sought to insulate communities where possible from its highly variable climate, initially through construction of large water storages. Australia has the world's highest per capita water storage capacity of any country (Australian Bureau of Statistics (ABS), 2010). Water

doi: 10.2166/wp.2016.012

policy, through dam construction, management of storages, regulations and licences structuring access to available water, provided a set of tools to ameliorate some of the impacts of drought for those users where water policy could affect access to water.

This paper reviews water policy responses to drought in Australia, focusing on the Murray-Darling Basin (MDB) during the two decades from 1997, a period that includes the Millennium drought (Bond *et al.*, 2008; Chiew *et al.*, 2011; van Dijk *et al.*, 2013). The year 1997 marks the permanent introduction of a cap on surface water diversions in the MDB, or the beginning of the period where water scarcity at some level was recognised as a permanent feature of the landscape, and not just a feature of droughts.

Water policy in the MDB focuses on three major user groups: irrigation businesses (the dominant user group); urban users; and the environment.

A key point to note at the outset is that the focus of water policy is to manage water availability, use and quality. Dry land agriculture in the MDB, where water policy has only a modest impact, accounts for over 60% of the gross value of agricultural product in the Basin. Irrigated agriculture, where water policy has a significant impact, by contrast accounts for around 37% of the gross value of agricultural product, but it occupies only 1% of agricultural land.

Water policy has been but one area affecting risks and input costs facing rural Australia, in the MDB and elsewhere. Exchange rates, commodity prices and changes in demand for products have been additional and sometimes more important factors (Mallawaarachchi & Foster, 2009; Horne, 2014; Ashton & Oliver, 2015). Moreover, during periods of drought, drought policy per se has played a role in how on-farm decisions have been managed (Botterill, 2003; Botterill & Cockfield, 2013; Department of Agriculture and Water Resources, 2016).

The paper is structured as follows. First, the paper reviews key historical developments in water policy in the MDB prior to the period under consideration, to provide some understanding of the state of water policy as at the mid-1990s. Second, the paper examines water policy developments during the period under consideration and the role played by drought (Howard, 2007a; Wittwer, 2011; Horne, 2013b; Gale *et al.*, 2014; Loch & Adamson, 2015). Finally, the paper draws some conclusions from the Australian experience.

2. Setting the stage for scarcity

Water use in the MDB for most of the 1900s was largely determined by state government policy, within a federal constitutional structure. From 1915 a series of intergovernmental agreements between the national government and the states of New South Wales (NSW), Victoria and South Australia played a role in managing water use in the southern part of the basin (Connell *et al.*, 2005; Wheeler, 2014; Guest, forthcoming). In the late 1980s the adequacy of the traditional management approaches had come under increasing challenge. New institutional arrangements were put in place, with the formation of the Murray-Darling Basin Ministerial Council (MDBMC) (formed by Ministers of the national government and the three state governments involved in basin governance (NSW, Victoria and South Australia)) and the Murray-Darling Basin Commission (MDBC). In addition to overseeing surface water diversions, the new MDBC was given a remit to be more actively involved in areas of natural resource management of common interest, such as salinity (Brett, 1990; Powell, 1994; Stevenson, 1996).

Australian governments had come to realise that the push for further economic development was adversely affecting the ecological health of key catchments, including problems caused by increasing salinity and loss of ecosystem values (Emmery, 1994; Williams, 1995; Stevenson, 1996). There was also a widely held view within government, in the context of a broader nationwide debate on micro-economic reform (including national competition policy), that economic output could be boosted with a more efficient water management regime without increasing water use (COAG, 1994; COAG Working Group on Water Resource Policy, 1994). In 1994 the Council of Australian Governments (COAG) agreed in principle that statutory surface water rights from land rights should be separated. This process of unbundling of water rights from land started the development of water markets as we know them today, but it was to be a long convoluted process (Grafton & Horne, 2014). Fledgling trading of seasonal or allocation water within irrigation districts, particularly in periods of drought, had been going on for decades (Alvarez *et al.*, 1989), and small volumes of allocation water trading (water that is credited annually to owners of water entitlements) was taking place regularly in the early 1990s. But the substantial benefits from a much broader fully functioning market had not yet been realised.

As water scarcity became more apparent, both environmental and efficiency issues became central policy concerns in the MDB. In 1991, the MDBMC agreed a salinity management strategy, to ensure the health of the River Murray for both agriculture and environment alike. By 1995, the MDBMC (with the addition of Queensland, which had now joined) had determined that an interim cap should be placed on surface water diversions, in an attempt to arrest the decline in the environmental health in the MDB, and in an attempt to bolster water security for irrigators (MDBC, 1995; Scanlon, 2002; Martin, 2005). The member state governments had long imposed a moratorium on the issue of new surface water licences, meaning that new access to water could only occur through transfer of an existing entitlement. NSW also privatised its irrigation districts in the southern MDB in 1995, starting a long process of putting irrigation infrastructure operators onto a commercial footing.

The climate-adjusted cap on surface diversions in the MDB did not seek to achieve sustainable ecosystem outcomes (Cox & Baxter, 1996; MDBC, 2004). It was, instead, a first step to constrain growth in surface water extractions but not development, if that could be achieved, by increasing efficiency of water use and trade. Essentially, the cap determined the maximum quantity of water a state should allow to be diverted in a particular year. How much water would be provided to urban users, irrigators and the environment was a decision for each state (MDBC, 1999). Each state had its own classes of entitlements, with different rules around how much water would be allocated to each entitlement class each year. Unused water allocations could not be held in the dam and carried over to the following water year as a matter of course. Water management was at this point largely a state-by-state matter. Prior to 1997 there were only two occasions when NSW Murray irrigators received less than 100% allocation on their entitlement (in 1982–83 and 1994–95). In parallel to basin-wide developments, individual states were introducing reforms of their own to implement their 1994 COAG commitments. The cap applied to the whole basin, requiring its acceptance by Queensland. Queensland was reluctant to agree immediately as it sought expansion in some of its fledgling irrigation districts – it wanted to be allowed to catch up with southern states.

Water policy directed at the MDB by both the national and state governments up to the late 1990s can be characterised as a mechanism promoting ongoing irrigation development and longevity, by enhancing water security to existing entitlement holders (mainly irrigators), while attempting to arrest the ongoing decline in environmental outcomes. While there was an increase in scientific study and much talk about sustainable outcomes, the era of putting in place policies to achieve those outcomes had yet to arrive.

3. Policy responses to the Millennium drought 1997–2009

Figure 1 provides an overview of rainfall for the periods 1997–2001, 2001–09 and 1997–2009. Figure 2 provides aggregate cap details while Table 1 provides some examples of allocations to entitlements in key valleys in the southern part of the Basin.

The data describe a lengthy, extraordinarily dry period over much of the Basin, although the most pronounced water availability shortfall was in the southern MDB. Taken together the data illustrate a number of broad points:

- Rainfall declines of 11% in the southern MDB translated into inflow declines of around 35%. In the far south of the Basin, both figures were higher (Chiew *et al.*, 2011).
- Diversions for the basin as a whole ratcheted down from 2000–01, with widespread impacts on water allocations from the 2002–03 water year. Up to that point trendline use had grown slightly during the previous decade.
- Impacts were felt unevenly between valleys, entitlement types and therefore crops:
  - Victoria's conservative approach to water allocation reflected its higher proportion of perennial crops and dairying – its focus was on ensuring high security entitlements received full allocation in most years.
  - NSW's approach reflected more opportunistic annual crops – its general security Murrumbidgee entitlements illustrated the greater variability in allocation.
  - Up until 2006–07 allocations to South Australian irrigators had been largely unaffected by lower Basin inflows and declines in storage levels, although the impact was already being felt in lower water levels of the Lower Lakes, at the end of the system.

3.1. 1997–2001: implementing the cap and considering its shortcomings

In the years immediately following the 1995 MDBMC agreement to put the cap in place permanently, governments and the MDBC were very focused on practical implementation of policy and explaining its detail to the irrigation community, alongside continuing implementation of the 1994 COAG water reforms to achieve full cost recovery for provision of rural water services, greater access to water trading and broader programmes such as salinity management (MDBC, 1999; Quiggin, 2000; Martin, 2005; National Water Commission, 2011; Guest, forthcoming). The 2000 review of the operation of the cap confirmed the importance of the cap, but noted there was no certainty that the cap itself was sustainable (Cox & Baxter, 1996; MDBMC, 2000). A 2001 independent scientific review questioned whether existing approaches to water management would achieve sustainable environmental outcomes (Jones *et al.*, 2002). It was clear to all involved that despite the significance of the cap in the evolution of water management, much more needed to be done to put Basin water management onto a sustainable path (MDBMC, 2001a; Connell, 2007; Quiggin, 2007).

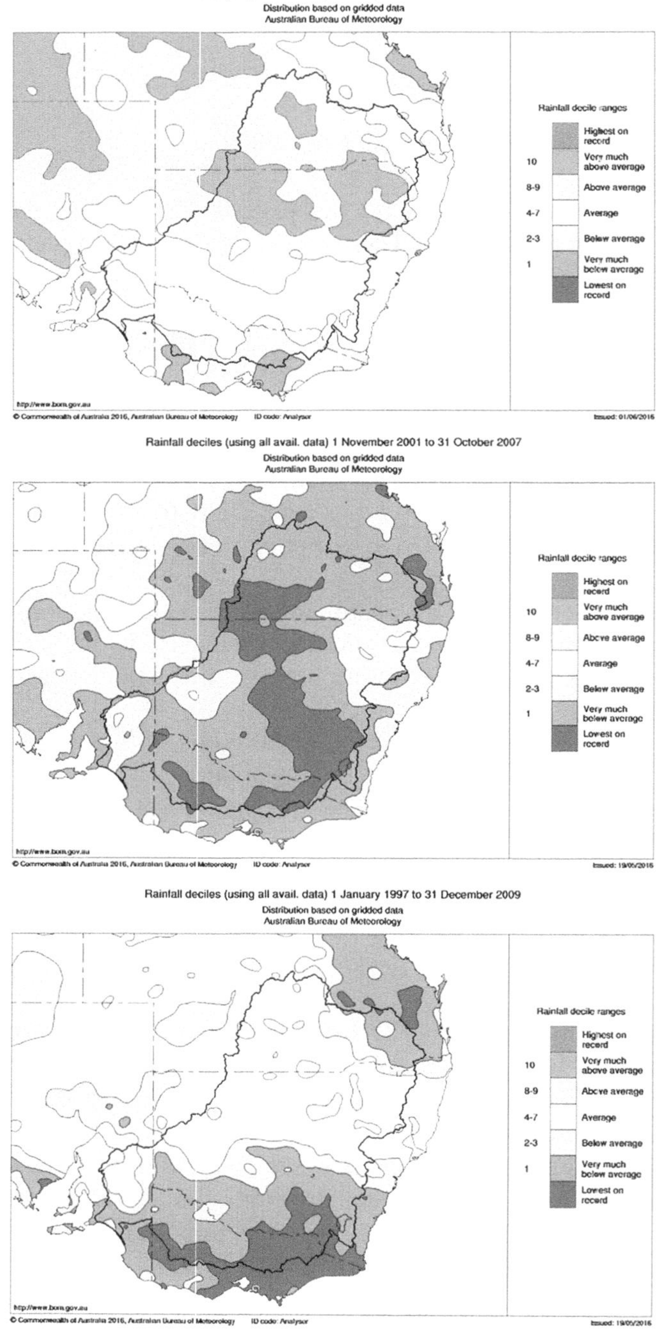

Fig. 1. Rainfall deciles 1997–2009. *Source:* Australian Bureau of Meteorology (BOM), personal communication with the author, 2016.

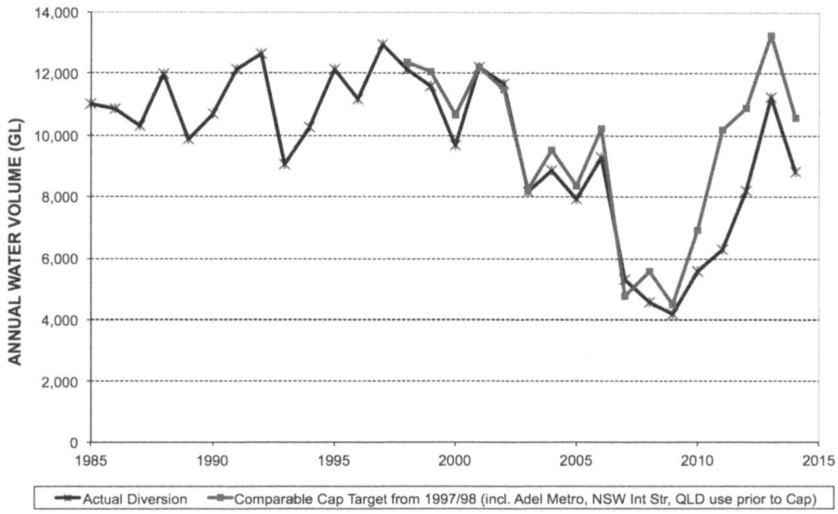

Fig. 2. Surface water diversions in the MDB under the cap[1]. *Source:* MDBA, personal communication with the author, 2016.

The early period of the Millennium drought (1997–2001) was not as severe as the extreme drought that was to follow, but water scarcity was being felt in allocation decisions in some valleys. Coming directly after the introduction of the cap, it was a period when the scientific community was questioning the adequacy of action in the Basin (Cox & Baxter, 1996; Blackmore, 2002; Jones *et al.*, 2002; Arthington & Pusey, 2003). At the same time, implementation of the COAG reforms was proceeding. For example, from 2000 NSW moved to put in place valley-by-valley water sharing plans under new legislation and increased cost recovery for water services (Martin, 2005). Much of its water management was preoccupied with implementing unfinished reforms, and considering their adequacy. The pressure from the drought underscored that the new cap was just a starting point, paving the way for two significant basin-wide initiatives.

3.2. Salinity and the 10-year Basin Salinity Management Strategy

Salinity issues became a renewed focus of attention, off the back of a 1999 basin salinity audit that concluded vast tracts of land would be unusable in the absence of concerted intervention, with salt damage estimated to cost around AUD$1 billion ($10^9$) per year (MDBMC, 2015). Action came through a Basin Salinity Management Strategy (MDBMC, 2001b, 2015) and as a part of the National Plan on Water Quality and Salinity (COAG, 2001; Pannell & Roberts, 2010), which was designed to address salinity in the worst affected Australian landscapes. Rather than being caused by drought, it was more a reflection of land and water use patterns, but together these national and basin-wide responses can be seen as reflecting drought and pressure on water resources more broadly, and pressure on governments to do something about it. At a very regional scale, the ongoing drought was seen in 2002 as a serious threat to environmental outcomes in the Coorong (Manning, nd). At a political level, the salinity

[1] 1 gigalitre (GL) = 10^6 cubic metres.

Table 1. End of season allocations to selected high and low/general reliability entitlements, major systems in the southern MDB (%)*.

Year	Vic Murray (high)	Vic Murray (low)	Goulburn (high)	NSW Murray (high)	NSW Murray (general)	NSW Murrumbidgee (general)	SA (high)
1997–98	100	30	100		84	90	
1998–99	100	100	100		93	85	
1999–00	100	90	100		35	78	
2000–01	100	100	100		95	90	
2001–02	100	100	100		105	72	100
2002–03	100	29	57		10	38	100
2003–04	100	0	100		55	41	95
2004–05	100	0	100	97	49	40	95
2005–06	100	44	100	97	63	54	100
2006–07	95	0	29	97	0	10	60
2007–08	43	0	57	28	0	13	32
2008–09	35	0	33	94	10	21	18
2009–10	100	0	71	97	28	27	62
2010–11	100	0	100	90	72	75	67
2011–12	100	0	100	55	42	74	100
2012–13	100	0	100	100	65	73	100
2013–14	100	0	100	100	110	63	100
2014–15	100	0	100	97	61	53	100

*Shading indicates change in series/policy.

Sources: http://www.water.nsw.gov.au/water-management/water-availability/water-accounting#historic; http://nvrm.net.au/resources http://nvrm.net.au/determinations/history; http://www.environment.sa.gov.au/managing-natural-resources/river-murray/water-allocations/historical-water-allocations; http://www.environment.sa.gov.au/managing-natural-resources/river-murray/water-allocations/historical-water-allocations.

strategy was sufficiently practical and anchored in the traditional agricultural base to be supported. It would turn out to be a very successful joint government intervention (Doolan *et al.*, 2015).

3.3. The Living Murray initiative: a first step recognition of environmental needs

Governments and the MDBMC had great difficulty dealing with the growing tension between environmental and irrigation needs. Following a review of the science in 2000, which indicated clearly that current water management practices were not sustainable, governments eventually agreed to look at options under the banner of The Living Murray (TLM), in 2002, designed to provide additional water to six icon sites along the River Murray (MDBA, 2011). In 2003 MDB governments agreed to implement an action plan to purchase 500 GL long-term average annual yield (LTAAY), even in the face of advice that a minimum additional pool of 1,500 GL was needed (Jones *et al.*, 2002; MDBMC, 2002, 2003; COAG, 2003).

Governments agreed to accumulate an environmental water portfolio under the banner of TLM, but then only proceeded very slowly to implement it, even though in 2006 the national government offered to spend an additional AUD$500 million to speed up the purchase (Connell *et al.*, 2005; MDBMC, 2006). The early post agreement years of TLM illustrated how drought conditions can bring an issue to the fore, but what happens then is determined by an array of factors, including institutional and agricultural politics (Connell, 2002, 2007; Horne, 2016a; Guest, forthcoming).

3.4. The National Water Initiative

On a parallel course, the persisting drought resulted in a renewed interest in national water reform. In 2003, at the same COAG meeting that discussed funding for TLM, governments agreed to develop a renewed water agenda focusing on implementing nationally compatible water access entitlements that encouraged investment and maximised the economic value created from water use, while ensuring rivers and aquifers remained healthy. The new framework would cover: nationally functioning water markets (an efficient water market enabling increased returns from water use and in the MDB, a review of entitlement products); best practice water pricing (user pays and full cost recovery); integrated management of environmental water (in the MDB, a basin-wide approach to enable environmental water management, including through the market); better measuring, monitoring and information; urban water reform (to reinforce the need for urban users to use water efficiently including by promoting water reuse and recycling); and new institutional monitoring of actions by government, to be undertaken by the National Water Commission (NWC) (COAG, 2003).

A year later governments agreed a negotiated compact, without any financial inducement, reflecting the worsening water outlook (Connell, 2007; Quiggin, 2007; Horne, 2013a). The National Water Initiative (NWI) framed the water problem in terms of triple bottom line outcomes, encouraging governments to address issues of overallocation and overuse in irrigation and the pressing water needs of the environment, as well as focusing on efficiency and markets in both urban and rural water use. Climate change was explicitly recognised as an issue of water policy (COAG, 2004; Khan, 2008).

While there are many analyses of the substance of the NWI and progress in implementation (Connell *et al.*, 2005; Crase, 2011; NWC, 2011), little is said of the role of drought, which assisted in elevating water policy issues to national political attention, giving it a place on a limited and crowded agenda in eight out of 10 meetings between 2000 and 2006 (COAG, various).

But state actions agreed under the NWI, as under TLM a few years earlier, were again implemented slowly relative to commitments, resulting in COAG having to 'renew' its commitment to implementation on several occasions (COAG, 2007). Tardy implementation of these water policy commitments by states was an ongoing reflection of difficult politics, for example deep-seated tension between irrigation and environmental objectives, and of course of the difficult economic conditions being experienced in many regional areas, which were still grappling with ongoing drought.

In a modest way, through TLM, governments were continuing the process of building a small portfolio of water to be set aside for the environment, but it was questionable whether substantial progress was being made, as increases in irrigation efficiency meant less water was being returned to the rivers, and with the growth of markets an increasing percentage of allocation water was being diverted (Loch & Adamson, 2015). As the drought deepened in 2006, this tension remained unresolved.

4. 2006–09: policy disruption: national insistence on MDB basin-wide objectives and action

The severity of the Millennium drought in the southern MDB increased in 2006–07. Inflows for the 2006–07 water year (July 2006 to June 2007) were tracking at lowest on record, at only around 10% of historical averages (Dreverman, 2013). In October 2006, the NSW government suspended some water allocations in the southern MDB that had already been credited to entitlement holder accounts, an indication that water management in the southern MDB was entering uncharted territory.

4.1. 2006–08 cooperative dry inflow contingency planning

First Ministers of the national government and MDB state and territory governments attended a 'crisis' meeting called at short notice by the Prime Minister in early November 2006. There were two key outcomes from this meeting. It commissioned the Commonwealth Scientific and Industrial Research Organisation (CSIRO) to prepare a water availability study for the basin as a whole, to provide an up-to-date assessment of water resources in the Basin, and how they were being used. The study would explicitly look at the impact of climate change out to 2030 (Howard, 2006). This was an outcome with positive long-term consequences, not only in the MDB, but nationally (CSIRO, 2008).

In the short-term, however, First Ministers were focused very much on ensuring all governments understood the problems at hand, and options to address them. They commissioned a senior officers group (SOG) to provide regular updates on the state of water resources in the southern MDB, and to recommend actions that went beyond the business as usual framework provided by the Murray-Darling Agreement (Howard, 2006).

The guiding principle for 'dry inflow contingency planning' was that where possible the operation of the market should be maintained, but if existing systems and rules failed, 'urban areas and towns should be provided sufficient survival water' (SOG, 2006). The focus of planning and policy actions was to secure urban supplies during 2007 and beyond, but also to ensure available water beyond urban provision was used to best effect. Initial scenario planning indicated that if inflows continued on their trajectory, critical demands could not be met for the River Murray (SOG, 2006). Previous assumptions about worst-case inflows based on the historical record were re-examined, as the basis for them proved to be hopelessly optimistic (Dreverman, 2013). This was a wake-up call on stationarity suggesting planning needed to be forward looking rather than anchored in history (Milly *et al.*, 2008; Quiggin *et al.*, 2010).

The nine reports from the SOG over a 2-year period set out immediate actions, actions that needed further consideration and actions that should be avoided (SOG, 2007–9). Many of these intergovernmental planning and policy discussions took place in circumstances where scenario planning suggested that there was a substantial risk that even 'critical human water needs' (CHWN) (a new term coined at the time) may not be met, which would have meant a collapse of irrigation in the southern part of the MDB.

Contingency planning continued for nearly 2 years, in parallel with major reforms being developed by the national government independently and in concert with state and territory governments, and developments being progressed by state governments individually. This period of extreme drought provided significant action and lessons on water policy, including strengthening of the water governance framework in the MDB.

Governments agreed that in managing issues around dry inflow contingency planning, markets should be used to the maximum extent possible (Howard, 2007b). There was an enhanced realisation that a fully functioning market required careful specification of all rights as far as possible, hence the development of policy around CHWN and carryover.

The 2008 amendments to the Water Act 2007 (discussed below) included formal arrangements to manage water sharing for CHWN being made a part of the MDB Basin Plan (Commonwealth of Australia, 2008). This policy development can be directly attributable to the 2006–07 dry inflow contingency experiences.

The nine SOG reports going to First Ministers over a 2-year period gives some sense of the depth of the crisis in the southern MDB, and the time that governments at the national and state level were required to set aside for its consideration. Overall, short-term policy responses had some positive impact on improving the management of the very small stock of available water (Dreverman, 2013; Ballard *et al.*, 2014). But some of these policy measures, notably carryover and water trading, had a much longer-term positive impact by encouraging policymakers and water users to place more emphasis on risk management within water years and between them, and for rights to be specified more clearly. The transition of carryover policy in each state from a drought response to a permanent feature of managing water allocations is illustrative of this (National Water Commission, 2014).

Combined with broader national deliberations related to the drafting of the national Water Bill and new water governance arrangements for the MDB, these developments give some sense of the direct policy focus brought about by the drought, and the long-term outcomes from it.

4.2. The National Plan for Water Security and the Water Act 2007

The Howard government's 2007 National Plan for Water Security (NPWS) (subsequently rebadged and fine-tuned under the Rudd government, as Water for the Future) had significant long-term intent. Rather than simply responding to acute but short-term issues brought into focus by the drought, it incorporated long-term policy and political responses made possible by the drought (Howard, 2007a; Wong, 2008; Horne, 2013a).

The major initiatives that formed the core of this well-resourced 10-year national thrust into MDB water governance and management included the following:

- Commissioning the long-term basin-wide water availability study noted above, and replicating it in three other areas where water resource use was under the microscope (CSIRO, 2013).
- Establishing the Murray-Darling Basin Authority (MDBA), a new independent national institution to oversee water management in the MDB (Commonwealth of Australia, 2008).
- Putting in place a Basin Plan, including establishing mandatory requirements for catchment level water planning for all water resources under a sustainable diversion limit (SDL), rather than just surface water, a basin-wide environmental watering plan, and regulations and rules to strengthen the operation of the water markets within the MDB (Commonwealth of Australia, 2012; Horne, 2014; Hart, 2015a, 2015b).
- Giving the Australian Competition and Consumer Commission a key role in overseeing irrigation infrastructure operators, monitoring and enforcing MDB water market rules, and making recommendations to the MDBA on water trading rules (Australian Competition & Consumer Commission (ACCC), 2016).
- Establishing the Commonwealth Environmental Water Holder, an environmental water holder with a strong basin-wide perspective, and an operating philosophy of developing close cooperation with states and communities (Docker & Robinson, 2013; Banks & Docker, 2014).
- Substantially increasing the volume of water entitlements set aside for the environment, to halt the decline in environmental outcomes, through national government entitlement purchases in the water market, and through investments in infrastructure (Australian Government, 2016).
- Providing better, more water transparent information to facilitate improved decision making and risk management by water users and policy decision-makers (Vertessy, 2013; Horne, 2015a; Bureau of Meteorology, 2016b).

These initiatives formed the basis for policy implementation over the period beyond 2009, well after the crisis had passed. The Water Act for its part put to bed once and for all the notion that water was simply the responsibility of the states, and it has become very much a shared responsibility between national government and the states, in conjunction with the community (Briese *et al.*, 2009; Kildea & Williams, 2010).

4.3. Putting drought into perspective

The scale of new action foreshadowed under NPWS/Water for the Future and the Water Act showed that as far as Australia had come during the decade to end 2006, much further work was needed over the following decade. Drought provided the central impetus or crisis for these initiatives to get started. An October 2005 public opinion poll ranked water as the most important issue for the national government to become more involved in (Roy Morgan Research, 2005). Three other factors were also important. These were:

- political leadership in the form of a national government led by an established long-serving Prime Minister, bolstered by an activist parliamentary secretary (later to become Water Minister and Prime Minister), combined with bipartisan political support at the national level;
- a game changing strategy (Christensen, 2013) in the form of a national Water Act seeking more determined water governance action across a spectrum of key water issues; and
- a strong national budgetary position making it easier to earmark the significant level of resourcing required to make a substantial difference.

Up to the end of 2006, the national government had played a subsidiary role to the states in managing water issues, although it had played important facilitating roles through the previous 100 years as successive River Murray and Murray-Darling Agreements were put in place (Clark, 1971; Kildea & Williams, 2010; Guest, forthcoming). In particular, the national government played an instrumental role in funding and putting in place the Snowy Mountain Scheme, and other major construction projects during the development phase in the MDB, and in the 1995 COAG Water Reforms, which sought to arrest the decline in the environment, and put a greater focus on recognising the value of the water resources. The 2004 NWI framework agreement again exhorted states to take action across many fronts but progress was slow. By 2007, with water shortages reaching crisis point not only in the MDB, but in key urban areas also, the national government moved decisively to become more involved, reflecting its view that state inaction was hurting the MDB regional economy, the environment and communities. Most, but not all states welcomed this approach. The Victorian state government in particular and from the outset steadfastly argued against much of what the national government sought to do, resulting in the Water Act 2007 being national legislation, based on the national government's own constitutional powers, including its external affairs power (its obligations in relation to biodiversity, Ramsar wetlands and migratory birds), its corporations power (in relation to regulation of market activities, and interstate trade of water entitlements) and powers relating to meteorological observations and census and statistics (Horne, 2007; Briese *et al.*, 2009). A change of national government in late 2007, bringing alignment between the political persuasion of national and Victorian government saw the legislation being revisited, and some modifications made, including importantly referral of some state powers to the national government (Briese *et al.*, 2009). But the thrust remained the same: the institutional framework and the thrust of water policy in the MDB had been changed decisively. Drought was necessary but not a sufficient condition for the reforms proceeding. Political will across the spectrum of views at the national level, constitutional powers, and resources provided the package of conditions needed to proceed.

5. Water markets and water trading through the drought

The MDB water markets account for over 95% of all water trading in Australia, and are the world's largest. The central, very well established notion behind the water market and water trading policies is that they enable water to migrate to higher value uses, and encourage more efficient use. However, as is also well recognised, their use has been contentious to some, and their development drawn out over a substantial period of time (NWC, 2011; Grafton & Horne, 2014; Wheeler, 2014; Grafton, *et al.*, 2015). The Millennium drought, and the period after the drought, saw the MDB's water markets coming of age, and in the process put to bed many of the myths about negative impacts of water market operations when there are sound governance arrangements in place (Grafton *et al.*, 2015). Figures 3 and 4 show the growth of both entitlement water trading (effectively buying and selling a share of the consumptive pool) and allocation water trading (effectively buying access to a specific volume of water in a specific water year) through the drought, and its continuation even after the drought had broken.

For governments, the development in the water markets during the drought years focused on strengthening property rights and removing differences between state systems that made trading difficult, particularly in entitlements. This was possible because the parties involved could see the benefits, notwithstanding specific issues. For irrigation infrastructure operators, there were issues around the impact of trading and their business models (ACCC, 2010). For irrigators, there was much to assimilate about how the markets operated, and how they could contribute to business profitability. Even in recent years, information availability continues to improve, as do options for participating in water trading (Schirmer *et al.*, 2015). Improvements in settlement times are another positive development (Morey *et al.*, 2015). There were also issues around integrating environmental water portfolios based on entitlements previously owned by irrigators, but purchased by the national government in the market. Consistent rules were needed that respected hydrology but also applied to basin-wide trading.

The increased acceptance and use of water markets can in large part be attributed to the drought creating irrigator demands for change. The extended and deep drought placed pressure on irrigators to reconsider how they did business, and how they worked their assets. The value and cost of water became very apparent and

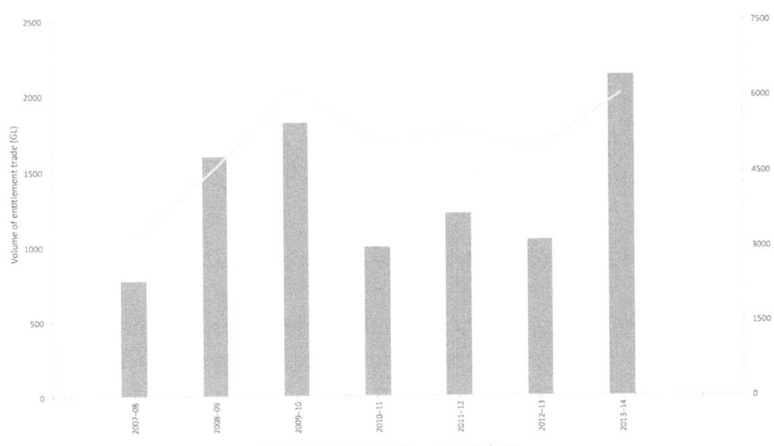

Fig. 3. Water entitlement trade in MDB. *Source:* Australian Bureau of Agricultural and Resource Economics and Sciences (ABARES), personal communication with the author, 2016.

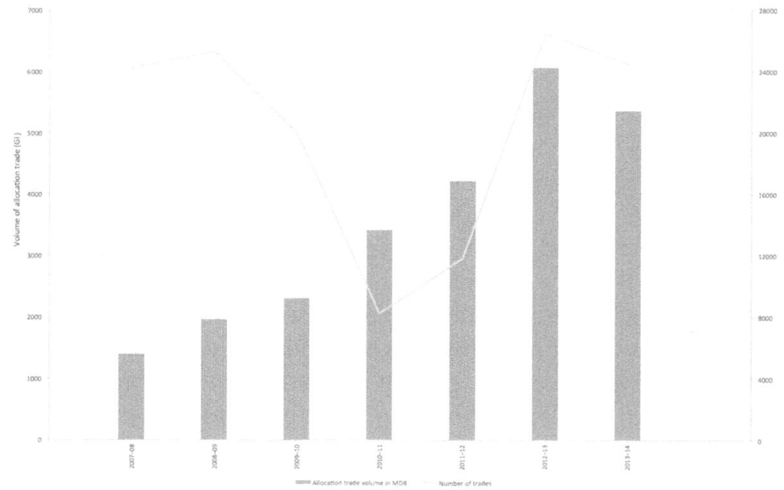

Fig. 4. MDB water allocation trade. *Source:* ABARES personal communication with the author, 2016.

central. But it took concerted effort to break administrative logjams and reduce transaction costs, to put in place new approaches and to test new trading models. Political scrutiny increased, and state and national institutions increased efforts to resolve ongoing problems, such as those around interstate trading and the level of transaction costs (Australian Government, 2008; COAG Standing Council on Environment and Water, 2013). The concept of water having value was and is very well accepted, so much so that 'water trading has become a vital business tool for many irrigators, providing flexibility to respond to variable water availability' (MDBA, 2016a). Water moved from being a silent part of balance sheets and input costs, to a very active part.

The benefit from well-functioning water markets is assessed in numerous studies. Wheeler summarises eight of these, concluding the economic benefits have been considerable (Wheeler, 2014). Another study notes *inter alia* the contribution of markets and trading to reducing risks and uncertainty (Grafton *et al.*, 2015); one survey reports the positive contribution of water trading to irrigator well-being in a non-drought year (Schirmer *et al.*, 2015). This seems likely to suggest greater peace of mind to MDB households and communities in managing drought. Another study suggests that irrigator adaption through the drought years resulted in a two-thirds reduction in water being associated with only a 20% decline in irrigated production gross value. Irrigators employed many drought adaption strategies (including crop mix, source of inputs and water trading) (Kirby *et al.*, 2014).

5.1. Integration of carryover policy into water market infrastructure

A less discussed but important example of water market development during the Millennium drought relates to carryover policy. Over a 10-year period, carryover policy went from being a drought contingency measure in some regions to a widely accepted part of the mainstream water allocation market.

As benefits of carryover became appreciated, it became an important tool for providing additional flexibility in using water assets, to all water users, with each state eventually introducing permanent carryover rules, and upstream states (NSW and Victoria) agreeing to the downstream state (South Australia) being allocated storage rights in upstream dams (NWC, 2011; Ballard *et al.*, 2014). Policy development resulted in better managing risks from extreme events that had implications for normal market operations. In a real sense, carryover policy represented a movement of risk management from government to the entitlement and allocation water owner. Subsequent flooding events laid bare some weaknesses in the new carryover policies, particularly in Victoria, suggesting this is one area where future refinements might be expected (Hughes *et al.*, 2013; Department of Environment and Primary Industries, 2016).

Figures 5 and 6 provide a metric for the significance of this policy development in the southern MDB. Over 25% of available water is now regularly carried over.

5.2. *National competition policy and markets*

The injection of national competition policy oversight in 2007 increased competition within the MDB water market, allowing irrigators to make choices involving buying and selling water entitlements and

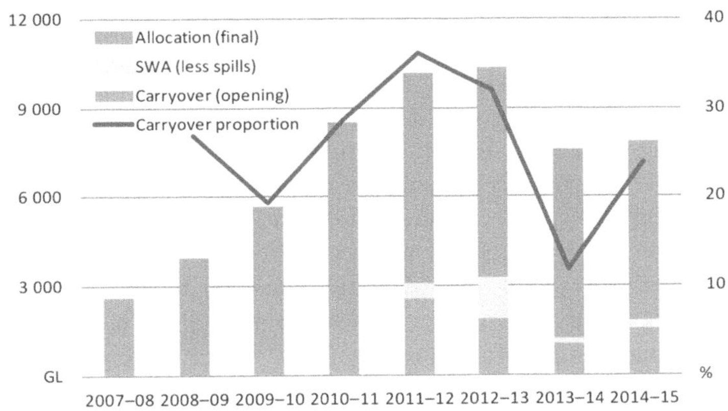

Fig. 5. Southern Murray-Darling Basin annual carryover volumes and allocations. Spillable water account (SWA). *Source:* Reproduced from Morey *et al.* (2015, p. 36).

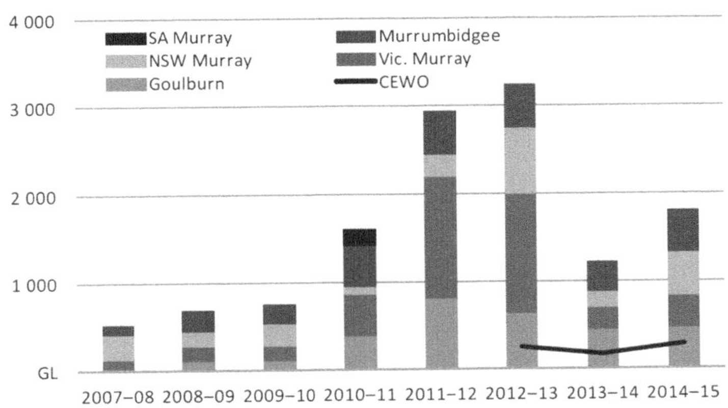

Fig. 6. Southern Murray-Darling Basin annual carryover volumes, by region. *Note:* Total carryover includes spillable water account volumes in the Goulburn and VIC Murray net of deductions due to spills. CEWO Commonwealth Environmental Water Office. *Source:* Reproduced from Morey *et al.* (2015, p. 35).

allocations in a friendlier, more neutral environment, without threats and in cases unconscionable action from irrigation infrastructure operators. These businesses themselves had to adjust how they did business as the needs and wants of their irrigators changed, and as the regulatory framework became much more transparent and monitoring more demanding (ACCC, 2010).

5.3. Information and markets

The quality of information available to water users was given a massive boost under the Water for the Future banner, by a water information programme led by the BOM to boost the quality and timeliness of water information. New systems had to be put in place, and irrigators and others had to learn how to use this information and integrate it into their businesses in the midst of drought (Morey *et al.*, 2015). Ten years later, this is having positive benefits throughout the water sector, assisting irrigators, urban water service providers, environmental water managers and policymakers better manage risks related to water in the immediate future and in the medium term (The Centre for International Economics, 2015; Bureau of Meteorology, 2016b). It has made much more transparent, information that was previously either not widely available or not available at all. Having implemented the programme and established its worth, the issue was whether its 10-year funding would be renewed in 2017. In the absence of a crisis and with a straitened national fiscal environment, the programme's future is uncertain.

In some parts of the MDB economic pressure on communities was enormous, as not only was a drought in progress, but commodity prices were falling and the exchange rate was rising. Part of the successful development of water markets can also be attributable to the decline in transaction costs as governments and the private sector improved their use of internet-based trading programmes. Taken together these developments, and a gradual increase in actual experience in the markets, altered significantly market participant perceptions around intrinsic value. None of this was instantaneous, but required persistent effort, and the building of market infrastructure to deliver positive outcomes from water trading. Concerns by some that water might move out of agriculture, or that some regions might be adversely affected have proved groundless. Research suggests water trading contributed substantially to regional domestic product or substantially ameliorated the effects of reduced water availability on irrigation output (National Water Commission, 2012; Kirby *et al.*, 2014).

Overall, as the drought unfolded irrigators were actioning on their own volition what economic studies later made clear, that the benefits from water trading were considerable (Mallawaarachchi & Foster, 2009; Productivity Commission, 2010; Wheeler, 2014; Grafton *et al.*, 2015). Simply put, markets enabled users to better manage scarcity (Young, 2013), and policy change enabled this behaviour.

6. Drought impact on urban and regional communities

One concern often expressed is that water trading would lead to increased water transfer to urban areas. While economically efficient, political concern and action during the drought years limited this occurrence. A 'north-south' pipeline to add to Melbourne's water security costing $750 m was to provide 75 GL of water, effectively subsidising irrigation modernisation in northern Victoria. For political reasons the water designated for Melbourne consumption (Melbourne is outside the MDB) has never been used there. The Victorian government's current policy introduced after the decision to build the pipeline is to only use the pipeline in a CHWN emergency (Productivity Commission, 2011).

In contrast, two pipelines (of 47 km and 87 km length, with an aggregate capacity of 38 GL) were constructed to increase water availability for two Victorian regional urban centres of 80,000, which had nearly exhausted their available supplies (Victorian Auditor General, 2008). No political opposition was raised on this occasion. Here the quantity of water involved in the context of overall water resources was very small (one centre had budgeted to purchase an additional 5 GL of entitlements), and the pipelines significantly enhanced water security to the two regional centres.

Some further constraints on water market activity (for example, a Victorian policy rule limiting entitlement trade out of an irrigation district to 4% annually) lasted for political reasons well past its use by date (Water Services Association of Australia (WSAA) & NWC, 2009; Productivity Commission, 2010). Politics, driven by a state-based farmers' group, kept in place a policy position that without doubt was damaging the welfare of many Victorian irrigators. Most of these state-based constraints have now been eliminated, in the MDB at least, by national rules.

The north-south pipe use rule and the 4% rule involved political decisions, and little attention to NWI principles. Moreover, a key element of government responses to drought affecting urban areas in Australia as a whole, and in areas that might draw on the MDB, was to invest heavily and, in part, unwisely in overly large urban desalination plants and in water reuse plants. This was accompanied by very large increases in the price of urban water. Despite it being inefficient, this policy approach increased urban water security in Adelaide and Melbourne, and took much pressure off water transfers from rural to urban areas. Drought also saw much greater attention being given to a range of demand management policies in major urban areas, again increasing long-term water security, and embedding a stronger appreciation of the value of water (Horne, 2016b).

7. Investments in infrastructure to increase efficiency of water use

Increasing efficiency of water use through investments in infrastructure has been a feature of increasing awareness of the value of water and acknowledging ever-present issues of water security in the face of drought and potentially longer-term climate change impacts. Significant public funding (many billions of dollars) was provided to drive this process. The funds underpinned irrigation renewal, community and regional projects to pipe previously open channels, and infrastructure to deliver environmental outcomes using less water (Victorian Auditor General, 2008; Australian National Audit Office (ANAO), 2010, 2012). Generally speaking, using public funding in this way has been heavily criticised (for example, Grafton, 2010; Loch & Adamson, 2015), but it is salutary to consider whether the thrust of reform would have proceeded without subsidising the private irrigation and community projects to increase water use efficiency. This author considers it is very problematic: in a sense these projects result from groups of private individuals and communities using the political process to reduce transition costs. Droughts provide a vehicle where the difficulties of these businesses and communities are most apparent. Policy action reflected this strongly political dimension. One such infrastructure project underscores the limitation of general rules such as the NWI pricing principles (Department of the Environment, 2010), and project assessment of large projects that offer water savings for the environment in return for capital funding. The Wimmera-Mallee water pipeline was a huge project: 17,500 km of open earthen channels replaced by an 8,800 km piped water distribution system. The new system covers 10% of Victoria. It provides water to 2,500 rural customers, 35,000 urban customers and 36 towns in the region (Victorian Auditor General, 2008). The $688 m project was over 75% funded by national and state governments, with a very small payback of environmental water, very little of which has been delivered. To regional communities, it provided much enhanced water security, but a more critical analysis suggests it never had any hope of passing any accepted benefit/cost test for public infrastructure, nor did it meet NWI pricing principles. Politics determined that it should be built.

8. Policies toward environmental water

The Millennium drought brought into sharp focus the damage being done to the environment. By late 2006 an environmental crisis was beginning to unfold in the Coorong and the Lower Lakes, the end point of the MDB. The Basin had effectively become a closed system, and salinity levels in the Lower Lakes had risen dramatically. As discussed above, MDB dry inflow contingency planning was focusing first and foremost on CHWN, and provided few avenues to provide any additional environmental water flows. It was clear by this stage, following a series of dry years, that Basin planning under the MDBC aegis had either underestimated system variability or that climate change was playing a role, which planning was not addressing. Later it became clear that state decisions about the use of water shares favoured irrigators over the environment (Young *et al.*, 2011). In the short to medium term it mattered little because there was not enough water to worry about anything but CHWN. But in the longer term it was significant.

Notwithstanding TLM had earlier been put in place as a first step to address declining environmental outcomes, the national government's Water for the Future provided resourcing for direct purchases from the Basin water markets of around 1,500 GL LTAAY of entitlements (or around 14% of MDB surface water extractions LTAAY), and funds to accumulate entitlements through on and off farm infrastructure investments (Horne, 2013b). Governance arrangements were established for this pool of environmental water to be managed by independent professionals within the national government, with close cooperative arrangements with state and local bodies (Docker & Robinson, 2013; Banks & Docker, 2014). State governments also established their own environmental water holders on a much smaller scale. For the first time, on a large scale, the environment would have access to water flows in its own right. The drought brought the plight of the environment to a head, but a crisis was required to generate new governance arrangements and funding programmes. These actions are, perhaps, helping to bring the MDB back from irreversible ecosystem decline, but some questioned whether the agreed volumes that were to be set aside would be sufficient to do that (Young *et al.*, 2011; Wentworth Group of Concerned Scientists, 2012).

9. After the dry: what happened?

The course of events after the Millennium drought broke in the MDB in 2010 provides some insight into how drought affected the development and implementation of water policy.

State governments continued to transform their water planning arrangements and their web-based water trading regimes, at least in the MDB. While national government spending on infrastructure in the MDB continued apace, state government general and financial commitment to joint operations in the MDB declined substantially (MDBA, 2015b).

From a national perspective, the MDB Basin Plan and the growth and use of environmental water portfolios stand out.

The national government put in place the first MDB plan at the end of 2012, cementing in place a key water planning document for the MDB, but really starting a process of further reforms that would take until 2019 to implement (Commonwealth of Australia, 2012; Hart, 2015b; Horne, 2015b; Pittock *et al.*, 2015; MDBA, 2016a). The plan was a product of extensive consultation with communities, and considerable interest group and state government pressure to dilute the timing and impact of the plan, in part reflecting the dissipating impact of the drought. Despite these concerns the plan represented a major step forward in putting in place a sustainable water resource management framework in the MDB. The new institutional arrangements, involving the MDBA, have been well bedded down, and the MDB water market rules are in place. But in the absence of drought or another crisis, two key, yet to be implemented, initiatives in the Basin Plan will test governments. The first is the new SDLs that govern the balance between irrigated diversions and water being set aside for the environment (Horne, 2015b; MDBA, 2016b). Before they are implemented, voices of opposition are likely to become louder, as evidenced by a recent parliamentary inquiry (Parliament of Australia, 2015).

The second is the related SDL adjustment mechanism. It is proposed that selected water efficiency projects in irrigation and environment could be used to reduce or increase the overall SDL limits respectively (MDBA, 2015a). The mechanism is currently being trialled. The robustness of projects selected under this mechanism will enable observers to tease out much more fully the extent to which governments remain committed to reform processes. The adjustment mechanism was at least in part an element of the compromise with the states to secure their support for the Basin Plan SDLs. Drought per se has nothing to do with the mechanism, but the issue of scarcity is at its core. In a design sense, states' interests are keen to increase the SDL as far as possible, so it will come as no surprise if proposed projects focus on increasing the efficiency of environmental watering.

By 2016 the Commonwealth Environmental Water Holder's portfolio of MDB environmental water entitlements had reached 1,600 GL LTAAY, larger than the TLM and all state environmental water portfolios combined. It accounted for flows of over 4,200 GL over 4 years, or about 10% of water diverted under the cap over this period. Annual basin wide environmental water plans had been put in place for the first time, and were now a part of established practice.

10. Summary and conclusion

Drought plays a significant role in the Australian landscape. By the early 1990s the limits on water use became increasingly apparent as the ecological health of key Australian catchments, including in the MDB, declined. Whereas the response to scarcity in the 1960s and 1970s had been largely to address supply, including building more dams, by the mid-1990s a new approach was beginning to take hold. It recognised that water was a scarce resource, and that the management paradigm needed to reflect it. Understanding the value of water was required if agricultural output was to continue to expand, and if ecological outcomes were not to be further compromised. National and state governments accepted these core notions, and began a long process of putting in place supportive water governance arrangements, including separating surface water rights from land rights, and in the MDB, capping water use and strengthening the operation of markets.

The Millennium drought brought a much sharper focus to discussions of scarcity and value of water, in irrigation districts and urban areas, and towards the environment. In addition to slowly implementing changes in water governance arising from policy changes that preceded the Millennium drought, the drought initially focused attention on rising salinity and environmental water availability, as action on both was supported by strong science and resonated politically. As the drought continued, nationally agreed reform attracted renewed attention because earlier efforts remained unfinished. The NWI was agreed and well regarded, including for its recognition of the risks from climate change. But implementation of reform continued slowly.

The most severe phase of drought brought concerted action. In the short term, the key issue was how to make sure communities did not run out of water. For the longer term, the national government responded with a 'big bang' package, which would only have been politically possible in an extreme drought. Over time, it gained uneasy community support across the MDB. The package strengthened many elements of MDB water governance infrastructure – including new legislation, new institutions, enhanced water resource planning and regulatory oversight, enhanced transparency and upgraded information systems and greater community involvement. Its completeness as a package can be attributed not only to the severity of drought, but also to political leadership, a disrupting strategy in the form of national legislation, and a strong national budget that provided resources. The drought provided a crisis, but other ingredients were necessary to ensure effective action.

Governance reforms over the past decade stand out in the overall sweep of reform in the MDB over the past 100 years, with the national government and communities playing much more central roles than previously, alongside state governments whose roles were, while still very important, much more shared. As noted above, with the Millennium drought having receded, agricultural interests are voicing concerns with some elements of the Basin Plan that affect their interests. Whether the reforms will be fully sustained will be an issue to watch over coming years.

References

ABS (2010). *1370.0 – Measures of Australia's Progress, 2010: Inland Waters*. Retrieved from http://www.abs.gov.au/ausstats/abs@.nsf/Lookup/bySubject/1370.0~2010~Chapter~Waterstorage (6.3.6.2) (accessed 31 May 2016).

ACCC (2010). *ACCC Water Monitoring 2009–10 Update to the Minister for Sustainability, Environment, Water, Population and Communities September 2010*. ACCC, Canberra. Retrieved from http://accc.gov.au/system/files/ACCCwatermonitoringupdate2009-10.pdf (accessed 31 May 2016).

ACCC (2016). ACCC role in water. Retrieved from https://www.accc.gov.au/regulated-infrastructure/water/accc-role-in-water (accessed 25 May 2016).

Alvarez, K., Cleary, D. & Wood, J. F. (1989). Rural water supply management in New South Wales. In: *Proceedings of the National Workshop on Planning and Management of Water Resource Systems: Risk and Reliability*. Dandy, G. & Simpson, A. R. (eds). Australian Water Resources Council Series No. 17, AGPS, Canberra, pp. 245–262.

ANAO (2010). *ANAO Audit Report No. 21 2009–10: Administration of the Water Smart Australia Program*. Retrieved from https://www.anao.gov.au/sites/g/files/net616/f/ANAO_Report_2009-2010_21.pdf.

ANAO (2012). *Auditor-General Audit Report No. 38 2011–12 Performance Audit: Administration of the Private Irrigation Infrastructure Operators Program in New South Wales*. Retrieved from https://www.anao.gov.au/sites/g/files/net616/f/201112AuditReportNo38.pdf (accessed 2 June 2016).

Arthington, A. H. & Pusey, B. J. (2003). Flow restoration and protection in Australian rivers. *River Research and Applications 19*, 377–395.

Ashton, D. & Oliver, M. (2015). *Irrigated Agriculture in the Murray–Darling Basin: an Economic Survey of Irrigators, 2012–13 to 2014–15. ABARES Research Report 15.13*. Retrieved from Canberra: http://data.daff.gov.au/data/warehouse/9aas/2015/IrrigatedAgMDB/IrrigatedAgMDBEconSrvy_v1.0.0.pdf (accessed 26 July 2016).

Australian Government (2008). *COAG Water Reform Work Program*. Retrieved from http://www.environment.gov.au/topics/water/australian-government-water-leadership/council-australian-governments-water-reform/coag (accessed 7 April 2016).

Australian Government (2016). *About Commonwealth Environmental Water.* Retrieved from http://www.environment.gov.au/water/cewo/about-commonwealth-environmental-water (accessed 30 May 2016).

Ballard, C., Garland, N. & Foreman, J. (2014). Management of drought in the southern Murray-Darling Basin, Australia, from 1996/1997 to the present. *Irrigation and Drainage 63*, 254–262.

Banks, S. A. & Docker, B. B. (2014). Delivering environmental flows in the Murray-Darling Basin (Australia) – legal and governance aspects. *Hydrological Sciences Journal 59*(3–4), 688–699.

Blackmore, D. (2002). Protecting the future. In: *Uncharted Waters.* Connell, D. (ed.). MDBC, Canberra, pp. 1–8.

Bond, N. R., Lake, P. S. & Arthington, A. H. (2008). The impacts of drought on freshwater ecosystems: an Australian perspective. *Hydrobiologia 600*(1), 3–16.

Botterill, L. C. (2003). Uncertain climate: the recent history of drought policy in Australia. *Australian Journal of Politics and History 49*(1), 61–74.

Botterill, L. C. & Cockfield, G. (eds) (2013). *Drought, Risk Management, and Policy: Decision Making Under Uncertainty.* CRC Press, Boca Raton, Florida.

Brett, D. (1990). *Restoring the Murray-Darling Basin, Ecos, 64, Winter 1990*, pp. 4–9.

Briese, R., Kingsland, A. & Orr, R. (2009). *Swimming in New Waters: Recent Reforms to Australian Water Law.* AGS Legal Briefing 21 July 2009. Retrieved from http://www.ags.gov.au/publications/legal-briefing/br90.htm (accessed 2 October 2015).

Bureau of Meteorology (2016a). What is drought? Retrieved from http://www.bom.gov.au/climate/drought/-tabs=About-drought (accessed 29 April 2016).

Bureau of Meteorology (2016b). *Improving Water Information Programme Progress Report: Advances in Water Information made by the Bureau of Meteorology in 2015.* BOM, Melbourne.

Chiew, F. H. S., Young, W. J., Cai, W. & Teng, J. (2011). Current drought and future hydroclimate projections in southeast Australia and implications for water resources management. *Stochastic Environmental Research and Risk Assessment 25*, 601–612.

Christensen, C. M. (2013). *The Innovator's Dilemma: When New Technologies Cause Great Firms to Fail.* Harvard Business Review Press, Harvard, USA.

Clark, S. (1971). The River Murray question: part II – federation, agreement and future alternatives. *Melbourne University Law Review 8*, 215–253.

COAG (1994). *Council of Australian Governments' Communique, 25 February 1994, Attachment A – Water Resource Policy.* Retrieved from http://www.coag.gov.au/coag_meeting_outcomes/1994-02-25/docs/attachment-_a.cfm (accessed 20 October 2015).

COAG (2001). *Council of Australian Governments' Meeting 8 June 2001 Canberra Communique.* Retrieved from http://www.pc.gov.au/inquiries/completed/gas/documents/coag080601.pdf (accessed 20 October 2015).

COAG (2003). *Council of Australian Governments' Communique 29 August 2003 Attachment A – Proposed National Water Initiative.*

COAG (2004). *Intergovernmental Agreement on the National Water Initiative.* Canberra. Retrieved from http://archive.coag.gov.au/coag_meeting_outcomes/2004-06-25/index.cfm–nwi (accessed 11 March 2015).

COAG (2007). COAG meeting outcomes. Retrieved from http://webarchive.nla.gov.au/gov/20070829161008/http://coag.gov.au/meetings/archive.htm (accessed 25 May 2016).

COAG (various). *Council of Australian Governments' Communique 2000–2006* [Press release].

COAG Standing Council on Environment and Water (2013). *Next Steps in National Water Reform: Preparation for the Future: A report to the Council of Australian Governments.* Retrieved from https://www.coag.gov.au/sites/default/files/NextStepsinNationalWaterReform.pdf (accessed 7 April 2016).

COAG Working Group on Water Resource Policy (1994). *Report of the Working Group on Water Resource Policy to the Council of Australian Government*, mimeo.

Commonwealth of Australia (2008). *Water Amendment Bill 2008 Explanatory Memorandum.* Retrieved from http://www.austlii.edu.au/au/legis/cth/bill_em/watmbb2012434/memo_0.html.

Commonwealth of Australia (2012). *Water Act 2007 Basin Plan 2012. Extracted from the Federal Register of Legislative Instruments on 28 November 2012 - F2012L02240.* Retrieved from https://www.comlaw.gov.au/Details/F2012L02240.

Connell, D. (ed.) (2002). *Uncharted Waters.* Murray-Darling Basin Commission, Canberra.

Connell, D. (2007). *Water Politics in the Murray-Darling Basin.* Federation Press, Sydney.

Connell, D., Dovers, S. & Grafton, R. Q. (2005). A critical analysis of the National Water Initiative. *Australasian Journal of Natural Resources Law and Policy 10*(1), 81–107.

Connor, J. D. & Kaczan, D. (2013). Principles for economically efficient and environmentally sustainable water markets: The Australian experience. In: *Drought in Arid and Semi-Arid Regions: A Multi-Disciplinary and Cross-Country Perspective.* Schwabe, K., Albiac, J., Connor, J., Hassan, R. & González, L. M. (eds). Springer Science, Dordrecht, Chapter 20.

Cox, W. & Baxter, P. (1996). *Setting the Cap – Report of the Independent Audit Group November 1996.* Retrieved from http://www.mdba.gov.au/sites/default/files/archived/cap/SETTING_THE_CAP.pdf.

Crase, L. (ed.) (2011). *Water Policy in Australia: The Impact of Change and Uncertainty.* RFP Press, Washington, DC.

CSIRO (2008). *Water availability in the Murray-Darling Basin. A report to the Australian Government from the CSIRO Murray-Darling Basin Sustainable Yields Project.* Retrieved from Australia: https://publications.csiro.au/rpr/download?pid=legacy:530&dsid=DS1 (accessed 14 October 2015).

CSIRO (2013). *The Sustainable Yields Projects.* Retrieved from http://www.clw.csiro.au/publications/waterforahealthycountry/sustainable-yields.html (accessed 27 May 2016).

Department of Agriculture and Water Resources (2016). *Drought Policy.* Retrieved from http://www.agriculture.gov.au/ag-farm-food/drought/drought-policy (accessed 13 May 2016).

Department of Environment and Primary Industries (2016). *Victorian Water Register: Carryover.* Retrieved from http://water-register.vic.gov.au/water-entitlements/carryover (accessed 18 May 2016).

Department of the Environment (2010). *National Water Initiative pricing principles.* Retrieved from https://www.environment.gov.au/water/australian-government-water-leadership/nwi/pricing-principles (accessed 2 June 2016).

Docker, B. & Robinson, I. (2013). Environmental water management in Australia: experience from the Murray-Darling Basin. *International Journal of Water Resources Development 29*(4), 164–177.

Doolan, J., Merrick, N. & Podger, G. (2015). *Report of the Independent Audit Group for Salinity 2013–14.* MDBA, Canberra.

Dreverman, D. (2013). Responding to Extreme Drought in the Murray-Darling Basin, Australia. In: *Drought in Arid and Semi-Arid Regions: A Multi-Disciplinary and Cross-Country Perspective.* Schwabe, K., Albiac, J., Connor, J., Hassan, R. & González, L. M. (eds). Springer, Dordrecht, Chapter 24.

Emmery, M. (1994). *Ecologically Sustainable Development Processes in Australia 1990–92.* Commonwealth Department of the Parliamentary Library, Australia.

Gale, M., Edwards, M., Wilson, L. & Greig, A. (2014). The boomerang effect: a case study of the Murray-Darling Basin Plan. *Australian Journal of Public Administration 73*(2), 153–163.

Grafton, R. Q. (2010). How to increase the cost-effectiveness of water reform and environmental flows in the Murray-Darling basin. Agenda : *A Journal of Policy Analysis and Reform 17*(2), 17–40. Retrieved from http://search.proquest.com.virtual.anu.edu.au/docview/817303303?accountid=8330.

Grafton, R. Q. & Horne, J. (2014). Water markets in the Murray-Darling Basin. *Agricultural Water Management.* doi:10.1016/j.agwat.2013.12.001.

Grafton, R. Q., Horne, J. & Wheeler, S. A. (2015). On the marketisation of water: evidence from the Murray-Darling Basin, Australia. *Water Resources Management.* doi:10.1007/s11269-015-1199-0.

Guest, C. R. (forthcoming). *Sharing the Water: One Hundred Years of River Murray Politics.* MDBA, Canberra.

Hart, B. (2015a). The Australian Murray-Darling Basin Plan: challenges in its implementation (Part 1). *International Journal of Water Resources Development.* doi:10.1080/07900627.2015.1083847.

Hart, B. T. (2015b). The Australian Murray–Darling Basin Plan: challenges in its implementation (Part 2). *International Journal of Water Resources Development.* doi:10.1080/07900627.2015.1084494.

Horne, J. (2007). *The State of Federalism – Is it working? Paper presented at the Constitutional Law Forum 2007*, Canberra.

Horne, J. (2013a). Australian water policy in a climate change context: some reflections. *International Journal of Water Resources Development 29*(2), 137–151.

Horne, J. (2013b). Economic approaches to water management in Australia. *International Journal of Water Resources Development 29*(4), 526–543.

Horne, J. (2014). The 2012 Murray-Darling Basin Plan – issues to watch. *International Journal of Water Resources Development 30*(1), 152–163.

Horne, J. (2015a). Water information as a tool to enhance sustainable water management – the Australian experience. *Water 7*, 2161–2183.

Horne, J. (2015b). As drought looms, the Murray-Darling is in much healthier shape – just don't get complacent. The Conversation. Retrieved from https://theconversation.com/as-drought-looms-the-murray-darling-is-in-much-healthier-shape-just-dont-get-complacent-50063 (accessed 1 June 2016).

Horne, J. (2016a). Achieving 'Sustainable Environmental Outcomes' Through Water Reform: the Case of the Murray Darling Basin (Unpublished).

Horne, J. (2016b). Policy issues confronting Australian urban water reuse. *International Journal of Water Resources Development 32*(4), 573–589.

Howard, J. (2006). Transcript of the Prime Minister the Hon John Howard MP Joint Press Conference with New South Wales Premier Morris Iemma. In: *Victorian Premier Steve Bracks, South Australian Premier Mike Rann and Acting Queensland Premier Anna Bligh*, Parliament House, Canberra. 7 November 2006. Retrieved from http://pandora.nla.gov.au/pan/10052/20061221-0000/www.pm.gov.au/news/interviews/Interview2235.html.

Howard, J. (2007a). *Address to the National Press Club, Great Hall*, Parliament House, 25 January 2007. Retrieved from http://pandora.nla.gov.au/pan/10052/20070321-0000/www.pm.gov.au/media/Speech/2007/speech2341.html.

Howard, J. (2007b). Prime Minister Media Release: Dry inflow contingency planning in the Murray-Darling Basin: fourth report [Press release]. Retrieved from https://web.archive.org/web/20081203232327/http://www.environment.gov.au/water/publications/mdb/dry-inflow-planning-sep07.html.

Hughes, N., Gibbs, C., Dahl, A., Tregeagle, D. & Sanders, O. (2013). *Storage Rights and Water Allocation Arrangements in the Murray–Darling Basin, ABARES Technical Report 13.07.* Canberra. Retrieved from http://www.agriculture.gov.au/abares/publications/display?url=http://143.188.17.20/anrdl/DAFFService/display.php?fid=pb_srwaad9abnw20131212_11a.xml (accessed 17 May 2016).

Jones, G., Hillman, T., Kingsford, R., McMahon, T., Walker, K., Arthington, A., Whittington, J. & Cartwright, S. (2002). *Independent Report of the Expert Reference Panel on Environmental Flows and Water Quality Requirements for the River Murray System – Report to the Murray-Darling Basin Ministerial Council*, February. MDBMC.

Khan, S. (2008). Managing climate risks in Australia: options for water policy and irrigation management. *Australian Journal of Experimental Agriculture 48*(3), 265–273.

Kildea, P. & Williams, G. (2010). The constitution and the management of water in Australia's rivers. *Sydney Law Review 32*, 595–616.

Kirby, M., Park, R., Connor, J., Qureshi, M. E. & Keyworth, S. (2014). Sustainable irrigation: how did irrigated agriculture in Australia's Murray–Darling Basin adapt in the Millennium Drought? *Agricultural Water Management 145*, 154–162.

Loch, A. & Adamson, D. (2015). Drought and the rebound effect: a Murray–Darling Basin example. *Natural Hazards 79*, 1429–1449.

Mallawaarachchi, T. & Foster, A. (2009). *Dealing with Irrigation Drought: the Role of Water Trading in Adapting to Water Shortages in 2007–08 in the Southern Murray-Darling Basin, ABARE Research Report 09.6 to the Department of the Environment*, Water, Heritage and the Arts, Canberra. Retrieved from http://data.daff.gov.au/data/warehouse/pe_abarebrs99001593/rr09.6_irrigation_drought_report.pdf (accessed 29 April 2016).

Manning, H. (nd). *Overview of Key Political and Policy Themes South Australia January to December 2002*. Retrieved from http://dpc.sa.gov.au/sites/default/files/pubimages/documents/proactive/Overview-Political-Themes2002.pdf (accessed 25 May 2016).

Martin, W. (2005). *Water Policy History on the River Murray*. SRI, Deniliquin.

MDBA (2011). *The Living Murray story*. MDBA, Canberra. Retrieved from http://www.mdba.gov.au/media-pubs/publications/living-murray-story/contents/brief-history-of-tlm (accessed 13 August 2015).

MDBA (2015a). *The SDL Adjustment Assessment Framework for Supply Measures. MDBA Publication No. 6/2015*. MDBA, Canberra. Retrieved from http://www.mdba.gov.au/sites/default/files/pubs/SDL-adjustment-assessment-framework.pdf (accessed 2 June 2016).

MDBA (2015b). *MDBA Annual Report 2014–15*. Retrieved from http://www.mdba.gov.au/sites/default/files/pubs/mdba-annual-report-2014-15.pdf (accessed 2 June 2016).

MDBA (2016a). *Towards a healthy, working Murray–Darling Basin – Basin Plan annual report 2014–15*. Retrieved from http://www.mdba.gov.au/publications/mdba-reports/basin-plan-annual-report-2014-15-download (accessed 3 March 2016).

MDBA (2016b). *Sustainable Diversion Limits*. Retrieved from http://www.mdba.gov.au/basin-plan-roll-out/sustainable-diversion-limits (accessed 2 June 2016).

MDBC (1995). *An Audit of Water Use in the Murray-Darling Basin. Water use and Healthy Rivers – Working towards Striking a Balance.* Retrieved from http://www.southwestnrm.org.au/sites/default/files/uploads/ihub/audit-water-use-murray-darling-basin-june-1995-1-3.pdf (accessed 9 September 2015).

MDBC (1999). *River Murray Water.* Retrieved from https://web.archive.org/web/20010303164654/http://www.mdbc.gov.au/river_murray/river_murray_water/rmw.htm (accessed 20 May 2016).

MDBC (2004). *The Cap.* MDBC, Canberra. Retrieved from http://www.mdba.gov.au/sites/default/files/archived/cap/cap_brochure_0.pdf (accessed 12 May 2016).

MDBMC (2000). *Ministerial Council – Mildura Communique 25/08/2000 Salinity and River Care, the Priorities in the Murray-Darling Basin*, NSW, Mildura.

MDBMC (2001a). *New Policy Statement – Managing Murray-Darling Basin Catchments 6/6/2001.* Retrieved from https://web.archive.org/web/20020331225942/http://mdbc.rucc.net.au/mdbc/news_room/media_releases/user_view.cfm?ItemNo=62 (accessed 31 May 2016).

MDBMC (2001b). *Basin Salinity Management Strategy 2001–2015.* MDBC, Canberra.

MDBMC (2002). *Corowa Communique. Council Meeting 31*, 12 April 2002, NSW, Corowa.

MDBMC (2003). The Living Murray Communique, Friday, May 9, 2003.

MDBMC (2006). *Murray-Darling Basin Ministerial Council Communique.* Ministerial Council Meeting 41, 29 September 2006, Sydney.

MDBMC (2015). *Basin Salinity Management 2030 (BSM2030).* MDBA, Canberra.

Milly, P. C. D., Betancourt, J., Falkenmark, M., Hirsch, R. M., Kundzewicz, Z. W., Lettenmaier, D. P. & Stouffer, R. J. (2008). Climate change – stationarity is dead: whither water management? *Science 319*(5863), 573–574. http://dx.doi.org/10.1126/science.1151915 (accessed 16 April 2016).

Morey, K., Grinlinton, M. & Hughes, N. (2015). *Australian Water Market Report 2013–14, ABARES Report to Client Prepared for the Department of Agriculture and Water Resources.* Canberra. Retrieved from http://data.daff.gov.au/data/warehouse/9aaw/2015/awmr_d9aawr20151211/awmr2013-14_20151211_v1.1.0.pdf (accessed 2 May 2016).

National Academies of Sciences Engineering and Medicine (2016). *Attribution of Extreme Weather Events in the Context of Climate Change.* doi:10.17226/21852.

National Water Commission (2011). *Water Markets in Australia: a Short History.* NWC, Canberra.

National Water Commission (2012). *Impacts of Water Trading in the Southern Murray–Darling Basin between 2006–07 and 2010–11.* NWC, Canberra.

National Water Commission (2014). *Australian Water Markets: trends and drivers 2007–08 to 2012–13*, NWC, Canberra.

Pannell, D. J. & Roberts, A. M. (2010). Australia's National Action Plan for Salinity and Water Quality: a retrospective assessment. *Australian Journal of Agricultural and Resource Economics 54*(4), 437–456.

Parliament of Australia (2015). *Senate Standing Committees on Environment and Communications. Water Amendment Bill (2015) Provisions. Submissions.* Retrieved from http://www.aph.gov.au/Parliamentary_Business/Committees/Senate/Environment_and_Communications/Water_Amendment_Bill_2015/Submissions (accessed 9 October 2015).

Pittock, J., Williams, J. & Grafton, R. Q. (2015). The Murray-Darling Basin plan fails to deal adequately with climate change. *Water 42*(6), 26–30.

Powell, J. M. (1994). Environmental degradation, 'sustainability' and the Murray-Darling Basin. *People and Place 2*(3), 6–13.

Productivity Commission (2010). *Market Mechanisms for Recovering Water in the Murray-Darling Basin, Final Report*, March. Retrieved from http://www.pc.gov.au/inquiries/completed/murray-darling-water-recovery/report/water-recovery-report.pdf (accessed 16 May 2016).

Productivity Commission (2011). *Australia's Urban Water Sector: Final Inquiry Report.* Productivity Commission, Canberra.

Quiggin, J. (2000). *Environmental Economics and the Murray-Darling river system (version of 24 August 2000).* Article retrieved from http://www.uq.edu.au/economics/johnquiggin/JournalArticles01/Murray01.pdf (accessed 8 September 2015).

Quiggin, J. (2007). *Issues in Australian Water Policy Australian Chief Executive: Official Journal of the Committee for Economic Development of Australia*, Feb 2007, 38–47.

Quiggin, J., Adamson, D., Chambers, S. & Schrobback, P. (2010). Climate change, uncertainty, and adaptation: the case of irrigated agriculture in the Murray-Darling Basin in Australia. *Canadian Journal of Agricultural Economics/Revue canadienne d'agroeconomie 58*(4), 531–554.

Roy Morgan Research (2005). *Nine in Ten Australian Opinion Leaders Think Government Not Doing Enough About Water Conservation.* Retrieved from http://www.roymorgan.com/findings/finding-3916-201302270505 (accessed 17 May 2016).

Scanlon, J. (2002). *From Taking, to Capping to Returning: The Story of Restoring Environment Flows in the Murray Darling Basin in Australia.* Paper presented at the Stockholm International Water Institute Seminar 2002 Towards Upstream/Downstream Hydrosolidarity, Stockholm. http://cmsdata.iucn.org/downloads/mdbc.pdf (accessed 21 August 2015).

Schirmer, J., Peel, D. & Mylek, M. (2015). *People and Place in Australia: The 2014 Regional Wellbeing Survey Part 2: Farmers and Agriculture.* Retrieved from https://www.canberra.edu.au/research/faculty-research-centres/ceraph/regional-wellbeing/survey-results/2014-survey-results/Farmers-and-agriculture_LR.pdf (accessed 6 May 2016).

Senior Officers Group (2006). *Murray-Darling Basin Dry Inflow Contingency Planning Report.* Canberra.

Senior Officers Group (2007–9). *Murray-Darling Basin Dry Inflow Contingency Planning (various).* Retrieved from https://web.archive.org/web/20090225104944/http://environment.gov.au/water/publications/mdb/index.html; https://web.archive.org/web/20081203232327/http://www.environment.gov.au/water/publications/mdb/dry-inflow-planning-sep07.html (accessed 6 October 2016).

Stevenson, B. (1996). *Cooperative Catchment Management: The Murray-Darling Basin Agreement Bill 1996.* Queensland Parliamentary Library, Brisbane.

The Centre for International Economics (2015). *Economic Analysis of the Bureau of Meteorology's Water Information* (unpublished).

van Dijk, A. I. J. M., Beck, H. E., Crosbie, R. S., de Jeu, R. A. M., Liu, Y. Y., Podger, G. M., Timbal, B. & Viney, N. R. (2013). The Millennium Drought in southeast Australia (2001–2009): natural and human causes and implications for water resources, ecosystems, economy, and society. *Water Resources Research 49.* doi:10.1002/wrcr.20123.

Vertessy, R. A. (2013). Water information services for Australians. *Australian Journal of Water Resources 16*(2), 91–106.

Victorian Auditor General (2008). *Piping the System: Incorporating the Wimmera-Mallee Pipeline and the Goldfields Superpipe.* Melbourne. Retrieved from http://www.audit.vic.gov.au/publications/2008/20080528-Wimmera-Mallee-and-Goldfields-Pipeline-Project.pdf (accessed 1 June 2016).

Wentworth Group of Concerned Scientists (2012). *Statement on the 2011 Draft Murray-Darling Basin Plan.* Retrieved from http://wentworthgroup.org/wp-content/uploads/2013/10/Statement-on-the-2011-Draft-Basin-Plan.pdf (accessed 27 May 2016).

Wheeler, S. A. (2014). Insights, lessons and benefits from improved regional water security and integration in Australia. *Water Resources and Economics 8*, 57–78. http://dx.doi.org/10.1016/j.wre.2014.05.006 (accessed 31 May 2016).

Williams, B. (1995). *Murray-Darling Basin: Ecologically Sustainable Irrigation? Research Paper No. 30 1994/95.* Canberra. Retrieved from https://www.aph.gov.au/binaries/library/pubs/rp/1994-95/95rp30.pdf (accessed 2 June 2016).

Wittwer, G. (2011). Confusing policy and catastrophe: buybacks and drought in the Murray–Darling Basin. *Economic Papers 30*(3), 289–295.

Wong, P. (2008). Water for the future. Speech to the 4th Annual Australian Water Summit, Sydney Convention and Exhibition Centre, 29–30 April 2008. Retrieved from http://www.climatechange.gov.au/,/media/Files/minister/previousminister/wong/2008/major-speeches/April/sp20080429.pdf (accessed 4 June 2016).

WSAA & NWC (2009). *National Performance Report 2007–08: Urban water utilities*, March 2009. NWC, Canberra.

Young, M. (2013). Trading into and out of trouble: Australian water allocation and trading experience. In: *Water Trading and Global Water Scarcity: International Experiences.* Maestu, J. (ed.). RFF Press, New York, NY.

Young, W. J., Bond, N., Brookes, J., Gawne, B. & Jones, G. J. (2011). *Science Review of the Estimation of an Environmentally Sustainable Level of Take for the Murray–Darling Basin. A Report to the Murray–Darling Basin Authority from the CSIRO Water for a Healthy Country Flagship.* Australia. Retrieved from http://www.mdba.gov.au/sites/default/files/archived/proposed/CSIRO_ESLT_Science_Review.pdf (accessed 2 June 2016).

Facing the challenge of extreme climate: the case of Metropolitan São Paulo

B. Braga[a] and J. Kelman[b]

[a]*Corresponding author. Secretary of State for Sanitation and Water Resources of the State of São Paulo, São Paulo, Brazil. E-mail: benbraga@hotmail.com*

[b]*Director-President of SABESP – Water and Sanitation Company of the State of São Paulo, São Paulo, Brazil*

Abstract

Climate variability has always posed important challenges to water resources planners and engineers. Recent hydrologic data from different parts of the world show that the use of traditional stationary series statistical methods to assess risk and uncertainty may not produce reliable estimates. In particular, the 2014–2015 drought in Metropolitan São Paulo that was followed by major flooding in 2016 shows the intensification of climatic extremes in this part of Brazil. The probability of this drought event is only 0.004 when estimated with data available from 83 years of record, until 2013. Was the drought of 2014–2015 a statistical outlier or should one consider it as an event that can occur again along the planning horizon? The paper describes the hydrologic conditions that resulted in the most severe drought ever recorded and the structural and non-structural initiatives taken by water authorities to avoid social chaos in the Metropolitan Region of São Paulo (21 million inhabitants). Furthermore, it discusses how the drought affected the existing water resources development plan, especially the anticipation of investments in water security. Because financial resources are limited, other investments, equally important but less urgent, are being delayed.

Keywords: Cantareira system; Climate change; Demand management; Drought; Infrastructure; Resilience; SABESP; São Paulo Metropolitan Region; Water crisis; Water security

Introduction

Water availability, both in quantitative and qualitative terms, is today one of the major concerns of both the business community and government authorities worldwide. Water is a fundamental element in achieving food, energy and health security. Water is the engine for socio-economic development. In September 2015 the United Nations General Assembly passed the resolution 70/1 – Transforming our world: the 2030 Agenda for Sustainable Development (United Nations, 2015) – in which a standalone target on water was approved. A recent report issued by the US National Security Council (2012)

doi: 10.2166/wp.2016.113

claims that 'water may become a more significant source of contention than energy or minerals out to 2030 at both the intrastate and interstate levels'.

Given the importance of water in the global scene and understanding that people's concerns about climate are not about climate per se, but the consequences of excess or lack of water resources due to climate change, we must prepare ourselves to increase our resilience to climate variability. By and large, all the impacts of climate variability are manifested through, by and with water, whether you talk about impacts on ecosystems or hydrologic extremes.

Hydrologic extremes have recently manifested themselves in different parts of the planet. In the United States, the states of California, Texas and Arizona have been facing severe drought since 2013; Australia has gone through the millennium drought from 2000–2010; Singapore has faced unexpected climate during 2013, to list some of many situations around the world. In all these cases, we have noticed the importance of water infrastructure and demand management in facing the challenge.

The State of São Paulo in SE Brazil experienced its most severe water crisis in the historical hydrologic record of 125 years of data during 2014–2015. Should this extreme event be considered as an event that can occur again in the coming years or should it be disregarded in planning as an unlikely outlier?

The paper describes the structural and non-structural initiatives taken by water authorities to avoid social chaos and how the drought affected planning, especially the anticipation of investments in water security. Because financial resources are limited, other investments, equally important but less urgent, are being delayed.

The physical and hydrologic context

Figure 1 shows the 12 hydrographic regions of Brazil and their corresponding freshwater availability, in terms of average annual yield. Brazil has 12% of the world's freshwater resources but, as can be appreciated in the figure, most of this water is concentrated (70%) in the Amazon basin that hosts 7% of the Brazilian population. SE Brazil, which is the focus of this paper, is located in the Parana Basin.

There are three States in SE Brazil: São Paulo, Minas Gerais and Rio de Janeiro. The population of São Paulo is 43 million, which corresponds to 21% of the Brazilian population[1], and the gross domestic product (GDP) is equivalent to 32% of that of Brazil[2].

A recent report from the National Water Agency of Brazil – ANA (2015) indicates that in 2014 SE Brazil experienced anomalous precipitation patterns. Figure 2 shows the rainfall amounts and the corresponding return periods during the rainy season (January, February and March) for years 2012–2014. Warmer colors represent drier climate (red means a return interval of the seasonal precipitation larger than 100 years).

Figure 3 shows the average annual rainfall distribution in the State for the period 1979–2013.

Figure 4 shows the contrast with what happened during 2014. During the drought, most of the northern and eastern parts of the State experienced annual rainfall below 900 mm, while the average annual rainfall is above 1,200 mm. This situation led the Water Resources Department of the State of São Paulo

---

[1] http://www.bibliotecavirtual.sp.gov.br/temas/sao-paulo/sao-paulo-populacao-do-estado.php.

[2] https://pt.wikipedia.org/wiki/Lista_de_unidades_federativas_do_Brasil_por_participa%C3%A7%C3%A3o_no_PIB.

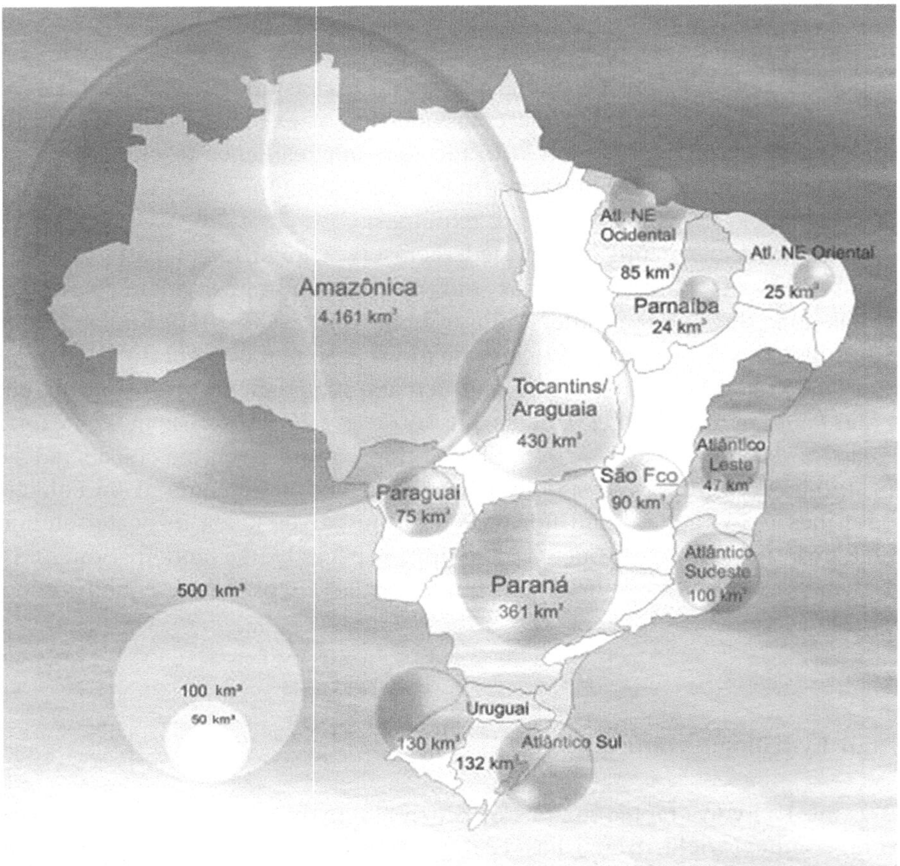

Fig. 1. Annual average runoff of the 12 Brazilian hydrographic regions.

(DAEE) to issue new water permits exclusively for the water supply of the population. Many municipalities in these areas passed municipal decrees of 'state of public calamity' by which they became entitled to receive state and federal funds and to use expedited bidding processes for emergency civil works. Figure 5 shows the water resources management regions affected and the respective municipalities involved.

The case of the São Paulo Metropolitan Region

São Paulo Metropolitan Region (SPMR) is located in the upstream reach of the Tiete River, a tributary of the Parana River (Figure 6). It is the most active economic region of Brazil (18% of Brazilian GDP), with a population of 21 million[3]. Eight water treatment plants (located at the circled areas in

[3] https://www.pdui.sp.gov.br/SPMR/https://www.pdui.sp.gov.br/SPMR/.

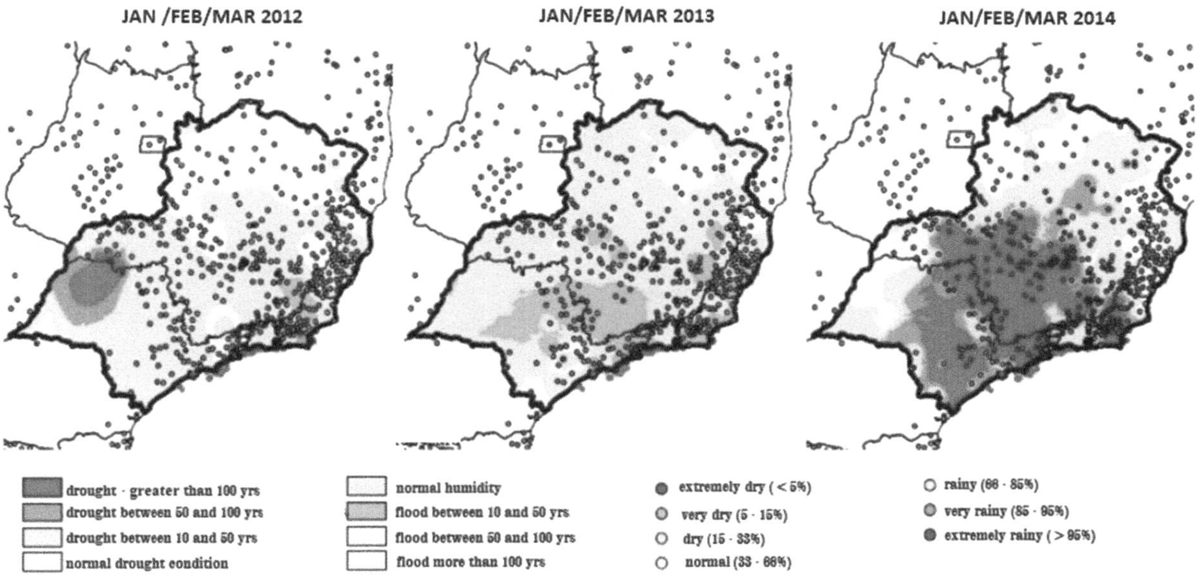

Fig. 2. Criticality of rainfall during rainy period (JFM) for 2012–2014 in SE Brazil (ANA, 2015). A color version of this figure is available in the online version of this paper: http://dx.doi.org/10.2166/wp.2016.113.

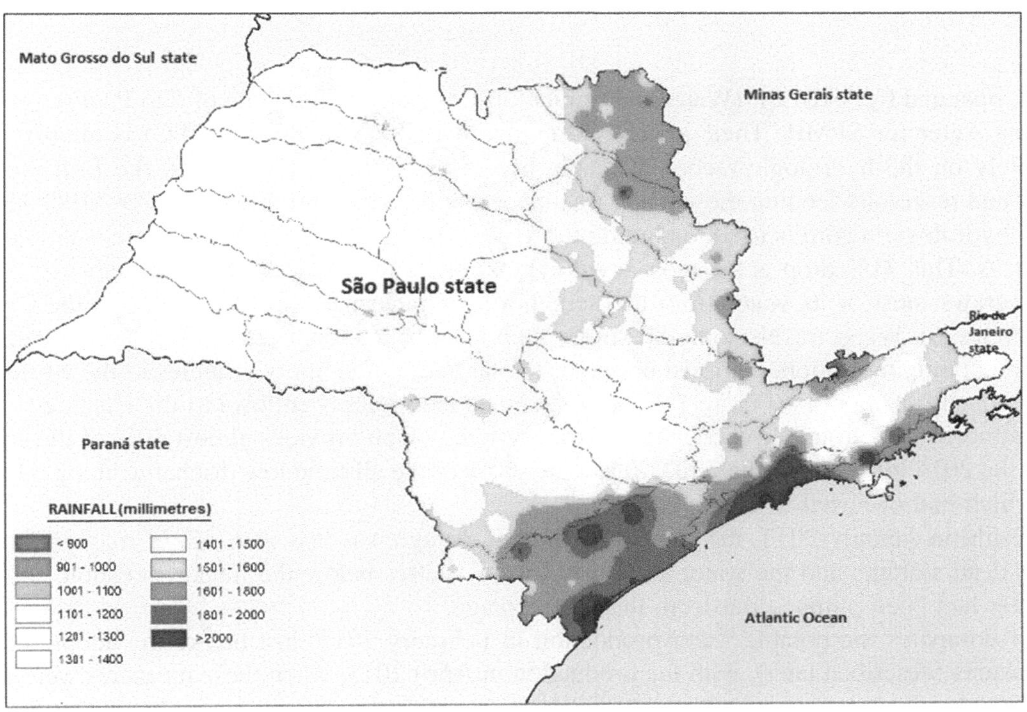

Fig. 3. Average annual precipitation in the State of São Paulo (1979–2013).

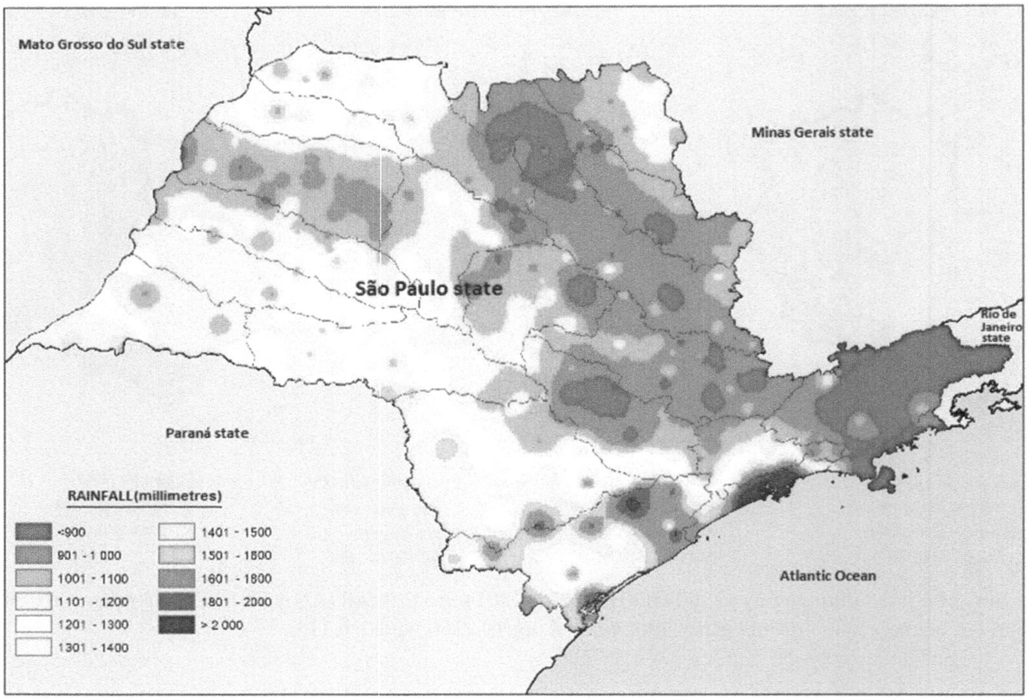

Fig. 4. Annual precipitation in the State of São Paulo in 2014.

Figure 6), operated by SABESP (Water and Sanitation Company of the State of São Paulo), provide all the potable water for SPMR. Their joint Maximum Production Capability (MPC) is almost 73 m^3/s. Based solely on the hydrologic record available before the 2014–2015 drought, the firm yield from the rivers and reservoirs feeding these water treatment plants would roughly match the MPC. However, if the full hydrologic record is used, including the years 2014–2015, the firm water yield shrinks to less than 52 m^3/s. This 31% drop is the most vivid expression of the severity of the drought.

SPMR draws most of its water from the neighboring Piracicaba river basin through the Cantareira system (Figure 6). Water travels some 80 km through a set of interconnected reservoirs, tunnels, channels and one pumping station, comprising one of the largest water supply schemes in the world, with a nominal capacity of 33 m^3/s (Figure 7). Figure 8 shows the monthly inflows to the Cantareira system. The magnitude of the drought experienced by this system, which provides almost 50% of the supply, is such that the 2014 inflow to Cantareira was roughly 50% of the all-time low discharge along 83 years of record, which had occurred in 1953.

As a result, in January 2015, the Cantareira system storage was less than 5% of maximum storage, including dead storage, and the water level was several metres below the intake elevation. Since May 2014, water had been pumped out from the dead storage.

Table 1 compares the potable water production in February 2014, just before the adoption of anti-crisis measures (described later), with the production in April 2015, when these measures were in force.

The water supply system operated by SABESP serves practically the entire population that lives in the formally urbanized area of SPMR, where one can find meters in virtually all households. However, this

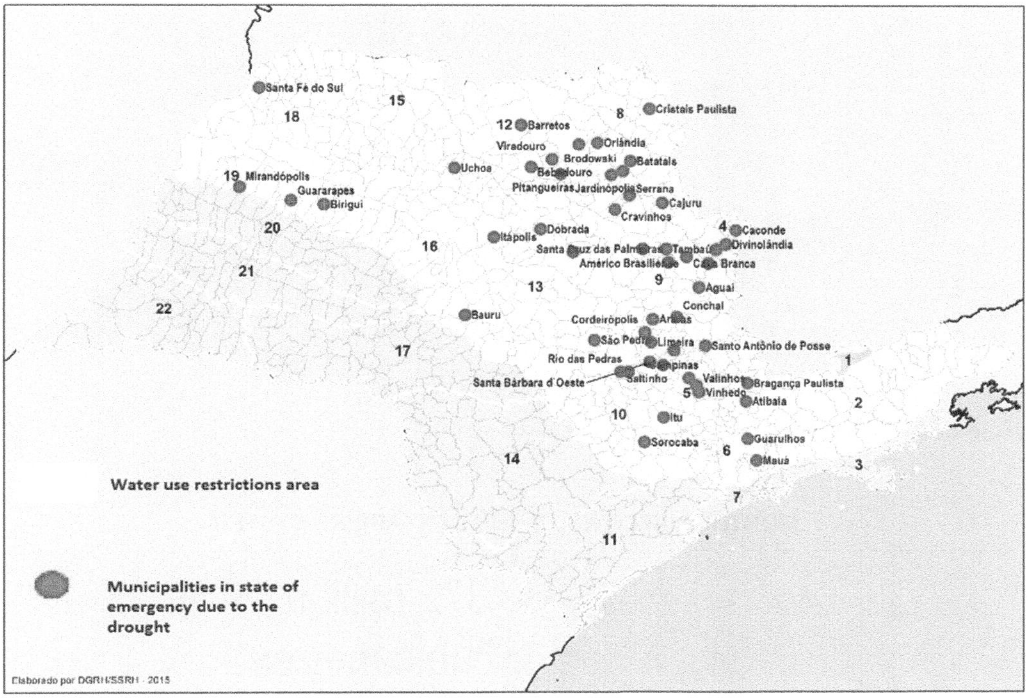

Fig. 5. Water resources management units and municipalities affected by the 2014 drought.

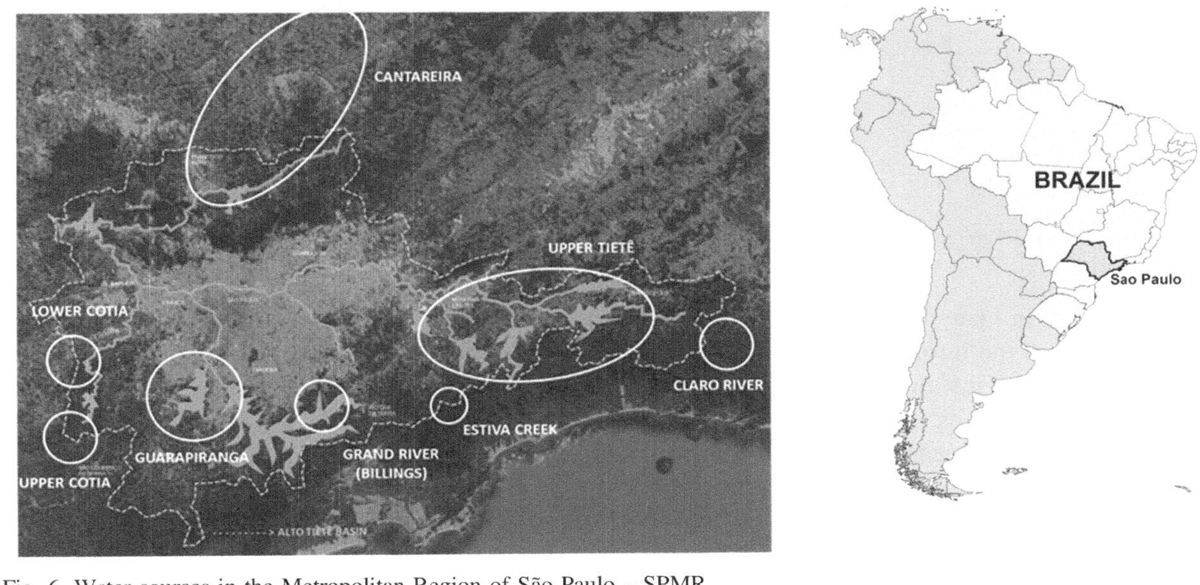

Fig. 6. Water sources in the Metropolitan Region of São Paulo – SPMR.

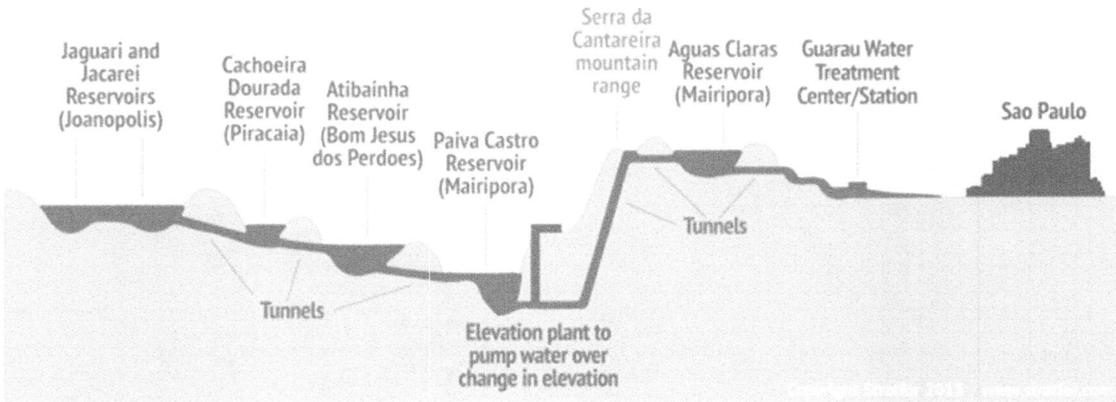

Fig. 7. Cross-sectional view of the Cantareira interbasin transfer system.

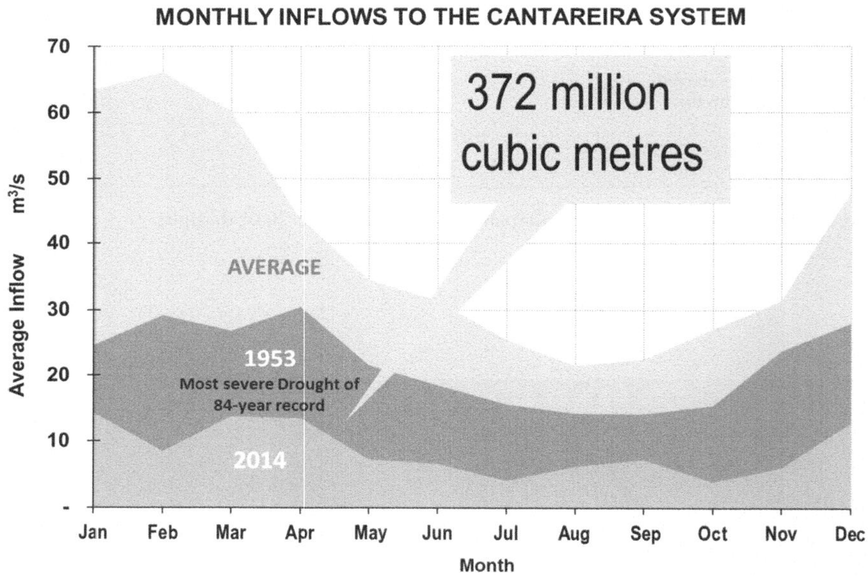

Fig. 8. Monthly hydrographs of the inflows to the Cantareira system.

is not the case in many of the informal settlements, where roughly 10% of the population lives[4]. Most people living in the informal settlements have illegal and unpaid access to potable water through precarious and wasteful distribution systems formed by a bundle of small-diameter plastic tubes connected to the mains (Figure 9).

Several barriers impede the water utility from entering these irregular settlements to provide regular services. Some relate to judicial disputes about the land ownership (people have invaded private

---

[4] http://www.ibge.gov.br/home/estatistica/populacao/estimativa2015/default.shtm.

Table 1. Potable water production before and during the water crisis (m^3/s).

SYSTEM	FEB/14	APR/15	DIFF
Cantareira	31.77	13.48	−18.29
Guarapiranga	13.77	15.05	+1.28
Alto Tiete UPPER TIETE	14.97	12.25	−2.72
Rio Grande GRANDE RIVER	4.94	5.09	+0.15
Rio Claro CLARO RIVER	3.83	3.87	+0.04
Alto Cotia UPPER COTIA	1.16	0.76	−0.40
Baixo Cotia LOWER COTIA	0.88	1.01	+0.13
Ribeirão Estiva ESTIVA CREEK	0.096	0.086	−0.010
TOTAL SPMR	**71.42**	**51.60**	**−19.82**

Fig. 9. Illegal connections in informal settlements.

property), others relate to environmental restrictions (people have invaded protected areas), and still others – the majority – relate to the impossibility of installing water supply and sewage collection systems in neighborhoods where there are no streets to bury the pipes. This unorganized urbanization process began in the 1950s, when only 30% of the Brazilian population lived in the cities, and has grown worse, especially in the early 1970s with the intensification of the industrialization process.

In a few years, the urban population swelled to 70% of the national population, but the infrastructure growth was much slower, resulting in several problems. Among them, a large quantity of untreated

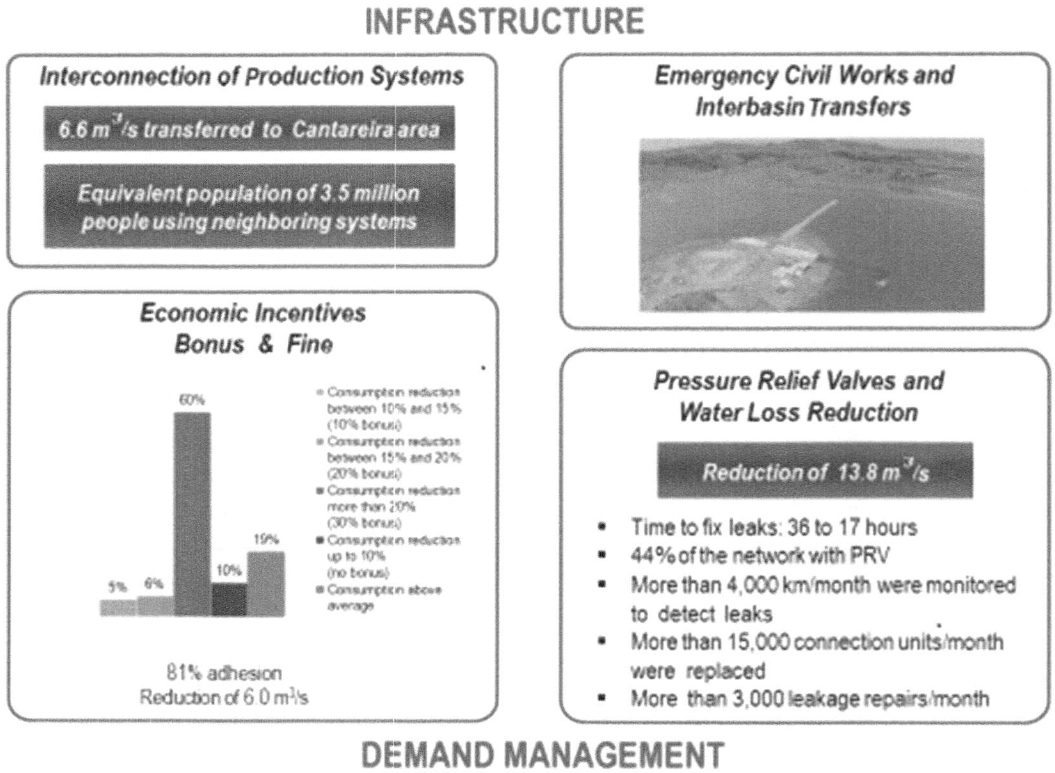

Fig. 10. Strategy to fight the drought of 2014 in Metropolitan São Paulo (PRV = pressure relief valve).

sewage is dumped into the local water bodies, impelling the search for new water supply sources far away from the demand center (Braga *et al.*, 2006). In spite of difficulties, practically all people living in Metropolitan São Paulo have access to potable water in their households, either through formal or informal connections.

Facing the challenge of the 2014–2015 drought

The strategy adopted by the State Government and SABESP to face a potential social disaster included the implementation of structural measures (building of emergency infrastructure) and non-structural measures (demand management via economic instruments and raising public awareness) (Figure 10).

Infrastructure development

Emergency civil works

When the water level was approaching the minimum elevation necessary for free flow through the intake of the upstream Jaguari Reservoir (Figure 11), SABESP installed a set of floating pumps and

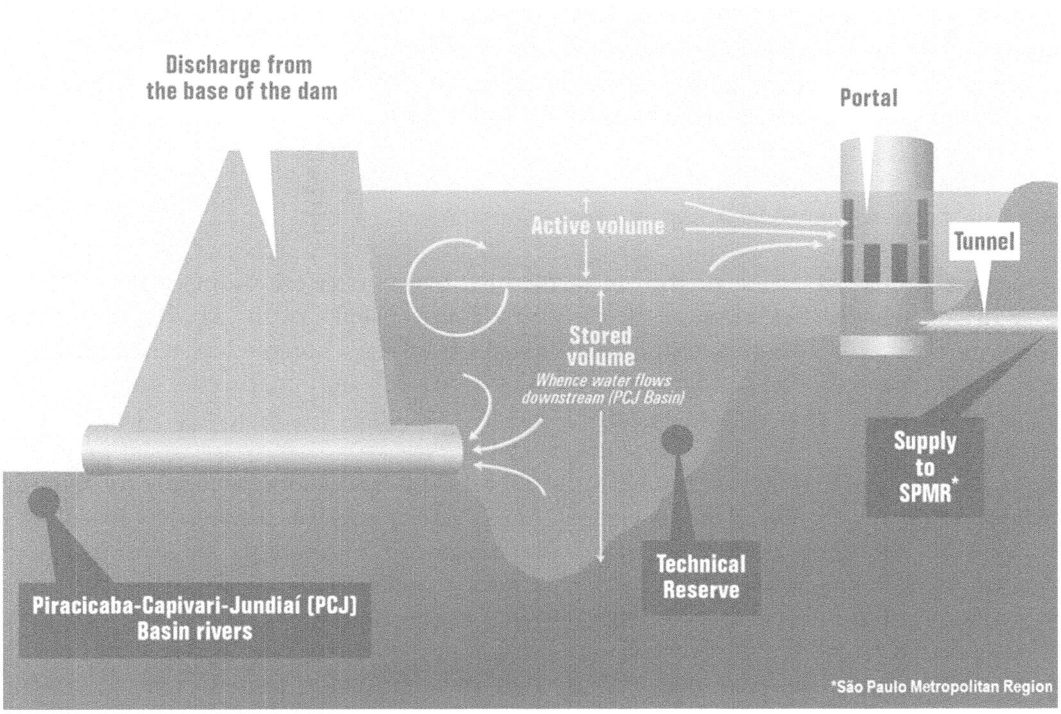

Fig. 11. Tapping the dead storage of Jaguari Reservoir.

Fig. 12. Floating pumps and cofferdam to tap dead storage.

built channels and cofferdams in order to pump up water from the dead storage to a provisional upper pond built adjacent to the intake (Figure 12).

This simple engineering solution stirred a public debate, fueled by the media, about alleged health hazards for the people who would drink water from the 'dead' storage. In fact, this was a false problem because a bottom gate had been, for many years, continuously releasing water from the reservoir to downstream in order to ensure the minimum environmental flow along the natural waterway. Nevertheless, an intense communication campaign was launched to explain that there was no risk to health, including the change of the name 'dead storage' to 'technical reserve'.

Interbasin transfers and interconnection of production systems

Many construction works were performed in the distribution network during 2014–2015 in order to increase the operational flexibility of the eight producing systems of the SPMR. In this way, sectors that before the crisis were only supplied with water from the Cantareira system began to receive water from other better-stocked producing systems. Some 3.5 million people stopped consuming water from Cantareira, and instead received water from the Guarapiranga, Upper Tiete and Grande river systems.

Since April 2015, the Grande river system has also become a new source of supply for some 250,000 people who were previously served by the Guarapiranga system. This was achieved by installing 2.1 km of water mains to carry the water to the southern part of the city. The surplus water from Guarapiranga was redirected to areas previously served by the Cantareira system. At the end of May 2015, it was the Claro river system's turn to boost its contribution to the area previously attended by the Cantareira, with the construction of a new link between two water mains in the eastern part of São Paulo.

Pumping stations and mains that had been decommissioned due to excessive leakage were refurbished. In the case of the Franca Pinto pumping station and main, this required the injection of high-density polyethylene piping into 550 m of the 6,600 m water main. This allowed an additional supply from Guarapiranga to help the sector served by the Cantareira system.

The steps taken during the 2-year period 2014–2015 to increase the operational flexibility of water production and distribution relieved the water sources most affected by the drought. The overall result was the increase of the capacity to transfer flows between the production systems by 6.6 m^3/s. This meant that 3.5 million people who previously received water exclusively from the Cantareira started to also receive water from other systems (Figure 13).

An important initiative undertaken in 2015 was the construction of the pipelines to interlink different bulk water sources. The Grande river and the Upper Tiete river systems were interconnected by two parallel water mains made of HDPE with a flow capacity of 4 m^3/s over a distance of 10.5 km (Figure 14). This construction work also stirred much public debate, mostly fueled by the press, and a litigation presently is in course in the judiciary. Some feared that this arrangement would jeopardize the quality of the drinking water. Others were concerned with the impacts on the environment. Still others used the debate as a political weapon to blame the government for alleged poor planning. It was necessary to launch a difficult battle at the public-opinion front.

Once the links were in place for redistributing areas of the city from one production system to another, the next step was to increase the water treatment capacity and to build new water storage within the distribution system. A module of ultrafiltration membranes was installed at the Guarapiranga, increasing its capacity production from 14 m^3/s to 16 m^3/s. Steel storage reservoirs were installed in different sectors of the city, to supplement supply safety. These reservoirs provide a reserve of treated water and uninterrupted supplies to districts that are located far from the principal water mains. During the crisis, 22 reservoirs with a storage capacity of 147,000 m^3 were installed and are now operating.

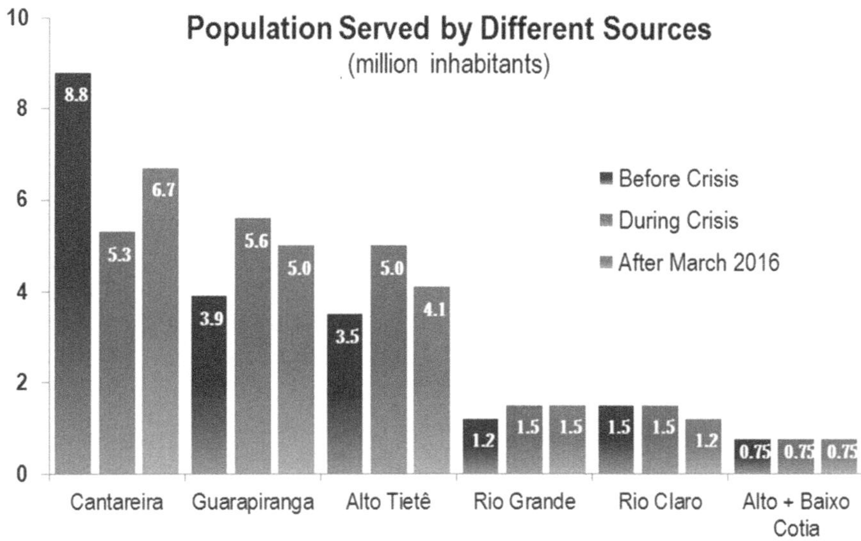

Fig. 13. Interconnection of sources and population served.

Fig. 14. Interconnection of water sources Grande river and Upper Tiete river systems.

Demand management

Pressure relief valves and water loss reduction

The SPMR's distribution system is divided into sectors for better management. Boosters pump the water to serve regions at a higher altitude and pressure relief valves (PRVs) installed at lower levels alleviate leaks and eventual damages to the pipes and other structures. PRV technology is used in the best systems worldwide and has been used in SPMR since the 1990s. PRVs are also used for remote control of pressure at times of low demand, such as the early hours of the morning, when there is more pressure in the networks. Starting in October 2014, with the worsening of the drought, the water loss due to leakages decreased by 7.6 m^3/s, which corresponds to around 42% of the total water savings during the crisis, thanks to the pressure reduction in the distribution system that was extended to time intervals beyond the early hours of the morning. This policy proved more effective than the traditional method of rotating shutdowns among sectors.

At the start of 2016, the periods of pressure reduction returned to normality, that is, they were restricted to night-time. Currently there are 1,303 PRVs operating in the SPMR. During the crisis, a large number of these PRVs were installed, representing today a coverage of 55% of the distribution network, which consists of 33,000 km of secondary pipes and 1,200 km of water mains.

Reducing water losses has been a priority of the water supply sector in the State of São Paulo for more than two decades. In 2009, SABESP introduced the Corporate Loss Reduction Program, in technical partnership with Japan International Cooperation Agency. The estimated total investment in the period 2009–2020 is greater than R$5 billion ($5 \times 10^9$) (US$1.5 billion) (1.5×10^9). The objective is to decrease the unaccounted-for water down to 18%. In 2015 the unaccounted-for water corresponded to 27.1%, which includes 17.8% due to leakages (Figure 15).

The reduction of leakages demands the installation of PRVs, substitution of mains, acoustic sweeps to locate leaks (so far, more than 87,000 repairs) and substitution of connection pipes to households (so far, more than 280,000 branches). The reduction of non-physical losses due to illegal connections or frauds, as well as inaccurate metering, demands respectively the action of anti-fraud patrols and the substitution

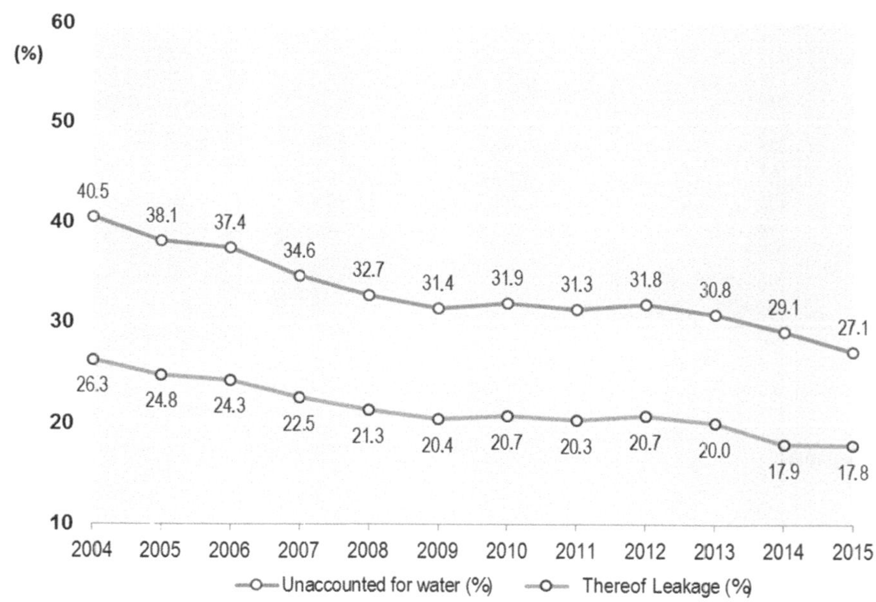

Fig. 15. Water losses in SPMR.

of meters. In 2015 alone, it was possible to identify a volume of 3.7 billion (3.7×10^9) litres illegally consumed and to replace more than 255,000 water meters.

Economic instruments – bonus and fines

The granting of bonuses to those who reduce water consumption and charging of a contingency tariff to those who waste water were strategic measures to restrain demand. Starting in February 2014, at the beginning of the crisis, a discount on the water bill (bonus) was granted to the water savers, that is, to those who decreased water consumption, compared to their own average, by at least 20%. This program got support from the majority of families. More than 80% of the consumers indeed reduced consumption, from which 49% received discounts in 2014 and 70% in 2015. The net effect was a lumped reduction of 3.5 m^3/s in the potable water production, which corresponds to around 19% of the total water savings during the crisis (Figure 16). Starting in January 2015, a penalty was imposed on those who did not understand the severity of the crisis and who increased their consumption. Some 19% of consumers were in this situation, but 8% were exempt from the charge because they did not exceed the minimum consumption level of 10 m^3 per month.

Communication campaigns

The government of the State of São Paulo and SABESP encouraged conscientious use of water through visits, lectures and the distribution of posters and leaflets at condominiums, schools, commercial establishments and homes, making people aware of the importance of economizing water. Concurrently with the growing press coverage, eight advertising campaigns were also carried out in 2014, with over 3,000 TV insertions and another 13,000 radio insertions, besides publications in newspapers and magazines, with an estimation that each São Paulo State citizen has been reached at least 40 times by water-saving messages.

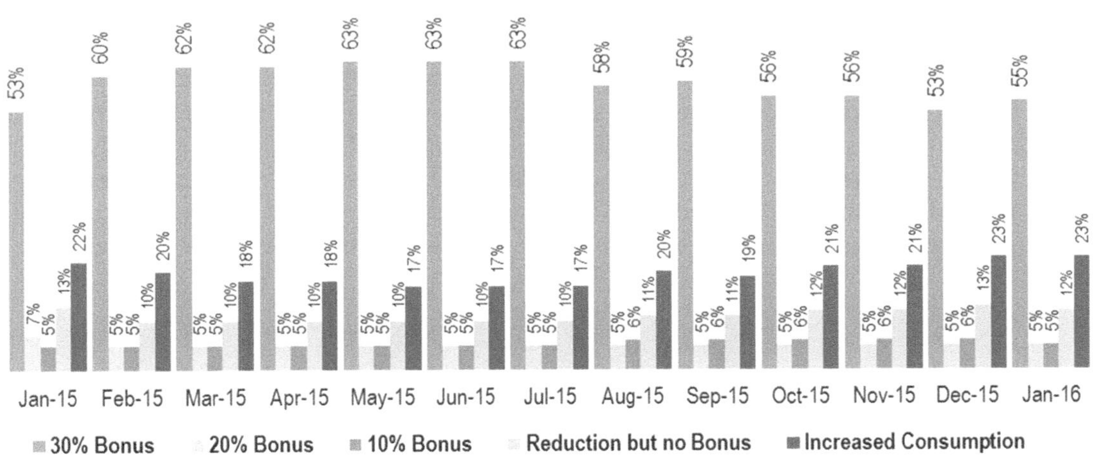

Fig. 16. Impact of economic instruments on water consumption in SPMR.

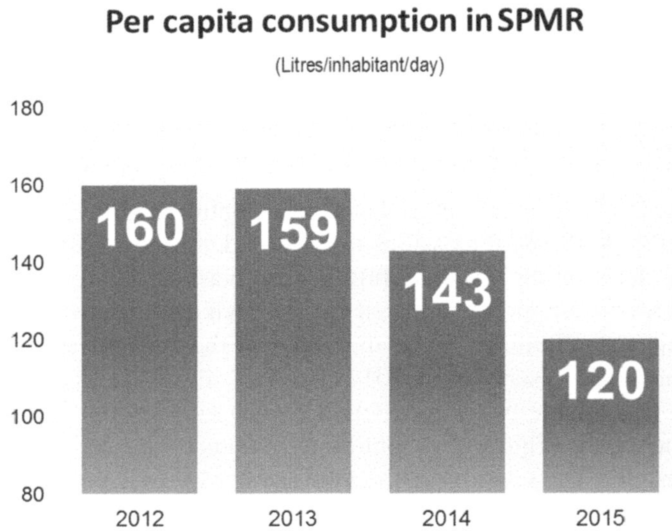

Fig. 17. Impact of non-structural measures on per capita consumption in SPMR.

In February 2015, the State government began a campaign called Each Drop Counts, reinforcing the relevance of the water rational use. More than 20 million printed materials were distributed by the 'Water Guardians' team. Campaign teams made almost 74,000 visits in 2015. Two million people were approached directly and almost 20 million indirectly. Additionally, it was necessary to launch a media campaign consisting of ads and radio publicity.

Figure 17 shows the compound effect of economic incentives and communication campaigns on the consumption pattern of the population of SPMR. These incentives have produced a reduction of 25% in prevailing per capita water consumption. It is expected that this lower per capita consumption will persist for many years to come.

Reduction in water supply production of the Cantareira system

Figure 18 synthesizes the overall result of all these actions: it was possible to reduce potable water production in the Cantareira system from approximately 32 m^3/s to 13 m^3/s. The breakdown from different actions shows that this enormous saving was due to: PRV/leakage reductions (7.7 m^3/s); interconnection of systems (6.6 m^3/s); economic incentives (3.4 m^3/s); and reduction in supply to partner utilities in SPMR (0.8 m^3/s). This reduction of almost 60% was essential to avoid Cantareira storage from hitting the bottom, which would mean a major collapse in the water supply system of SPMR (Figure 19).

Lessons learned

The worst drought in the history of SPMR could have led to social unrest. The storage level in the most important water source system reached the impressive mark of less than 5%. A combination of

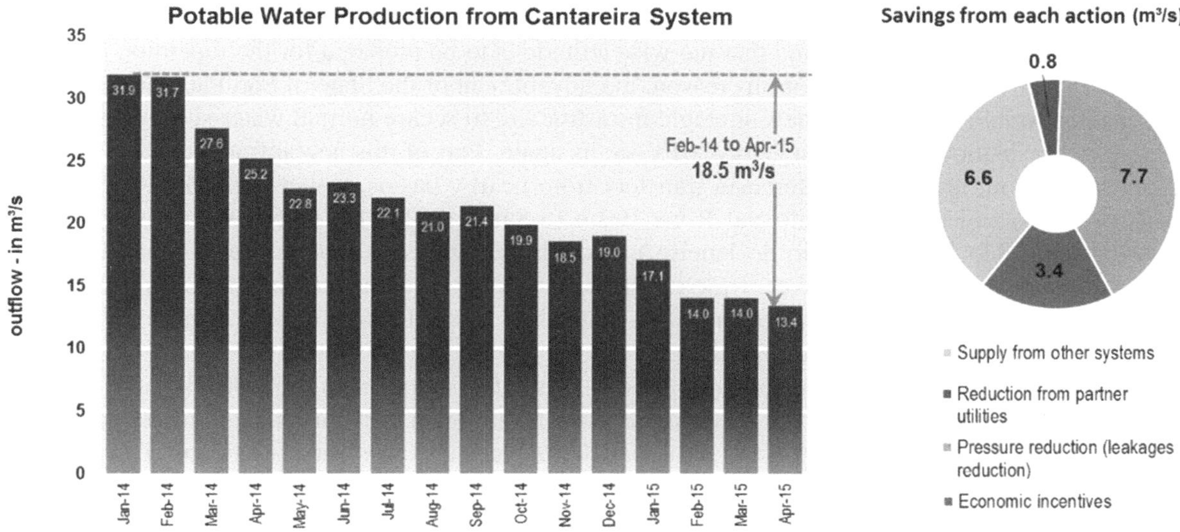

Fig. 18. Water supply production in SPMR during water crisis.

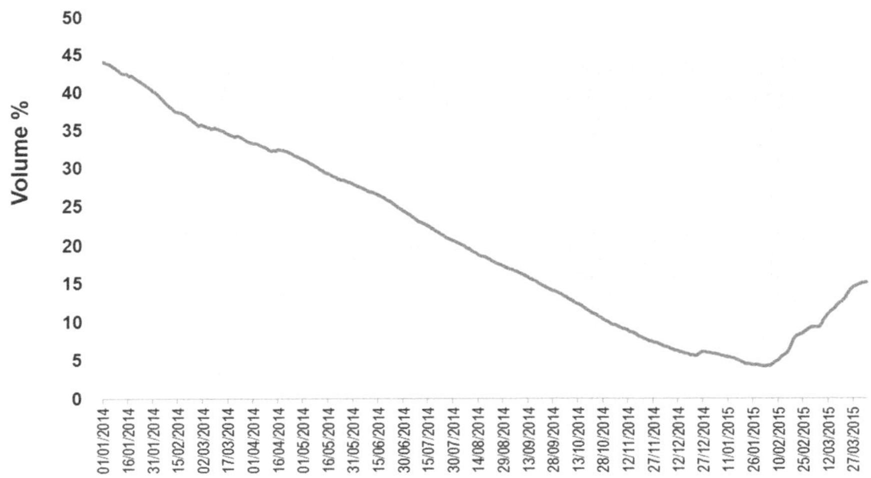

Fig. 19. Depletion of storage in the Cantareira system during the crisis.

structural and non-structural measures including an important collaboration from the population in using water more efficiently were essential in facing this challenge. A large number of emergency water infrastructures have been implemented in record time allowing a more flexible use of the eight reservoir systems in the region. Today, the SPMR is far better prepared to face much worse droughts than those foreseen in the various water resources plans produced since the 1960s.

Based on a long-time series and assuming stationary conditions, the 2014–2015 drought is such a rare event that one could be tempted to consider it as an outlier, with no relevant consequences for planning.

However, the prudent decision maker has to consider that the stationary hypothesis may be wrong, due to climate and land use changes, and that the wise attitude is to be prepared for the repetition of similar events (the 'firm yield' criteria). For this reason, the government of the State of São Paulo and SABESP have decided to invest heavily in new strategic infrastructure to secure normal water supply, even if the adverse water situation observed in 2014–2015 occurs again. Part of this new infrastructure is currently under construction, namely the interbasin transfers from nearby basins, respectively the Ribeira River Basin (6.4 m^3/s) and the Paraiba do Sul River Basin (8.5 m^3/s). This last interbasin transfer stirred a debate between São Paulo and Rio de Janeiro metropolitan regions. Although they are 400 km apart, both fetch water from the same river, which runs for roughly 1,000 km crossing three Brazilian states (Kelman, 2015). A third interbasin transfer, the Itapanhau River Basin (up to 2.5 m^3/s) is currently under environmental impact assessment.

Reducing wastage is another fundamental task to guarantee supplies. In the last 20 years, the State Government has made major efforts in this direction, resulting in the building of infrastructure that cost more than US$2.5 billion ($2.5 \times 10^9$), partly financed by the Inter-American Development Bank. However, much more needs to be done. A key initiative is being launched by the State Government for demand management including: subsidies for the poor to install domestic water-efficient equipment; low-interest rate loans for apartment owners to install individual metering; and regulation for non-potable water reuse. SABESP is working towards eliminating the maze of pipes scattered along the alleyways of the irregular but irreversibly established settlements. This of course requires articulation with city councils and the Public Attorney's Office. The Catholic Church, which included sanitation as a theme of its 2016 Fraternity Campaign, is another important partner.

Giving priority to works to underpin water security means, inescapably, that other equally important, but less urgent, investments have to be postponed. This situation, however, does not exclude the permanent goal of providing sewage collection and treatment services for everyone.

Another important lesson learned from the water crisis is that there is an urgent need to review how water supply and sanitation tariffs are established. When the population rightly becomes more parsimonious on water use, the income of water utilities decreases. SABESP relies solely on revenues from consumers for operation and expansion of the system supply. No subsidies from taxpayers are provided.

It is also necessary for society and the courts to understand that the purpose of sanitation is to maximize social well-being with the resources available, making water treatment and distribution the priority, followed by sewage collection and, lastly, by sewage treatment. In other words, Brazil has to follow the example of developed countries: put people's health above any other consideration. It makes little sense, therefore, to use the scarce financial resources of the water companies to pay the so-called 'environmental compensation' fees, which are penalties intended to punish companies for polluting the rivers or the ocean during the period that sewage is collected but not treated. Such impositions have the effect of eliminating resources that are extremely necessary for the expansion and improvement of the quality of these services.

Conclusions

The water crisis that SPMR suffered in the 2-year period 2014–2015 is an example of the impacts of climate variability and land use change that are felt in the water sector. These impacts were magnified by the fact that the drought hit a densely populated area. The consequences could have been catastrophic if structural and non-structural measures had not been timely implemented.

Tapping water from the dead volume of the largest reservoir in the Cantareira system through the construction of cofferdams and the use of floating pumps was an essential measure in such a crisis. The interlinking of storage reservoirs was fundamental to provide flexibility in the joint operation of the three major water sources in the region. Structural integration, in the form of extending the system to sectors belonging to other water sources, was made possible by a series of operations, such as refurbishing the water mains and the pumping stations, and replacing or adapting boosters. To increase water treatment capacity, the option selected took into account the advantages offered by membrane filtration technology. The stocks of treated water at strategic points were increased with the construction of sector reservoirs, minimizing interruptions in supply. Large investments were necessary to reduce physical and commercial water loss by replacing networks, branch lines and water meters, and by strict measures to control fraud.

Demand management is fundamental in facing the challenge of a water crisis. The economic incentives provided by the water utility allowed a significant reduction in the consumption rate. The contingency tariffs, with reductions for those saving water and increases for those using water in excess, worked admirably. However, the extension of these incentives for a long period of more than a year severely affected the income of the water utility. SABESP had to adjust its budget to anticipate investments in water infrastructure needed to increase water security at a time when there was a severe drop in revenues due to the combination of lower water sales and lower tariffs. The unavoidable result was a temporary slowdown in initiatives already under way to expand the sewage collection and treatment.

As the Chinese wisdom teaches, every crisis represents an opportunity – for innovating, for changing paradigms and, above all, for implementing sustainable long-term solutions. After 2 years of intense challenges and much hard work, the population and the political class are much more aware of the importance of water security. The lesson learned is that in the future we need to be more resilient to the vagaries of climate. This implies more infrastructure and good governance through strong public institutions, efficient regulatory agencies, adequate legal frameworks and the consciousness of the population to use water parsimoniously and efficiently.

References

Agência Nacional de Aguas (ANA) (Brazilian National Water Agency) (2015). *Encarte especial sobre a crise hídrica, Conjuntura dos Recursos Hídricos no Brasil (Water Resources Conjuncture in Brazil – Water Crisis special issue)*, 2014 Report. Brasília (in Portuguese), p. 13.

Braga, B., Porto, M. F. A. & Silva, R. T. (2006). Water management in Metropolitan São Paulo. In: *Water Management for Large Cities*. Tortajada, C., Varis, O., Lundqvist, J. & Biswas, A. K. (eds). Routledge – Taylor & Francis Group, London, pp. 151–166.

Kelman, J. (2015). Water supply to the two largest Brazilian metropolitan regions. *Science Direct, Aquatic Procedia 5*, 13–21.

United Nations (2015). *70th General Assembly*, Resolution A/70/L.1 September, 2015.

US National Security Council (2012). *Global Trends 2030: Alternative Worlds*, December 2012.

Integrated water resources management and drought risk management in Japan

Tomoyuki Okada

River Planning Division, Water and Disaster Management Bureau, Ministry of Land, Infrastructure, Transport and Tourism (MLIT), Tokyo, Japan. E-mail: okada-t2vd@mlit.go.jp

Abstract

Japan has experienced several severe drought disasters in the past and is still suffering from drought conditions in some parts of the country every year. To cope with the drought disasters, Japan takes three types of countermeasures: (i) water resources planning and development as supply side measures; (ii) promotion of efficient water use as demand side measures; and (iii) coordinated drought risk management as risk control measures. Owing to continuous efforts in water resources development, water saving, and water recycling, drought risk is currently mitigated to some extent. Japan further needs to improve the drought safety level and secure stable water supply.

Keywords: Dam reservoir; Drought risk management; Integrated Water Resources Management (IWRM); Japan; Water saving

1. Climatic and topographic condition

Japan appears to have sufficient water resources because it is located in East Asia and surrounded by the sea; however, Japan's available water resources are limited and drought issues are observed every year in some parts of the country.

In terms of climatic condition, Japan belongs to the monsoon Asia, which is well known for having one of the largest amounts of rainfall in the world. Based on the data from around 1,300 nationwide rainfall observation stations from 1981 to 2010, average annual precipitation of Japan is 1,690 mm (MLIT, 2015). On the other hand, average annual precipitation of the worldwide land area is 810 mm, which is calculated from AQUASTAT data of the Food and Agriculture Organization (FAO) from 2014. Japan's precipitation is approximately twice as large as that of the world. However, if we look at annual precipitation per person, which is defined as the average annual precipitation multiplied by total land area and divided by total population, Japan's value is about 5,000 m^3/person/year, which is only one-third of the world average, 15,000 m^3/person/year.

doi: 10.2166/wp.2016.214

In terms of topographic condition, nearly three-quarters of Japanese land is mountainous areas with steep and rugged landforms. Because its rivers are also steep and short, rainwater runs off quickly to the sea.

Heavy rainfall is observed during the rainy season, from June to July, and the typhoon season, mainly from August to September. The total capacity of dam reservoirs in Japan amounts to over 22×10^9 m^3, which is approximately only half the volume of the Hoover Dam in the USA (MLIT, 2007a). Moreover, half of the Japanese population lives in alluvial plains, which count for only 10% of Japanese land. Skewed population distribution over the land limits available water resources and makes Japan historically vulnerable to droughts. A large part of Japan has suffered from disruption or halt of domestic water supply due to droughts. Water supply areas affected by drought at least once within the past 30 years are shown on the map (see Figure 1).

1.1. Precipitation trend

Annual precipitation data, based on 51 nationwide rainfall observation stations, show a downward trend, from 1,650 mm in 1900 to 1,570 mm in 2014 (see Figure 2). From the middle of the 1960s,

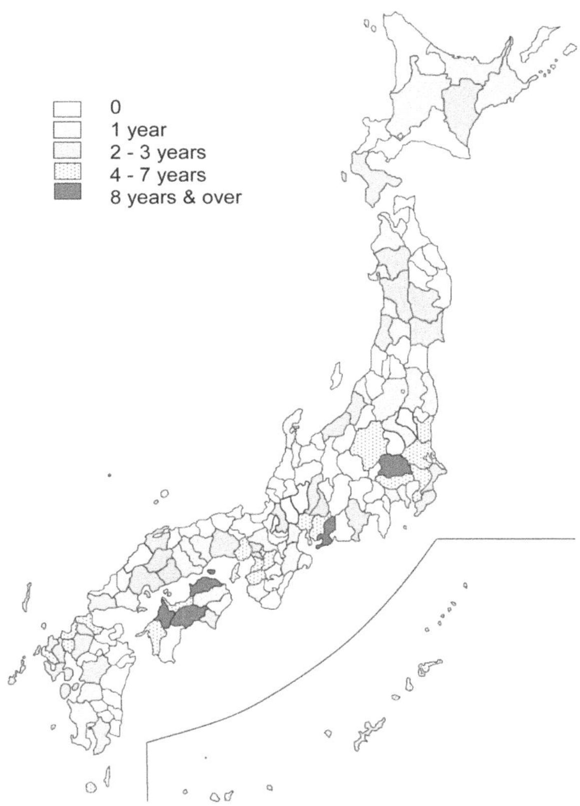

Fig. 1. Domestic water supply areas disrupted or halted due to droughts over the past 30 years.

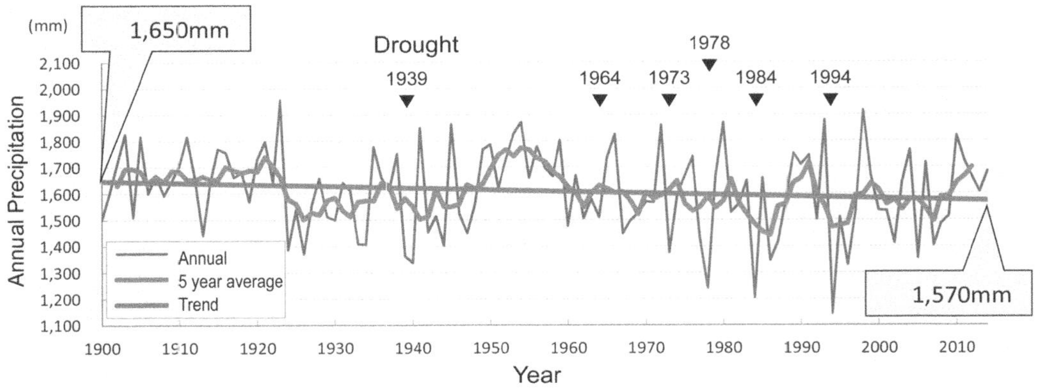

Fig. 2. Annual precipitation trend from 1900 to 2014.

the difference has widened between the smaller and larger annual precipitation. Especially, annual rainfall significantly decreased in 1973, 1978, 1984, 1994, 1996, and 2005, most of which resulted in historically severe droughts. The analysis of the data also shows increasing variances of precipitation in recent years compared to a century ago. The standard deviation of annual precipitation is 112 mm in the early 20th century and 152 mm in the early 21st century (see Table 1).

1.2. Drought conditions

The number of domestic, industrial, and agricultural water supply areas affected by droughts is summed up from 1965 to 2014 (see Figure 3). Although the total number of water supply areas varies from a minimum of 144 to a maximum of 175, a large part of Japan was under the influence of droughts, especially in 1967, 1973, 1978, 1984, 1985, and 1994. From the late 1990s, the number of drought-hit areas decreased but they still exist every year.

The definition of drought for each water use is as follows.

Domestic water: Water agency reduced water supply by lowering water pressure or temporarily halting supply.

Industrial water: Water agency reduced water supply by lowering water pressure or temporarily halting supply, or Water agency requested water users to save water with a certain target.

Agricultural water: Insufficient crop growth was caused by decreased river flow or intake restriction.

Table 1. Annual precipitation change for over a century.

Trend		Variance		
Year	Annual Precipitation	Period (10 Years)	Min. & Max. Differences from Average	Standard Deviation
1900	1,650 mm	1900–1909	−150 mm, +180 mm	112 mm
2014	1,570 mm	2005–2014	−220 mm, +250 mm	152 mm

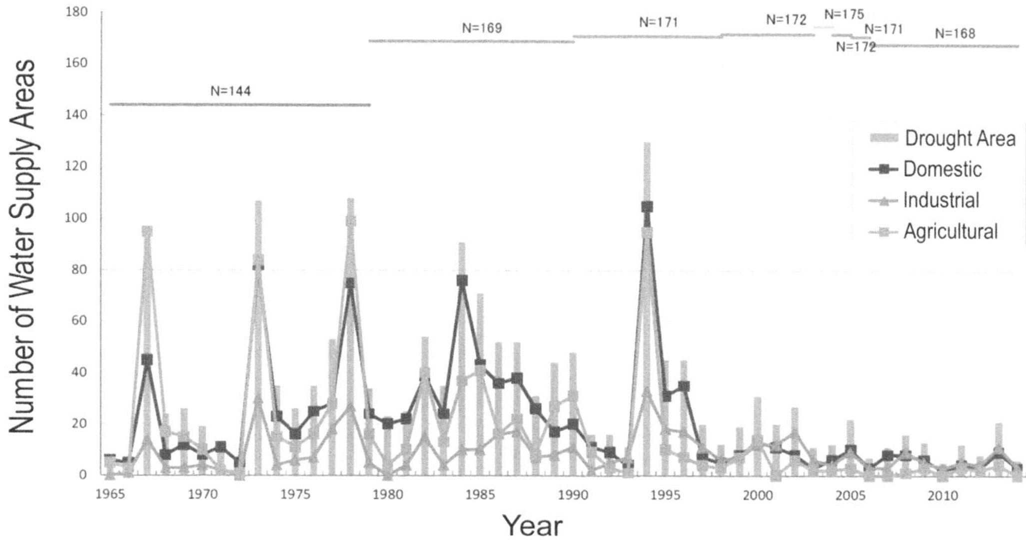

Fig. 3. Number of water supply areas affected by droughts.

2. Current water use

Water resources can be classified by the purpose of the water use, such as domestic water, urban activity water, industrial water, and agricultural water (see Figure 4). The total amount of nationwide water use measured at intake level in 2012 is 80.5 × 10^9 m^3/year, which comprises daily life water (15.1 × 10^9 m^3/year), industrial water (11.5 × 10^9 m^3/year), and agricultural water (53.9 × 10^9 m^3/year) (see Figure 5). The amount of urban water has been increasing since 1965; however, since the 1990s it has remained the same or gradually decreased. The industrial water is the summation of fresh water withdrawn from rivers or underground, excluding recycled water. The data are collected from industrial facilities with four employees or over. If recycled water is included, the total industrial water use is 48.9 × 10^9 m^3/year.

2.1. Daily life water use

The daily life water use measured at tap level peaked at 14.4 × 10^9 m^3/year in 2000 and gradually decreased to 13.2 × 10^9 m^3/year in 2012 (see Figure 6). At the same time, average daily life water use per person, which is defined as the total daily life water use divided by water supply population,

- Urban Water
  - Daily Life Water
    - Domestic Water: drinking, cooking, bath, toilet
    - Urban Activity Water: shop, office, public place, hydrant
  - Industrial Water: washing, cleaning, coolant
- Agricultural Water: irrigation, live stock

Fig. 4. Classification of water use and examples.

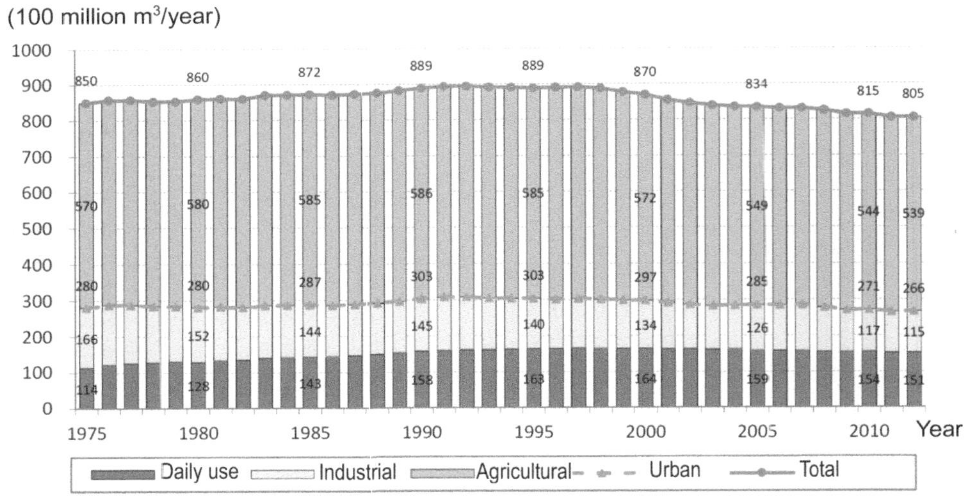

Fig. 5. Amount of nationwide water use.

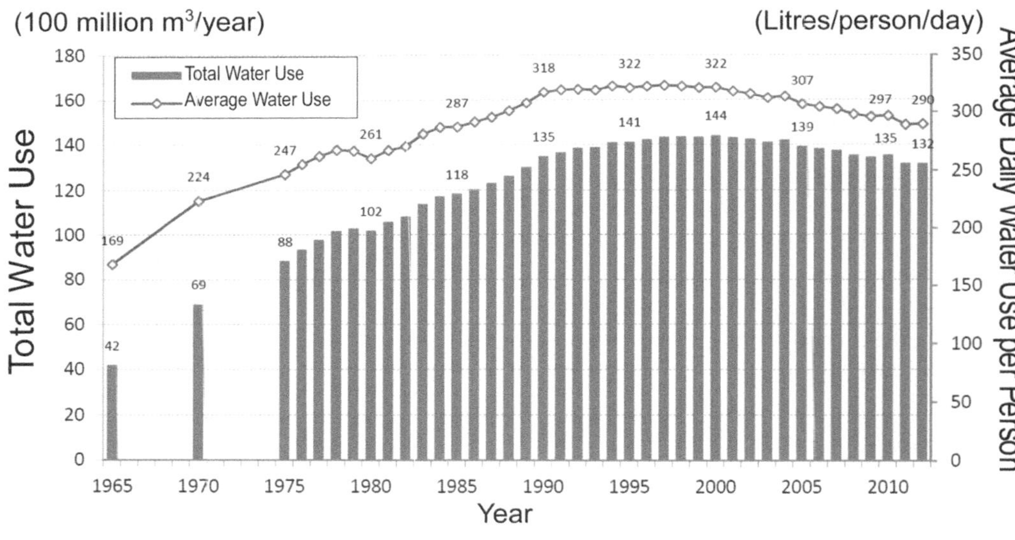

Fig. 6. Trend of daily life water use.

reached its peak (324 litres/person/day) in 1997. The declining trend of the water consumption unit is partly because of prevailing equipment with efficient water use. The Japanese population also reached its peak (128 million) in 2008 and has been gradually decreasing since then.

Tap water was rapidly distributed to the public between the middle of the 1950s and the early 1970s. The penetration rate of tap water exceeded 90% in 1978. As of 2012, the penetration rate reached 97.7% and water supply population was 124 million. Water supply tends to increase during the summer season and decrease during the winter season; however, the difference between the two seasons is decreasing these days.

2.2. *Industrial water use*

In 2012 the total volume of industrial water used by industrial facilities with 30 employees or over was 47.2×10^9 m^3/year, of which 79.1% was recycled water (see Figure 7). In parallel with increasing industrial water demand, the water recovery rate rapidly improved between the mid-1960s and mid-1970s to use limited water resources efficiently and meet wastewater discharge standards. From the early 1980s the improvement in the recovery rate slowed down. The top three water consumption industries, chemical, steel, and pulp and paper industries, amount to 71% of total industrial freshwater use. The water recovery rates of the chemical and steel industries are quite high (80 to 90%) but that of the pulp and paper industry remains low (40 to 50%).

2.3. *Agricultural water use*

The majority of agricultural water is used for irrigation for rice paddies, followed by irrigation for crop fields and water for animal husbandry of cows, pigs, chickens, etc. Because agricultural water is not measured in many areas, the amount of agricultural water is estimated by available data, such as irrigation areas, water seepage rate, and the number of livestock. Both paddy and crop field areas have been decreasing since 1975, while more irrigation water is required to increase the crop yield. As a result, irrigation water for paddies has been decreasing since the late 1990s and water for crop fields remains the same (see Figure 8).

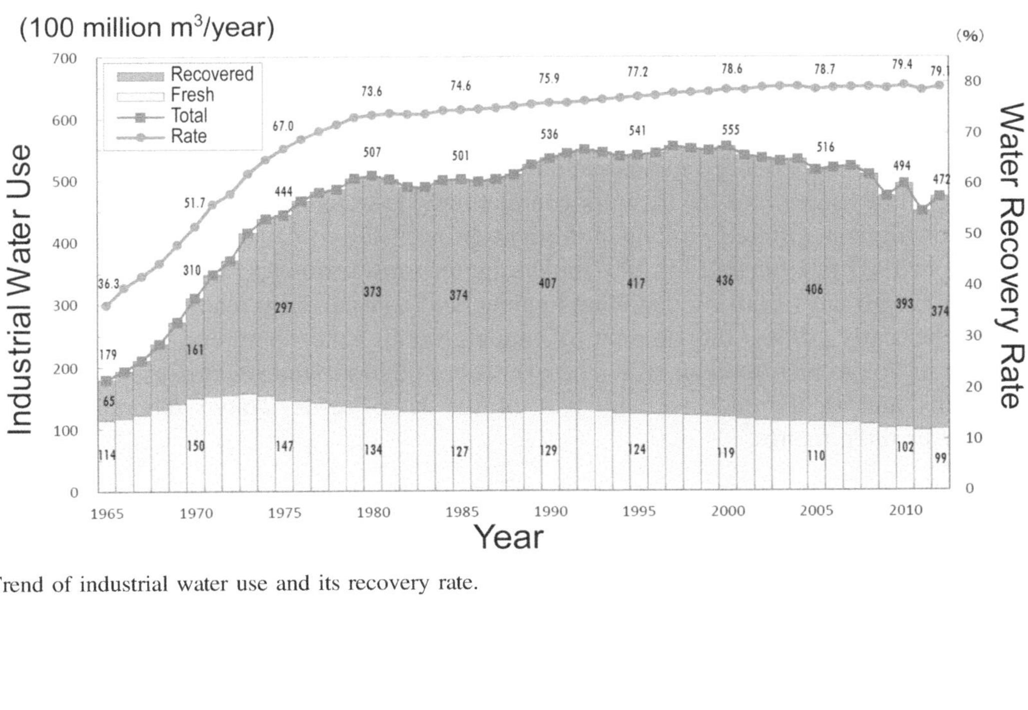

Fig. 7. Trend of industrial water use and its recovery rate.

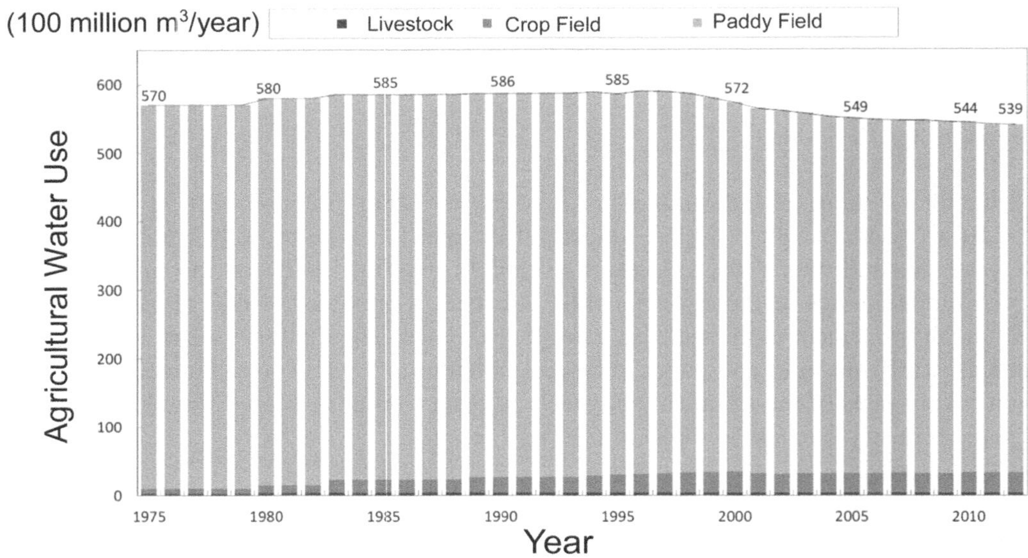

Fig. 8. Trend of agricultural water use.

3. Countermeasures against drought disasters

Japan takes three types of countermeasures to cope with drought disasters:

(i) Water resources planning and development – Supply side measures.
(ii) Promotion of efficient water use – Demand side measures.
(iii) Coordinated drought risk management – Risk control measures.

3.1. Water resources planning and development

From 1950 Japan has experienced rapid economic growth, urban population increase, and improved living standards, all of which have led to significant nationwide water demand. For example, the Tokyo Metropolitan Government suffered prolonged drought disaster, the so-called 'Tokyo desert', for a total of 42 months from 1961 to 1965 (MLIT, 2007b). During the most serious period, domestic water supply was cut by 50% for 10 days, causing degraded hygiene and consequently many food-poisoning cases. Moreover, in the early 1960s the amount of underground water pumping reached 500,000 to 600,000 m^3/day in Tokyo metropolitan area. The overuse of underground water caused severe land subsidence problems and in the most serious part of the Koto delta, Tokyo recorded accumulated subsidence of over three metres.

3.1.1. Legal framework for water resources development. In order to meet ever-growing water demand and mitigate serious water shortage, the Japanese Government in 1961 enacted two laws: the Water Resources Development Promotion Law, and Water Resources Development Public Corporation

Law (later amended to Japan Water Agency Law). The former aims to facilitate consensus building on water resources among concerned parties by establishing a basin-wide water plan (Water Resources Development Basic Plan). The latter aims to promote basin-wide comprehensive water development projects by mobilizing government's fiscal investment and loans through the newly created public corporation.

Water resources development often requires long-term projects for efficient allocation of limited resources. To this end, the central government made a first National Comprehensive Water Resources Plan in 1978. The Plan was later revised into Water Plan 2000 in 1987 and then Water Plan 21 in 1999. These national plans estimate long-term (10 to 15 years) water demand and supply, and establish basic national policies on water resources development, conservation, and utilization.

The government has also prepared separate Water Resources Development Basic Plans targeting seven designated river basins where Japan's major cosmopolitan areas are located. The designated areas play a significant role in the Japanese economy, comprising 66 million people, 52% of Japan's total population and JPY 118 trillion yield of manufacturing products (46% of national total). In addition, 23 prefectural governments out of a total 47 have their own local plans on long-term water demand and supply.

3.1.2. Water resources development. Under the above-mentioned legal framework and long-term plans, Japan has been developing water resources facilities including dam reservoirs, river weirs, lake storage development, canals, etc. in order to secure stable and sufficient water supply for all citizens. Water resources development within each river basin targets a drought safety level of one in ten years, that is, water supply is not disrupted under the condition of smallest river discharge for a 10-year period. The long-term water development plan considers future water demand based on projected demographic and economic growth in the river basin.

Water resources have been steadily developed for many decades. By March 2015, newly developed water resources in Japan for domestic use amounted to 12.6×10^9 m^3/year, and those for industrial use 6.0×10^9 m^3/year (see Figure 9). The sum of this domestic and industrial water, 18.6×10^9 m^3/year, covers about 70% of the total water use for both purposes. The remaining 30% is withdrawn from rivers and lakes without developing new storage facilities. The natural river flow, most of which is already appropriated to agricultural use, cannot cope with the growing water demand without constructing new water storage facilities. As of April 2015, 19 water resources development projects are being implemented, which will provide additional domestic and industrial water of 0.3×10^9 m^3/year and agricultural water of 0.3×10^9 m^3/year.

If water shortage is critical and necessary water resources development facilities are not yet completed, the planned amount of water is tentatively withdrawn from rivers. Such water withdrawal is allowed only when river runoff is more than sufficient to other prioritized users. As of the end of 2013, this unstable domestic and industrial water amounted to 0.9×10^9 m^3/year, which was 3.3% of the total river withdrawal, 26.6×10^9 m^3/year.

3.1.3. Construction of dam reservoirs. According to Dam Almanac, Japan, there are over 2,700 dams in Japan, where a dam is defined as an embankment with a height of 15 m or over (JDF, 2014). The Ministry of Land, Infrastructure, Transport and Tourism (MLIT) administers dams for flood control purposes. MLIT, Japan Water Agency (JWA), and prefectural governments construct such dams under relevant laws (see Table 2). As of August 2016, MLIT, JWA, and prefectural governments operate

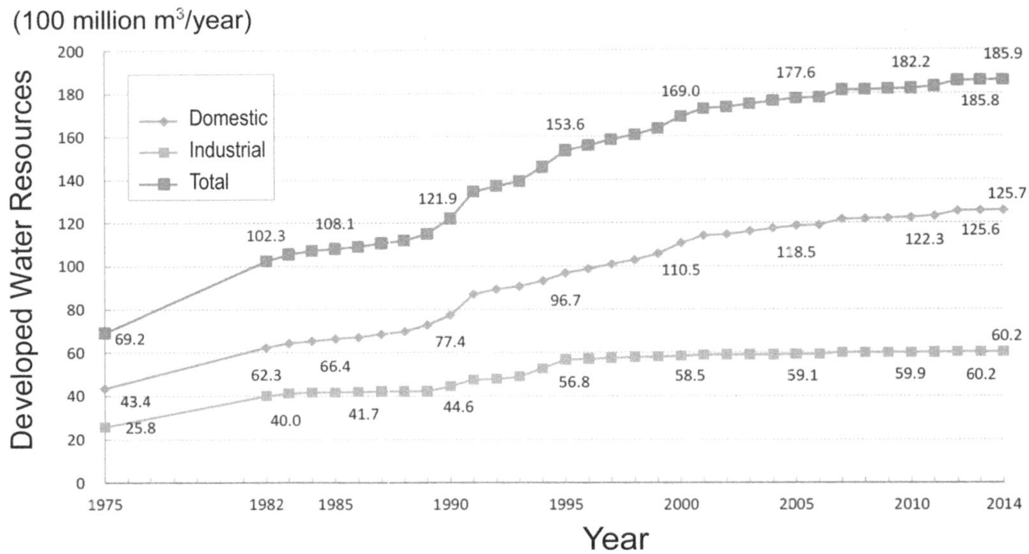

Fig. 9. Water resources amount developed by new facilities.

Table 2. Legal system and implementing organization of flood control facilities.

Organization	Facility Classification	Relevant Law
Ministry of Land, Infrastructure, Transport and Tourism	Designated multipurpose dam	Designated Multipurpose Dam Law
	Comprehensive river development project	River Law
	Runoff coordination river (Transboundary aqueduct)	
Japan Water Agency	Water resources development facility	Japan Water Agency Law
Prefectural Government	Multipurpose dam	River Law
	Flood control dam	

and maintain 99, 23, and 434 dams with flood control, respectively. The Designated Multipurpose Dam Law, enacted in 1957, has been the driving force to meet the growing needs of water resources by allowing the Minister of Construction (currently Minister of Land, Infrastructure, Transport and Tourism) to implement multipurpose dam projects.

3.1.4. Redevelopment of existing dam reservoirs. The Kinugawa River upstream dams coordination project connects two existing dam reservoirs by discharge tunnels and pumping stations for efficient use of reservoir volumes (MLIT, 2012) (see Figure 10). Both Ikari Dam and Kawaji Dam are located in the upstream of the Kinugawa River system in the Kanto region. Ikari Dam has large inflow into the reservoir but a small reservoir capacity. Consequently, the dam often wasted excess water when the reservoir was full. On the other hand, Kawaji Dam has small inflow but a large reservoir capacity (see Table 3). This JPY 13.8 billion (10^9) project constructed discharge tunnels totalling a length of

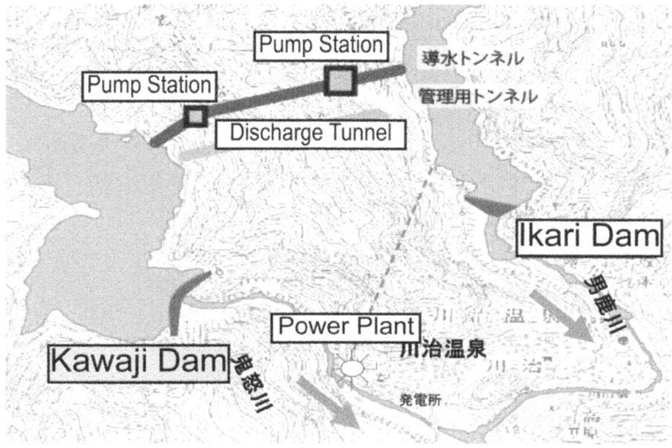

Fig. 10. Outline of Kinugawa River upstream dams coordination project.

Table 3. Comparison of Kawaji Dam and Ikari Dam.

	Kawaji Dam	Ikari Dam
Purpose	Flood control, Natural river runoff, Irrigation, Water supply	Flood control, Natural river runoff, Irrigation, Hydro power
Year of completion	1983	1956
Dam height (m)	140	112
Catchment area (km^2)	144	169
Inflow to reservoir (million m^3/year)[a]	140	230
Effective reservoir volume (million m^3)	76	46

[a]The inflow is the average from 2001 to 2010.

1.1 km over a period of 10 years (1997–2006). The tunnels convey a maximum 20 m^3/s of water from Ikari Dam to Kawaji Dam for water storage and release a maximum 1 m^3/s of water from Kawaji Dam to Ikari Dam for improving the downstream river environment. This return water annually averaged 3.7 million m^3/year from 2007 to 2010.

After the completion of Ikari Dam in 1956, all water was bypassed from the dam, causing a degraded downstream environment with murky water and odour. The downstream of the dam runs through Kawaji hot spring district, which later requested the improvement of river environment and scenery. Accordingly, reservoir users agreed to discharge 0.46 m^3/s of water (0.23 m^3/s in winter) in 1971. Moreover, the electric power company agreed to increase the discharge amount to 0.7 m^3/s in 1981. The Kinugawa River upstream dams coordination project further improved the condition to 1.0 m^3/s in 2006. Finally, Yunishigawa Dam, completed upstream of Ikari Dam, increased the discharge to 1.5 m^3/s in 2012.

During the 1950s and 1960s, when Japan experienced rapid economic growth, hydro power was a major energy source and had a priority among different river water uses. Because all or most of the river flow was diverted from dam reservoirs to hydro power plants, the dried-up river bed appeared between the intake and outlet of bypass tunnels. People's desire for a natural river environment gradually gained power and the Ministry of Construction and the Ministry of International Trade and Industry agreed to issue the guideline entitled 'Assurance of normal river runoff at the renewal of water rights for hydro power' in 1988. Increasing minimum river flow in the Kinugawa River coincides with this national trend. Thanks to the increased river discharge, the Kinugawa River recovered popular river activities for tourists as well as improved fish habitat.

3.1.5. Interlinked and coordinated rivers. The Saga Aqueduct project in northern Kyushu region connects two adjacent river systems, Chikugo River System and Kase River System, through two aqueducts for water resources development and flood control (MLIT, Chikugo River Office, n.d.) (see Figure 11). The project comprises West Saga Aqueduct 9.8 km, East Saga Aqueduct 13.2 km, nine pumping stations, and Kose River Retarding Basin. The JPY 99.5 billion (10^9) project started in 1974 and was completed in 2008, lasting for over 30 years.

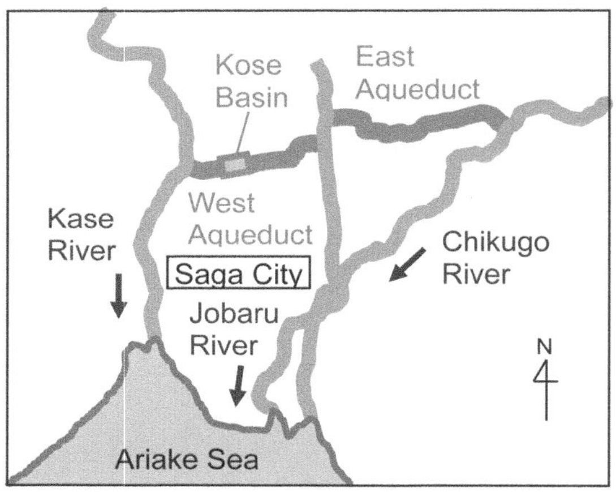

Fig. 11. Map of Saga Aqueduct project.

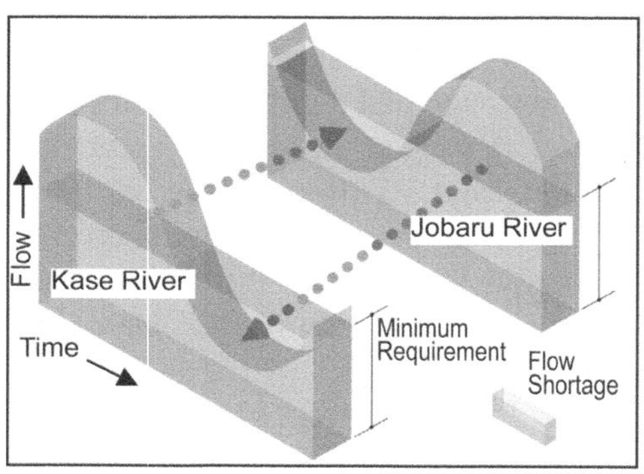

Fig. 12. Interlinked and coordinated rivers.

One of three purposes of this water resources development is to improve river runoff conditions in Kase River and Jobaru River. If Kase River's runoff is more than sufficient and Jobaru River's runoff is insufficient, river water is pumped from Kase River to Jobaru River (see Figure 12), and vice versa. If both rivers' runoff is insufficient and Chikugo River's runoff is more than sufficient, river water is pumped from Chikugo River to both rivers.

The second purpose is to deliver water to Saga West Water Supply Agency whose coverage area is the west side of Kase River. The Agency used to rely heavily on underground water as its water source, before the Saga Aqueduct project was completed. The underground water extraction for nearly four decades caused over 1 m of land subsidence in land reclamation areas of Chikugo and Saga plain. This area is one of three subsidence-regulated areas in Japan along with Nobi plain around Nagoya City and Northern Kanto plain to the north of Tokyo. Owing to the shift of water source from underground water to surface water, land subsidence is alleviated in a large part of Chikugo and Saga plain. Immediately after the water source shift, once-fabled spring water started to come out into Nuinoike pond after a 40-year dry spell.

The third purpose is to increase runoff of rivers and canals running through downtown Saga City, Capital of Saga Prefecture. Additional river discharge by the project is expected to cleanse river and canal networks and mitigate degraded river water quality.

3.2. Promotion of efficient water use

The water distribution rate to customers in drinking water service reached 92.8% in 2012 from 81.1% in 1975 by preventing water leakage from water pipelines. Japan's leading water utility agency, the Bureau of Waterworks in Tokyo Metropolitan Government, supplies water to more than 13 million people; however, its water leakage rate was merely 3.1% in 2014 (Tokyo, 2015). As a part of demand control, most water agencies adopt tiered water rates, or increasing block water rates, which promote rational water consumption and control user demand for water.

Rainwater and recycled water are increasingly utilized for relatively low-quality water usage, such as coolant, toilet flush, gardening, etc. Recycled water use is classified into four systems: individual circulation system within a building; district circulation system with large-scale residential buildings or office buildings; sewage treatment plant system; and rainwater harvesting system. Lessons learned by past mega-earthquakes promote the use of recycled water as a substitute water source in case of emergency for toilet flushing, moistening the ground for dust abatement, fire hydrants, and so on.

3.2.1. Rainwater harvesting. As of the end of 2012, 1,851 buildings (public facilities and offices) were using a total amount of rainwater of 7.9 million m^3/year, which is equivalent to 0.01% of the total water use in Japan (see Figure 13). Nearly half of these buildings are located in Tokyo, where rainwater harvesting has been promoted by municipal regulations since the middle of the 1970s. These days, around 100 new rainwater harvesting facilities are constructed annually. The purpose of rainwater use is mostly toilet flushing (42% of facilities) and sprinklers (32%), followed by fire hydrants, cleaning, scenery, coolant, car washing, washing, and air conditioning.

The Rainwater Use Promotion Law was enacted in 2014 aiming at both efficient use of water resources and prevention of excessive rainwater discharge into sewers and rivers. The national government provides subsidies to local governments to construct large-scale rainwater storage facilities for flood control. Tax incentives are also specified by the Law on Special Measures Concerning Taxation

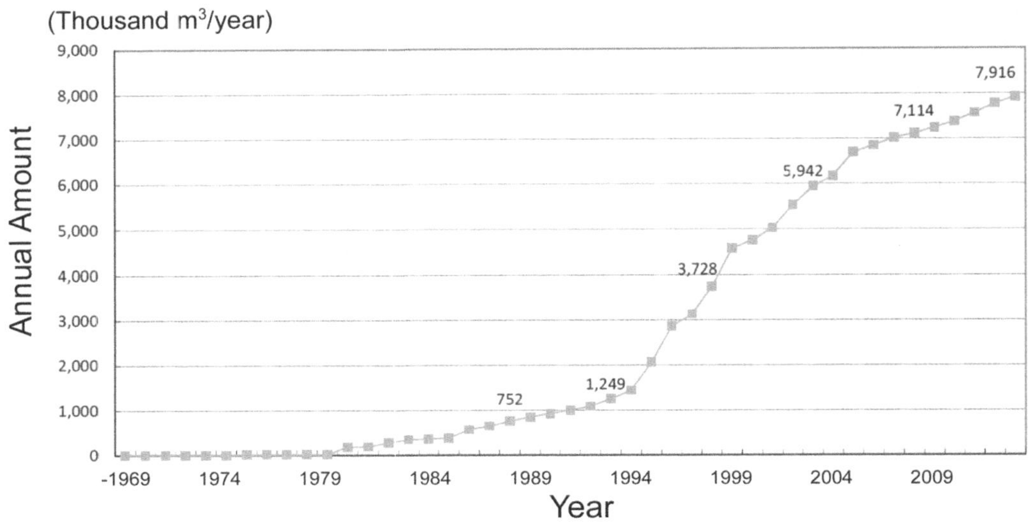

Fig. 13. Amount of annual rainwater harvesting.

and the Local Taxation Law. The former entitles facility owners to receive a 10% increase of depreciation on the assets of rainwater storage and infiltration for five years. The latter allows local governments to reduce the property tax base of these assets. In addition to the national government, 227 municipalities out of 1,742 have ordinances or guidelines to promote rainwater harvesting, as of 2013. Based on these ordinances, municipalities are responsible for propagating rain water harvesting to the public. The local governments also subsidize building owners to install rainwater facilities, such as rainwater tanks, recycled sewage tanks, infiltration pits and trenches, and permeable pavements.

3.2.2. Recycling of wastewater. In order to use water resources efficiently and conserve the water environment, wastewater is properly treated and recycled for various beneficial uses with consideration of economic feasibility. Approximately 2,100 sewage treatment plants produced 14.5×10^9 m^3 of treated wastewater in 2012 and rural area sewage facilities are estimated to have discharged 0.4×10^9 m^3 in 2009. The treated wastewater is reused in (1) closed water cycles where treated wastewater is separated from the natural water cycle and directly consumed and (2) open water cycles where treated wastewater is discharged into rivers and utilized all together. More than half of wastewater treatment plants reuse treated wastewater for their treatment process. In 2012, a total of 306 wastewater plants produced 211 million m^3 of recycled wastewater and distributed it outside the plants for the purpose of river discharge maintenance, snow melt, scenery, and so on (see Table 4).

3.2.3. Water saving equipment. People's awareness and behaviour toward water saving are steadily improving. The percentage of people who save water increased from 51% in 1986 to 77% in 2010. Water saving equipment is also prevailing in households, such as water saving washing machines (24%), dishwashers (19%), and so on. Because a washing machine and a toilet consume a lot of water in domestic water use, these two items with water saving functions are the equipment most desired

Table 4. Recycling of treated wastewater in 2012.

Purpose	Recycled Wastewater (million m^3)	Ratio	Number of Wastewater Treatment Plants
1. Toilet flushing	7.8	3.7%	54
2. Environmental use			
(1) Scenery	48.1	22.8%	101
(2) Recreational use	5.4	2.6%	29
(3) River flow maintenance	61.8	29.2%	15
3. Snow melt	52.7	24.9%	37
4. Gardening, sprinkler, etc.	0.6	0.3%	160
5. Agriculture	11.6	5.5%	28
6. Industrial water supply	2.5	1.2%	7
7. Office, factory	20.9	9.9%	61
Total	211.3	100%	306*

*The total of 306 treatment plants does not include double counting.

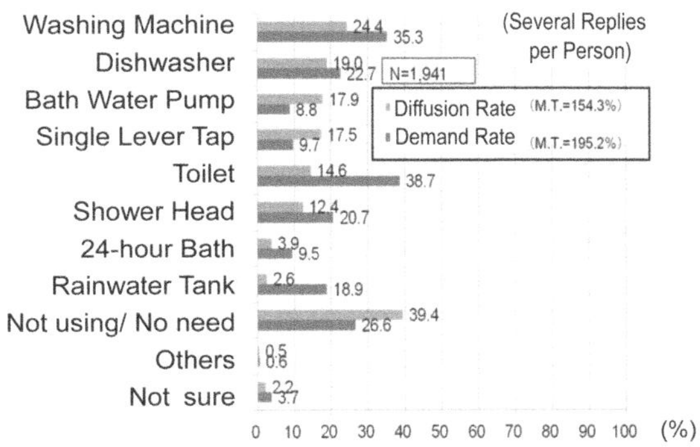

Fig. 14. Diffusion rate of water saving equipment in 2010.

by citizens (see Figure 14). Moreover, the water saving tap plug is a popular tool, which is even provided by some municipalities free of charge.

Technical innovation helps promote water efficiency and demand control as well. The water consumption of an automated washing machine per 1 kg of clothes decreased from 40 litres in 1990 to

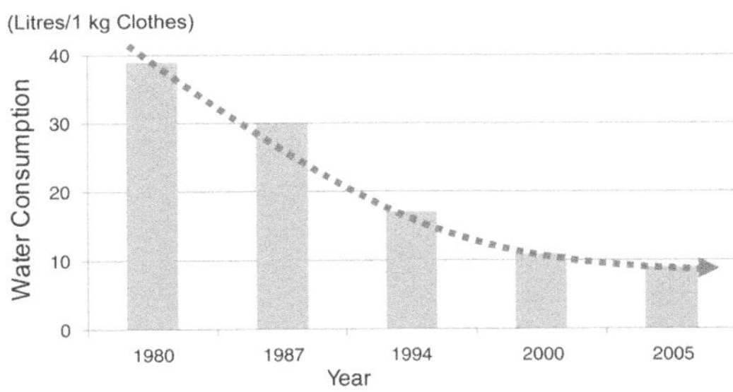

Fig. 15. Water consumption of washing machine.

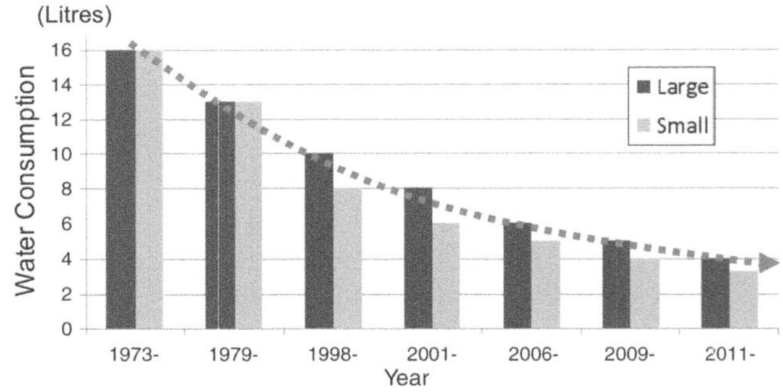

Fig. 16. Water consumption of toilet flush.

10 litres in 2005 (Toshiba, 2006) (see Figure 15). Similarly, that of a drum style washing machine decreased from 13 litres in 2001 to 8 litres in 2008 (Toshiba, 2008). The water used for a toilet flush shows the same decreasing trend from 16 litres in 1973 to 4 litres in 2011 (JSEIA, n.d.) (see Figure 16).

3.2.4. Municipal policy for water saving. Local governments subsidize the purchase of water saving appliances and the renovation of household equipment. Matsuyama City in Shikoku region started the subsidy system to raise citizen's awareness of water saving as well as to achieve the outcome of water saving (Matsuyama, 2012). Registered citizens are qualified to receive each subsidy once if an appliance is purchased at the shop located in the city. The subsidy for a household bath pump for a washing machine is half of the purchase price or JPY 2,000, whichever is higher. The subsidy for a water saving washing machine with a function of pumping bath water is JPY 5,000. The subsidy for a single lever cold and hot water mixing tap is JPY 3,000. The subsidy for a dish washer and drier, which completed the target number and finished in 2010, was half of the purchase price or JPY 20,000, whichever was higher.

Owing to comprehensive water saving policies in Matsuyama City, the amount of daily life water use per person per day decreased from 358 litres in 1993 to 281 litres in 2014 (see Figure 17). Because daily life water includes business water use, household water use is smaller than these amounts. Although the

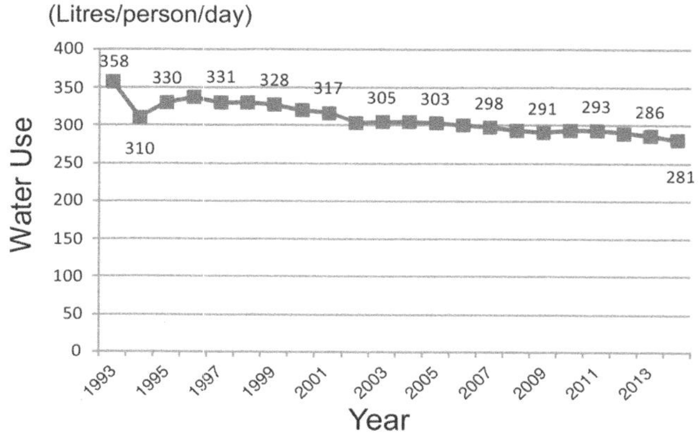

Fig. 17. Daily life water use per person per day in Matsuyama City.

city set up in 2004 the long-term water saving target, 300 litres/person/day in 2015, water saving progressed faster than the original plan. Out of 47 regional core cities in Japan, Matsuyama City records the second-smallest daily life water use per person per day next to Fukuoka City.

3.3. Coordinated drought risk management

River runoff and water storage volume of dam reservoirs are constantly monitored by river and dam administrators through real-time telemeter systems. When river runoff or reservoir level decreases and entitled water use is estimated to be affected, a drought coordination council is established in the river basin. The council is composed of river and dam administrators (Ministry of Land, Infrastructure, Transport and Tourism, local governments, Japan Water Agency, etc.) and water users (domestic and industrial water supply agencies, irrigation organizations, hydraulic power companies, etc.). The council regularly organizes drought coordination meetings to share river and reservoir observation data, drought forecast, and estimated impact on water users. If the drought conditions deteriorate, the council members discuss and agree on the step-by-step water withdrawal restriction. During the restriction, the members keep monitoring the impact or damage caused by the drought and encouraging citizens and end users to save water resources.

Article 53 of the River Law stipulates the coordination of water use in a drought condition by saying that water right holders (entitled water users) should respect other water usages and coordinate with other users. The river administrator should provide necessary information for this coordination and can mediate or arbitrate water users' negotiations if they cannot reach consensus.

The priority of water rights is not based on the purpose of water use, such as drinking, industry, and agriculture but on the order that the rights are granted. The water rights are usually based on the lowest river flow in 10 years. If the river flow decreases below this threshold, the designated water amount may not be withdrawn. The water right is dependent on river flow, which is a natural phenomenon and unstable in itself. Even if water users cannot withdraw water, they cannot claim for their rights against the river administrator. If a water user with high priority exercised their right during severe drought, social and economic disturbance would be caused. Japanese drought management follows the principle of giving way. Since 1972, the Ministry of Land, Infrastructure, Transport and Tourism has advised setting up a river basin drought coordination council for smooth arrangements among relevant water users.

In addition to the basin-level drought coordination, each water-related ministry and prefectural government sets up its drought response headquarters. In severe cases of drought, the government organizes a ministerial drought coordination meeting with 11 relevant ministries and agencies. The meeting aims to share the information on the drought conditions and responses to them. Several measures have been taken in the past as follows: reservoir water appropriated for hydraulic power was converted to water supply; underground water from private wells was temporarily used; drinking water was delivered to citizens by water tank; a public campaign for water saving was conducted; public and school swimming pools stopped operation.

3.3.1. Drought conditions in 2016. Several river basins suffered a shortage of reservoir water in 2016 and water users restricted their water intake. One of such basins is the Tone River basin, Japan's largest river basin, which covers an area of 16,840 km^2 and includes a population of 12.8 million (MLIT, Kanto Regional Development Bureau, n.d.). Connected to the Arakawa River basin with a 9.8 million population, the Tone River is the major water source of Tokyo metropolitan area and its surrounding Kanto

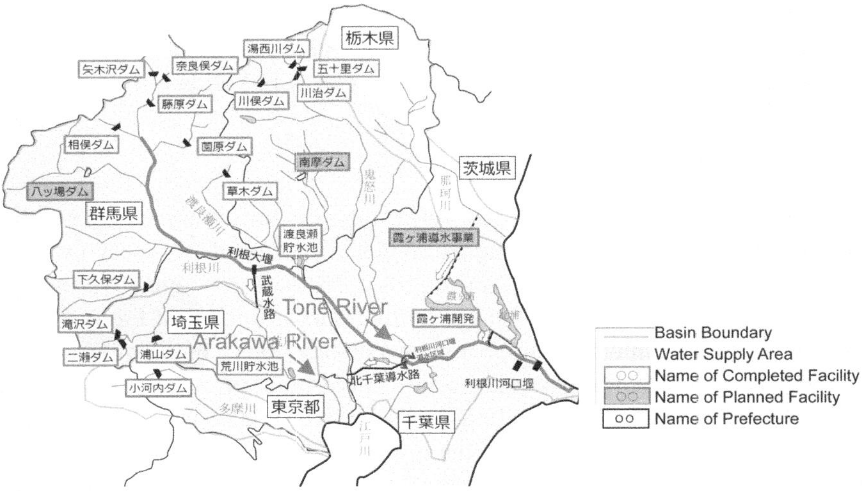

Fig. 18. Major water resources development facilities in the Tone and Arakawa River basins.

plain (see Figure 18). A total of 83% of urban water, including domestic water and industrial water, is provided by dams and other water resources development facilities in Kanto region. Of the 83% of urban water, 64% comes from the Tone River, 6% from the Arakawa River, and 13% from other rivers.

The water shortage of the Tone River was caused by the lack of both snowfall in winter and rainfall in May. Ozenuma observation station in the upper Tone River basin recorded a maximum snow depth of 172 cm in 2015–2016 winter season. This snow depth is the lowest for its 62-year observation period and only 60% of the annual average. The entire snow pack at the station melted away at the end of April, which is one month earlier than the average. Furthermore, the upper Tone River basin recorded only 48% of the average monthly rainfall in May 2016. Eight reservoirs in the area filled the deficit of river flow from early May to early June (see Figure 19). When the total volume of the eight reservoirs

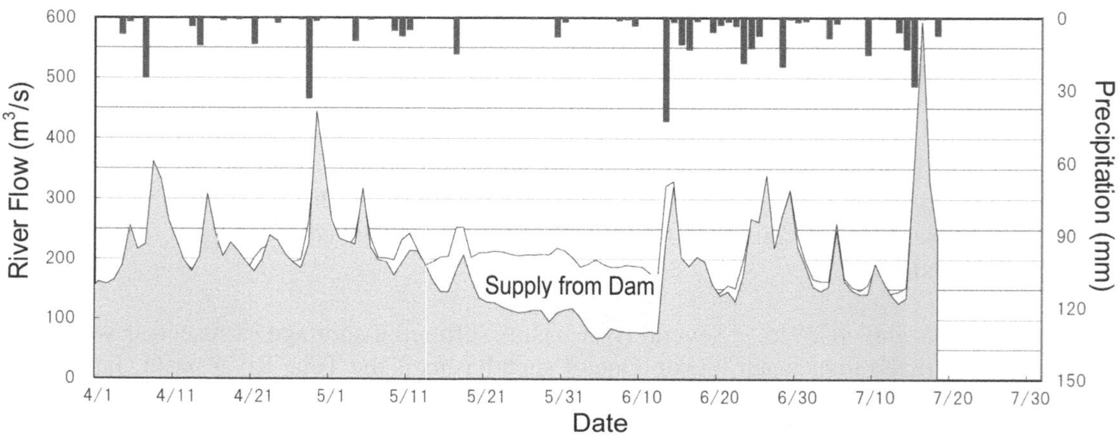

Fig. 19. Tone River flow and supply from dam reservoirs in 2016.

reached 45% of the annual average, Kanto Regional Development Bureau organized the first Tone River Drought Coordination Council on 14 June. The domestic, industrial, and agricultural water users started a 10% intake cut on 16 June and maintained this until 24 August when Typhoon No. 9 brought sufficient rainfall to the area.

4. Conclusion

Underpinned by legislation and long-term plans, water resources have been steadily developed for years. As a result, nationwide supply targets are more or less satisfied; however, some water resources facilities are still planned or being constructed to eliminate unstable water supply. According to the report of the Intergovernmental Panel on Climate Change (IPCC), climate-related extremes, such as floods and droughts, reveal vulnerability of human systems. The climate change is likely to reduce the safety level of future water supply. Even today the drought coordination council avoids serious drought damage by taking precautionary measures of intake restriction. In order to improve the safety level of water supply, we cannot rely solely on developing new water resources facilities as social and economic costs are increasing. Existing water resources facilities need to be utilized efficiently. Society should focus more on risk management for secured water supply.

References

Japan Dam Foundation (JDF) (2014). *Dam Almanac* (in Japanese). Available at: http://damnet.or.jp/cgi-bin/binranA/Syuukei.cgi?sy=sou (accessed 16 October 2016).

Japan Sanitary Equipment Industry Association (JSEIA) (n.d.). *History of Toilet* (in Japanese). Available at: http://www.sanitary-net.com/history/transition.html (accessed 16 October 2016).

Matsuyama City (2012). *Subsidy System for Water Saving Appliances* (in Japanese). Available at: https://www.city.matsuyama.ehime.jp/kurashi/kurashi/hojokin/sessuikiki.html (accessed 16 October 2016).

Ministry of Land, Infrastructure, Transport and Tourism (MLIT) (2007a). *Dam Projects Pamphlet* (in Japanese). Available at: http://www.mlit.go.jp/river/pamphlet_jirei/dam/gaiyou/panf/dam2007/index.html (accessed 16 October 2016).

Ministry of Land, Infrastructure, Transport and Tourism (MLIT) (2007b). *Japanese Water Resources - Current Situation and Issues* (in Japanese). Available at: http://jcsepa.mri-jma.go.jp/outreach/20070324/Presentations/P2_Yamamoto.pdf (accessed 16 October 2016).

Ministry of Land, Infrastructure, Transport and Tourism (MLIT) (2015). *Current Situation of Water Resources in Japan* (in Japanese). Available at: http://www.mlit.go.jp/mizukokudo/mizsei/mizukokudo_mizsei_fr2_000013.html (accessed 16 October 2016).

Ministry of Land, Infrastructure, Transport and Tourism (MLIT), Chikugo River Office (n.d.). *Saga Aqueduct* (in Japanese). Available at: http://www.qsr.mlit.go.jp/chikugo/gaiyou/dosui/ (accessed 16 October 2016).

Ministry of Land, Infrastructure, Transport and Tourism (MLIT), Kanto Regional Development Bureau (2012). *Ex-post Analysis on Kinugawa River Upstream Dams Coordination Project* (in Japanese). Available at: http://www.ktr.mlit.go.jp/ktr_content/content/000059170.pdf (accessed 16 October 2016).

Ministry of Land, Infrastructure, Transport and Tourism (MLIT), Kanto Regional Development Bureau (2016). 2016 *Drought in Tone River System* (in Japanese). Available at: http://www.ktr.mlit.go.jp/ktr_content/content/000655125.pdf (accessed 16 October 2016).

Ministry of Land, Infrastructure, Transport and Tourism (MLIT), Kanto Regional Development Bureau (n.d.). *Water Resources and Water Usage in Tone River* (in Japanese). Available at: http://www.ktr.mlit.go.jp/river/bousai/river_bousai00000065.html (accessed 16 October 2016).

Tokyo Metropolitan Government, Bureau of Waterworks (2015). *Water Supply in Tokyo*. Available at: https://www.waterworks.metro.tokyo.jp/eng/supply/ (accessed 16 October 2016).
Toshiba (2006). *Toshiba Review Vol. 61, No. 10* (in Japanese). Available at: https://www.toshiba.co.jp/tech/review/2006/10/61_10pdf/a02.pdf (accessed 16 October 2016).
Toshiba (2008). *Toshiba Review Vol. 63, No. 10* (in Japanese). Available at: https://www.toshiba.co.jp/tech/review/2008/10/63_10pdf/a05.pdf (accessed 16 October 2016).
Water Resources Development Subcommittee, Land Council (2015). *Advisory Report on Water Resources Policy in the Future* (in Japanese). Available at: http://www.mlit.go.jp/common/001084369.pdf (accessed 16 October 2016).

National drought management framework for drought preparedness in Korea (lessons from the 2014–2015 drought)

Ilpyo Hong[a], Joo-Heon Lee[b] and Hyo-Seob Cho[c]

[a]*Hydro Science & Engineering Division, Korea Institute of Civil Engineering & Building Technology (KICT), Goyang, Korea*
[b]*Corresponding author. Department of Civil Engineering, Joongbu University, Goyang, Korea. E-mail: leejh@joongbu.ac.kr*
[c]*Water Resources Information Center, Ministry of Land, Infrastructure and Transport (MLIT), Seoul, Korea*

Abstract

The scale of damage caused by drought is on the rise in South Korea. The country has been experiencing a 4–6-year cycle of extreme droughts at a nationwide scale. From 2014–2015, South Korea suffered from its worst drought in the past 50 years. This study aims to provide an effective drought management policy by analyzing the Korean government's response to the 2014–2015 drought under the National Drought Management Framework, which is composed of four stages: prevention, preparedness, response, and recovery. The findings indicate that the Korean government effectively addressed the drought, yet there are no measures currently in place to cope with megadroughts that last for more than 5 consecutive years. Immediate attention is required to conduct research and introduce policies that will help in managing megadroughts. This paper takes an interdisciplinary approach to drought preparedness research in the context of megadroughts and proposes an efficient national drought management framework that involves engagement of relevant line ministries.

Keywords: Drought; Drought management policy; Drought preparedness; National disaster management framework

1. Introduction

Since 2015, droughts that follow abnormal weather patterns have been seriously impacting Asian countries, including India, Vietnam, Thailand, and Korea. Approximately one-quarter of India's population is faced with inaccessibility of safe drinking water and with water shortage due to extreme drought. River levels in the Mekong Delta are at their lowest in the past century due to severe drought. The resulting intrusion of seawater has ruined almost 700,000 hectares of farmland in the region. Korea has been experiencing its worst drought in 50 years. Many farmers in Australia and New Zealand have been forced to sell their livestock due to water and pasture shortages, while approximately 32 million people in Africa are suffering from hunger as a result of drought-induced agricultural losses. Drought

doi: 10.2166/wp.2016.015

has serious socioeconomic repercussions and is regarded as a serious issue (Wilhite, 2000). Many predict that the negative impacts of droughts will continue to grow worldwide, and that solutions are urgently needed (Trenberth *et al.*, 2004).

South Korea has been experiencing a 5–6-year cycle of extreme droughts. Recent historical extreme droughts have occurred during the periods of 1967–1968, 1994–1995, and 2014–2015. Each drought has caused severe damage and has seriously impacted the Korean economy. If the current drought, which started in 2014 and persisted throughout 2015, continues in 2016, it is likely to become the most severe drought in Korea's history. The 2014–2015 drought alone is considered the worst in recorded history from the standpoint of severity and duration. As a result of the drought, South Chungcheong Province has introduced restrictive water rationing and restriction of in-stream flow and agricultural water for a number of multi-purpose dams in the Han River and Geum River Basin. This drought prompted in-depth discussions between relevant line ministries (Ministry of Public Safety and Security (MPSS); Ministry of Land, Infrastructure, and Transport (MLIT); Ministry of Agriculture, Food, and Rural Affairs (MAFRA); Ministry of Environment (MOE); and the Korea Meteorological Administration (KMA)) and water-related public companies (Korea Water Resources Corporation and Korea Rural Community Corporation), as well as academics with the goal of implementing cooperative mechanisms and reviewing Korea's current drought management system.

Drought management involves phases of monitoring, response, and recovery, and encompasses various disciplines, such as hydrology, meteorology, agriculture, and disaster prevention and mitigation. Thus drought management necessitates a holistic and integrated approach from relevant government authorities. As such, the Korean government implements a comprehensive drought management plan led by the MPSS. This ministry is in charge of disaster management through cooperation with the relevant line ministries, which include the MLIT, MAFRA, MOE, and the KMA.

The current study reviews Korea's natural disaster management policy, which focuses on flood and drought, and identifies the major issues at hand. The damage and impacts caused by droughts recurring every 4 to 6 years are examined along with the water-related ministries' preparedness and response plans for the different stages of drought. The actions of the central government, provincial government, and water-related public companies are analyzed to examine how the participants collaborated and responded coherently to the 2014–2015 drought. The findings are compared with responses to past droughts to draw lessons learned. The study concludes with recommendations to improve Korea's national drought management policy and framework.

2. 2014–2015 drought in historical perspective

2.1. Drought history in Korea

Climatologically, Korea is located in a monsoon area. The country experiences a wet summer season from June through August, when nearly half of the annual precipitation falls. The winter and spring are relatively dry, and, in times of absolute shortage of summer rainfall, the country has suffered droughts. This has been the fixed weather pattern in Korea for a very long time. The 'Annals of the Joseon Dynasty' record a total of 100 droughts occurring over a period of 519 years (1392–1910) of the dynasty, with severe droughts occurring every 4–6 years. The world's first rain gauge was invented in 1441 to observe rainfall in a quantitative manner, and as such, to monitor drought conditions.

Since agriculture was the key industry under the Joseon Dynasty, various agricultural technologies, such as irrigation and reservoir projects, were implemented to overcome droughts. These technologies were compiled into books. Table 1 presents details that were recorded in the annals during the reign of kings throughout the Joseon Dynasty, including the number of times that drought was mentioned. In the past, when a drought occurred, the king and his officials examined whether they had fulfilled their responsibilities and performed rituals and ceremonies praying for rain (Kim, 2001). The records show that the Korean peninsula has never been drought-free.

Figure 1 shows recent historical extreme drought years of Korea since 1960. Recent records kept since 1960 show that Korea has experienced an extreme drought every 4–6 years (Lee *et al.*, 2012). Multiyear droughts (>2 years' duration) have occurred in 1967–1968, 1976–1977, 1981–1982, and 1994–1995. According to modern scientific meteorological observations kept since 1910, the 2014–2015 droughts are considered to have been the most severe droughts around the Han River Basin in the past 100 years.

Table 2 shows the financial loss incurred per drought year. Multiyear drought hit South Korea in 1967–1968, 1976–1977, 1981–1982, and 1994–1995. The 1967–1968 drought stands as the drought incurring the greatest financial damage on record, with peak losses of 700,928 million Korean won in 1968. The financial loss due to drought damage was calculated for each year by converting the damaged area of paddy and upland fields into yield rates using the purchase price of rice. The annual gross domestic product (GDP) was used to estimate the relative scale of the damage costs. Lastly, the costs were adjusted to year 2000 prices to maintain comparability of the values. The results indicate a decreasing trend over time in both the damage costs and drought-affected area. When comparing the droughts in the 1960s with the 1980s and 1990s, the data show a decrease of about 50% in the

Table 1. Droughts recorded in the Annals of the Joseon Dynasty (Kim, 2001).

King	Regnal years	Length of reign (YYYY/MM)	Records of drought	King	Regnal years	Length of reign (YYYY/MM)	Records of drought
Taejo	6	1392/07–1398/09	21	Injo	26	1623/03–1649/05	168
Jeongjong	2	1398/09–1400/11	5	Hyojong	10	1649/05–1659/05	98
Taejong	18	1400/11–1418/08	158	Hyeonjong	15	1659/05–1674/08	105
Sejong	31	1418/08–1450/02	322	Hyeongae	–	–	154
Munjong	2	1450/02–1452/05	17	Sukjong	45	1674/08–1720/06	214
Danjong	3	1452/05–1455/06	25	Sukbo	–	–	5
Sejo	13	1455/06–1468/09	38	Gyeongjong	4	1720/06–1724/09	22
Yejong	1	1468/09–1469/11	7	Gyeongsu	–	–	5
Seongjong	25	1469/11–1494/12	442	Yeongjo	52	1724/08–1776/03	233
Yeonsan	12	1494/12–1506/09	53	Jeongjo	24	1776/03–1800/06	151
Jungjong	38	1506/09–1544/11	452	Sunjo	34	1800/07–1834/11	49
Injong	1	1544/11–1545/07	7	Heonjong	15	1834/11–1849/06	5
Myeongjong	22	1545/07–1567/06	145	Cheoljong	14	1849/06–1863/12	1
Seonjo	41	1567/07–1608/02	121	Gojong	44	1863/12–1907/07	–
Seonsu	–	–	14	Sunjong	3	1907/07–1910/08	–
Gwanghae	18	1608/02–1623/03	98				
Total no. of drought records							3,135

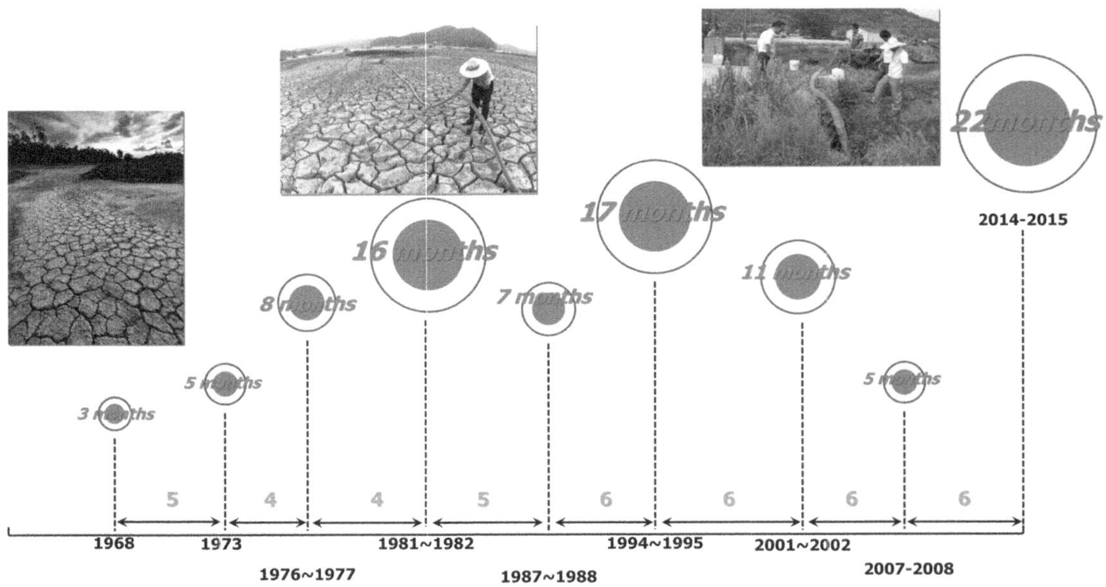

Fig. 1. Historical drought years in Korea since the 1960s. Korea's droughts have occurred every 4–6 years, with durations ranging from 3 to 22 months.

Table 2. Damage cost for historical drought years (MAFRA, 2001).

Drought year	Damaged area (ha)	Damage costs (million won)	Annual GDP (million won)	Damage costs adjusted to year 2000 prices (million won)
1967	420,547	626,615	1,313,260	822,244
1968	470,422	700,928	1,692,340	919,758
1976	28,218	42,044	14,413,200	55,170
1977	60,222	89,370	18,520,300	117,271
1981	145,457	216,730	49,324,000	284,393
1982	231,244	344,533	56,858,600	452,096
1992	31,523	46,969	273,267,400	61,633
1994	231,569	249,281	366,054,200	313,745

area damaged by droughts for similar volumes of annual rainfall. This is thought to be attributed to the continuous development of irrigation and agricultural water supplies.

2.2. *Spatiotemporal distribution of the 2014–2015 drought*

The 2014–2015 drought is assessed as the worst drought in Korea's history since the 1960s. The drought started in January 2014 and progressed through great summer rainfall shortages that were centered in the Gyeonggi and Gangwon Provinces (upper region of South Korea). The southern portion of Korea received a small amount of precipitation during the winter season, but the overall precipitation shortage continued in 2015, when the overall annual precipitation reached only half of the historical average. On June 19, 2015, the low-water level of the multi-purpose dam located on the Soyang River in Gangwon Province was recorded at 152.24 m, the second lowest since the 151.93 m record on June 24, 1978. The reservoir became nearly depleted as a result of the decreased inflow.

Figure 2 presents the meteorological drought conditions of the Han River Basin (the northern part of South Korea) during the 2014–2015 drought using the 6-month standardized precipitation index (SPI), a drought index that can be calculated from 6 months of accumulated precipitation. The SPI indicates that the drought, which started in June 2014, escalated to a severe drought when the annual summer typhoons and heavy rains did not take place. While a late typhoon in the fall did bring rainfall to the southern region after September, the drought continued into 2015, leading to a more severe drought from August to December 2015.

2.3. Hydrological impact: reservoir and stream flow

Table 3 presents drought disaster alerts released throughout 2014–2015 by nine main multi-purpose dams in South Korea. While most of the reservoirs managed to stay at a 'watch' level owing to flexible reservoir operation and water supply management, the Boryeong Dam reservoir level dropped very low and issued higher alert levels (i.e., 'emergency' alerts).

Table 4 and Figure 3 illustrate provincial areas that have adopted strict water rationing due to the 2014–2015 drought. Parts of Gangwon Province whose water access was via small-scale water works suffered the most from water shortage during the drought. In contrast to past records, drought preparedness and mitigation measures, such as emergency water supply and groundwater development, are now in place to significantly reduce the negative impacts of drought.

Table 5 shows the average storage rates of the major multi-purpose dams managed by the MLIT and K-Water (Korea Water Resources Corporation) during the drought. It shows that the majority of the storage rates dropped below 40%, which indicates severe hydrological drought conditions. The average volume of water in reservoirs in 2015 was 4,996 million m^3. The storage rate of the reservoirs was 39.4%, which is just 65% of the historical annual average rate. Seomjin River Dam and Boryeong Dam stored 42 million m^3 and 31 million m^3, respectively in 2015, which accounted for only 19% and 45%, respectively, of the storage rate compared to an average year.

2.4. Agricultural impact

Although the 2014–2015 drought was categorized as a meteorological drought that was serious in terms of both severity and duration, it caused relatively little agricultural damage due to the efficient utilization of approximately 3,000 agricultural reservoirs under the auspices of the Korea Rural Community Corporation. A survey on the crop damage caused by the 2014–2015 drought shows that drought ruined 5,810 ha of agricultural area. This total includes 2,451 ha of rice paddies, which is 0.3% of Korea's 799,000 ha total rice paddy area. In the case of fields, 3,359 ha was affected, equaling 0.4% of the country's 748,000 ha. The water release of some multi-purpose dams was limited due to low water levels, but most reservoirs minimized potential drought damage by applying optimal operation rules.

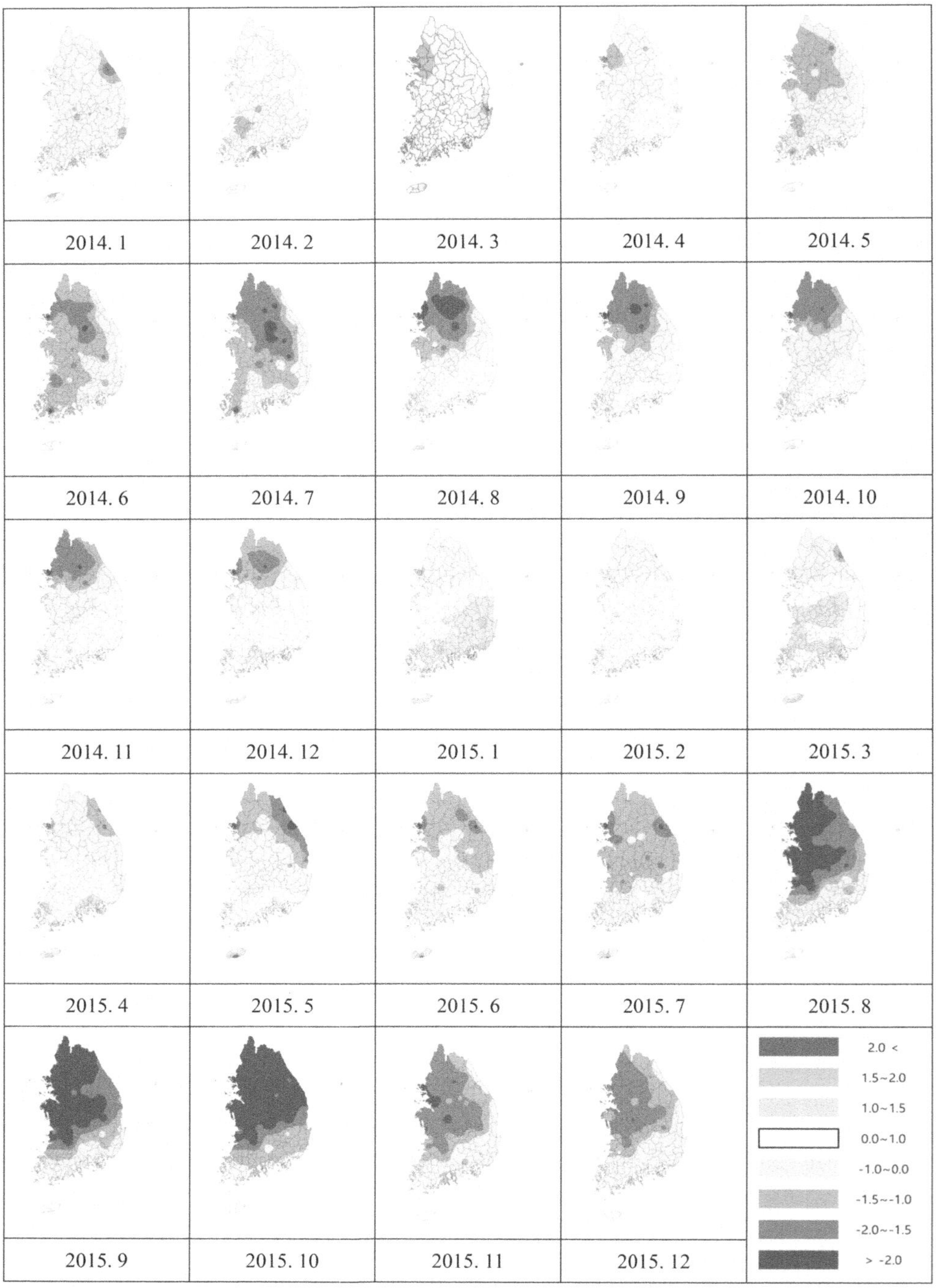

Fig. 2. Six-month SPI map for South Korea during 2014–2015 drought.

Table 3. Disaster alerts released by multi-purpose dams during the 2014–2015 drought, by alert classification and date range (yy/mm/dd).

Soyang River Dam/Choongju Dam	Andong Dam/ Imha Dam	Daecheong Dam	Yongdam Dam	Juam Dam	Boryeong Dam
(Watch) 15/3/25–15/12/31	(Watch) 14/8/20–14/10/22 (Watch) 15/7/8–15/12/31	(Watch) 15/8/14–15/9/22 (Warning) 15/9/23–15/12/31	(Watch) 15/3/27–15/12/31	(Watch) 15/9/21–15/12/31	(Watch) 15/8/5–15/8/14 (Warning) 15/8/15–15/8/17 (Emergency) 15/8/18–15/12/31

Table 4. Counties with water supply restrictions during the 2014–2015 drought.

Months	Counties	No. of persons
March 2015	Five areas, including Incheon City and Woongjin County	4,281
June 2015	19 areas, including Chuncheon City and Samcheok City	7,281
September 2015	Three areas, including Ongjin County and Ganghwa County	5,040
December 2015	Four areas, including Jecheon City and Shinan County	15,402

3. Natural disaster management policy of Korea

3.1. National framework for risk management

Figure 4 illustrates the national framework of governance for Korea's risk management strategy. A total of six authorities share the role of risk management, with the National Committee for Safety Management placed in charge of overall disaster and safety management related policies. The Central Disaster and Safety Countermeasures Headquarters are in charge of operations that oversee and manage large-scale disasters in the context of prevention, preparedness, response, and recovery. The Central Accident Response Headquarters is headed by the ministers from the MOE, MAFRA, and MLIT. The common role for all of these ministries is to form an early warning system for disaster risk reduction and to provide relevant technical assistance should disaster strike. Led by the heads of each region, the Regional Disaster Prevention and Countermeasure Headquarters administer and take the required measures for disaster prevention, preparedness, response, and recovery for all corresponding areas.

3.2. National plan for drought management

The national drought management policy is set and revised every 5 years based on the Framework Act on the Management of Disasters and Safety. It outlines Korea's national safety management plan in the case of natural disasters, including floods, droughts, and earthquakes. The National Safety Management Framework (Ministry of Public Safety and Security, 2015) outlines national drought management strategies as shown in Figure 5 and Table 6.

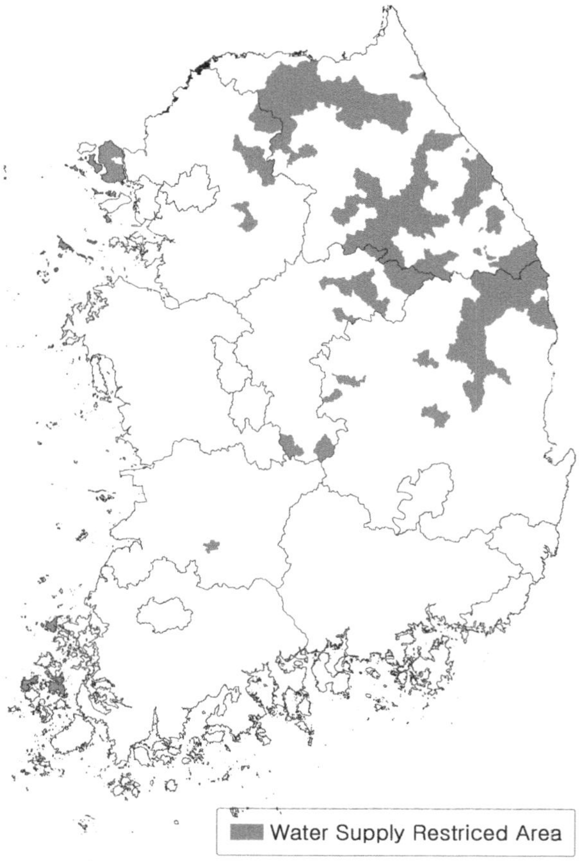

Fig. 3. Water supply restriction areas as a result of the 2014–2015 drought.

Table 5. Reservoir storage rate during the 2014–2015 drought.

Reservoirs		Stored volume (million m^3)	Rate (%) [% compared to a normal year]
Nationwide		4,996	39 [65]
Han River Basin	Soyang	1,275	44 [69]
	Chungjoo	1,135	41 [70]
	Hwengseung	27	31 [48]
Nakdong River Basin	Andong	422	34 [59]
	Imha	192	32 [65]
	Hapchoen	335	42 [72]
	Gunwei	15	31 [60]
Geum River Basin	Yongdam	266	33 [47]
	Daechung	577	39 [63]
Seomjin River Basin	Seomjin	42	9 [19]
	Juam	178	39 [58]
Others	Buan	26	51 [68]
	Boryeong	31	26 [45]

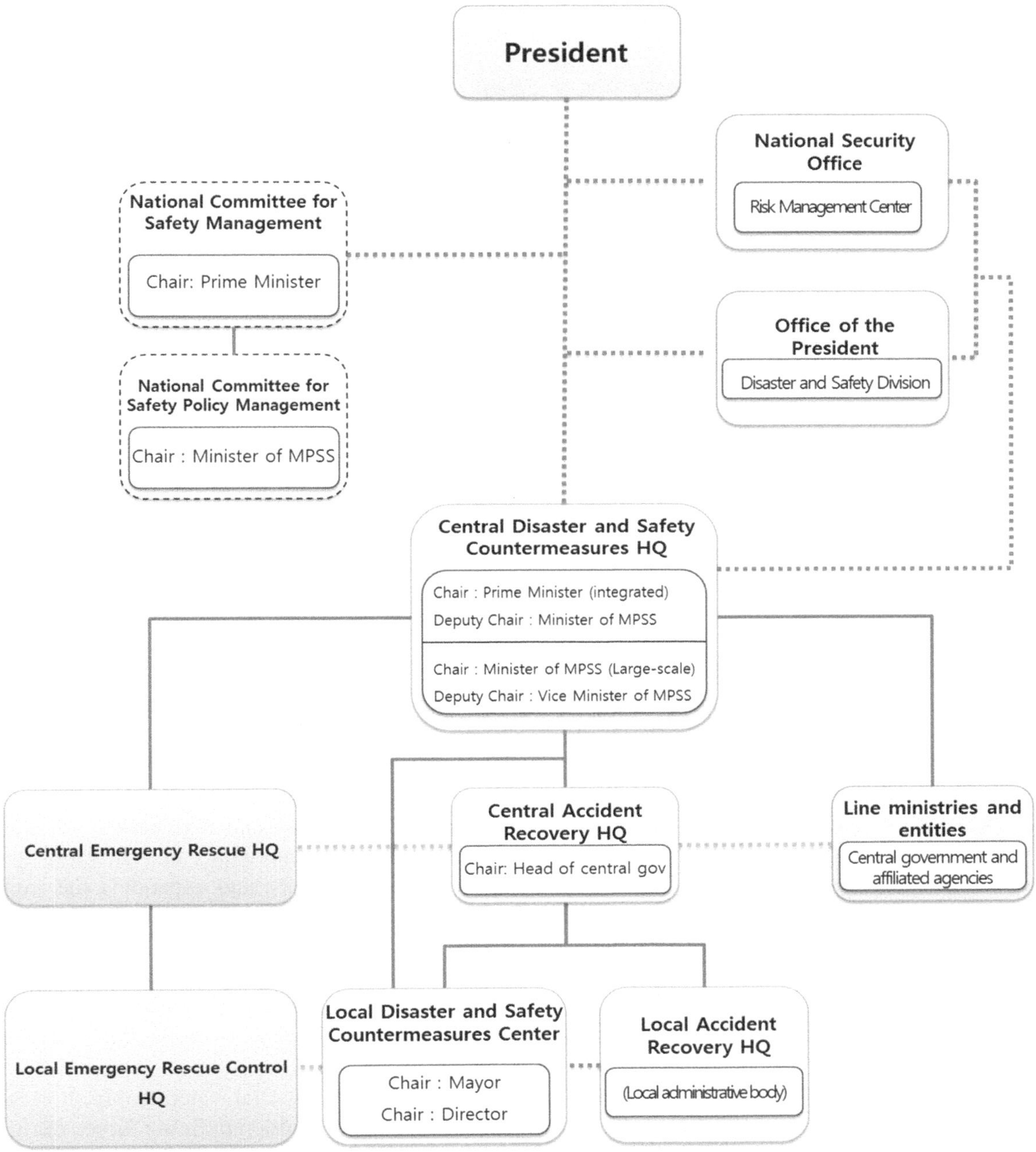

Fig. 4. National risk management governance of Korea (Multiple Ministries of Korea, 2015).

Drought management is composed of four steps: prevention (in normal circumstances), preparedness (when drought risk is expected), response (when drought has caused damage), and recovery (when drought is over). Relevant ministries participate in each step. The third step – response – is regarded

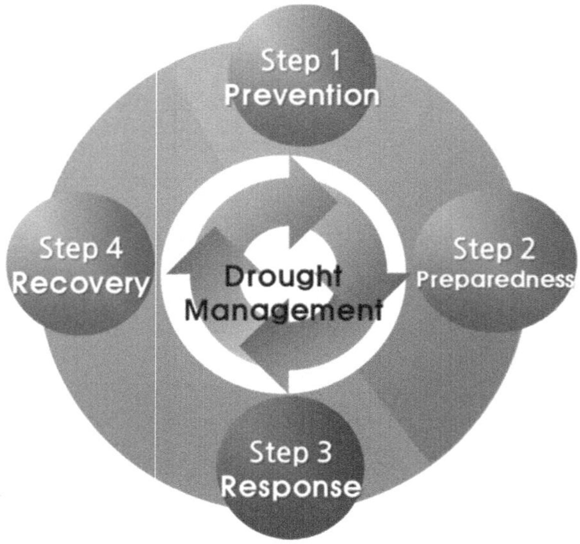

Fig. 5. The four stages of Korea's National Drought Management Framework.

as the most important, as it involves response to the drought risk at hand. Each ministry has its own response steps ranging from four to six stages depending on how the drought progresses.

3.3. Government responses and activities during the 2014–2015 drought

While national systems have not been established in South Korea for damage compensation or for casualty insurance due to droughts, the central government does currently provide support to local governments, which in turn address drought through special tax allocations and a reserve fund. These expenses are used to propel projects such as the construction of local waterworks in vulnerable areas, reservoir dredging, securing agricultural water in preparation for droughts, and expanding the supply of water-saving equipment. To minimize the agricultural damage during the 2014–2015 drought period, plans were established through the Four Major River Restoration Project to redirect fresh river water to agricultural reservoirs in advance of low water levels via multi-functional weirs. The project focused on filling more than 300 reservoirs that were expected to have insufficient volumes of water for agricultural use. Furthermore, financial support was provided to each local government to resolve the shortage of surface water by developing wells for pumping groundwater.

Various alternatives were also discussed to resolve municipal and industrial water supply shortages. For instance, purchasing or repurchasing vested water rights is now forbidden in South Korea during a drought. The MLIT and local governments also organized financial support for water use restriction in 2015, which encouraged municipal and industrial water conservation and attempted to prevent the wasting of emergency and temporary reservoir water. The monetary compensation of 1,240 Korean won per ton of water saved compared to the same period during the previous year was three times as high as the regional municipal water price (413 Korean won per ton). The financial support provided for 3 months (October to December of 2015) to the eight counties in the Chungcheong

Table 6. Drought management policies of the ministries based on Korea's National Safety Management Framework.

Step	MPSS	MAFRA	MOE	MLIT
Prevention	- Establish integrated drought management policy - Support establishment of local government's long-term plan	- Agricultural water planning - Set up medium and long-term plan for agricultural drought management - Promote water conservation policies - Monitor agricultural water status	- Promote water quality conservation policies - Prevent water contamination during shortage in rainfalls - Monitor water quality	- Manage multi-purpose dams and develop underground water resources - Secure additional water resources - Expand bulk water supply system and network - Monitor water resource system
Preparedness	- Raise awareness of drought preparedness - Collect rain water - Support research and development - Manage region where drought is forecasted	- Initiate research and development - Manage region where drought is forecasted	- Initiate research and development - Manage region where drought is forecasted	- Initiate research and development - Manage region where drought is forecasted
Response	- Lead national drought management governance - Support policies to limit water supply at each stage - Encourage emergency water resources development	- Take measures to counter drought depending on its severity	- Set concrete risk management structure - Limit water supply at each stage - Secure/supply emergency water resources - Undertake water conservation campaigns	- Set concrete risk management structure - Limit water supply at each stage - Secure/develop emergency water resources - Undertake water conservation campaigns
Recovery	- Provide financial support for restoration - Provide financial support to purchase equipment - Assist daily management activities	- Provide financial support to cover oil and electricity costs - Provide financial support to purchase equipment - Provide financial support to install tube walls	- Provide emergency water resources management plan - Reinforce water quality management	- Continue water resources development

Province, which was the most severely affected by the 2015 drought, amounted to approximately 3.41×10^9 Korean won.

A task force team (TFT) responsible for the 2014–2015 drought was organized under the management of the MPSS and consisted of the following five departments: The MPSS, MLIT, MAFRA, MOE, and the KMA. The TFT conducted drought forecasting and warning through KMA's drought forecasting system. The MLIT and K-Water determined restriction of municipal and industrial water use and in-stream flow from multi-purpose dams. Adjustments of the agricultural water supply from the agricultural reservoir were determined by MAFRA. The MOE managed the drought response of the local waterworks and low water supply areas concerning residential use.

The cooperation between the multiple ministries during the drought resulted in a total saving of 2.4×10^9 m^3 of water through: (1) the restriction of in-stream flow and agricultural water by multi-purpose dams; (2) the joint operation of hydroelectric dams; and (3) self-adjustments by residents in the drought-affected areas. This total water saving thereby prepared South Korea for the drought extending into 2016.

4. Establishment of the integrated drought risk management framework and the drought preparedness study for future extreme drought risk

4.1. Drought risk management framework

Drought should not be regarded as a mere weather phenomenon but as a natural disaster that can negatively affect various sectors, including climate, agriculture, environment, and water resources. Due to this widespread impact, addressing drought risk requires an integrated approach. A critical component of the approach is in raising awareness about drought disaster. Losses due to drought are often repeated because drought attracts only short-term attention during its presence and leads to a lack of countermeasures of a long-term perspective. Another way to improve drought risk management is to bolster entities in the national and local governments that specialize in drought management. The National Drought Management Framework calls for close coordination and cooperation among water-related government ministries, as is presented in Figure 6.

The National Disaster Management Institute has established a disaster alert system for instances of flood and drought in Korea. This system, illustrated in Table 7, is composed of four alert levels: attention (blue), watch (yellow), warning (orange), and emergency (red). Conditions must be met for each level before releasing the alert and vital measures must then be taken on each level. The MPSS is in charge of capturing the potential signs of drought risk and releasing the warning alerts. This ministry coordinates in advance with the line ministries involved in water resource management (KMA, MAFRA, MOE, and MLIT) in the case of an 'emergency (red)' alert stage to address the disaster in a holistic manner.

4.2. Multidisciplinary and multi-ministry research and development for drought preparedness

Understanding and addressing drought requires an interdisciplinary perspective because it is pertinent to many sectors, including water, climate, agriculture, and disaster prevention. Thus, drought research should be conducted via cooperation among experts from related fields. Current ongoing research includes finding ways to efficiently manage drought via support from interested line ministries, including the MPSS, MLIT, MAFRA, MOE, and the KMA.

In order to improve preparedness for future megadrought events, the TFT dedicated to drought is under the management of the National Committee on Water Resources Management. As shown in Table 8, this committee is now in the course of discussing 14 research areas under the following three categories: Drought Watch & Forecasting; Drought Impact Assessment & Drought Management; and Preemptive Drought Response. The 14 research areas will require teamwork amongst the

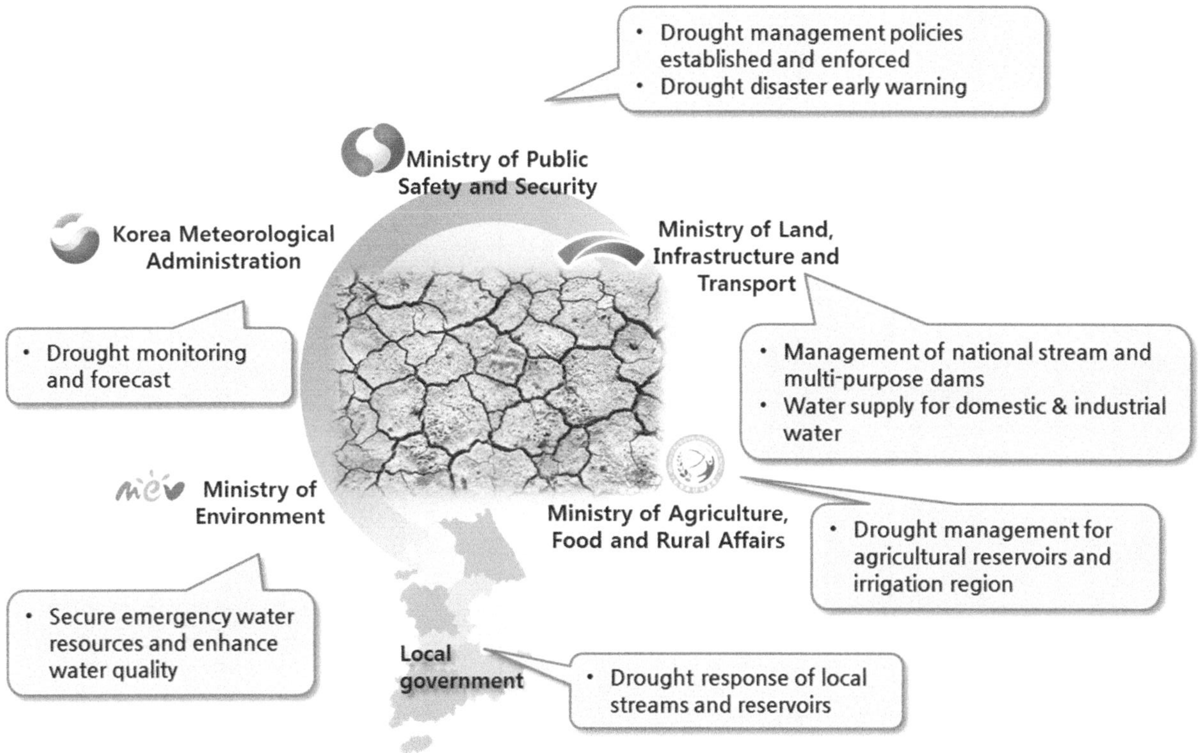

Fig. 6. Drought-related duties and responsibilities of the Korean government ministries.

Table 7. National drought risk alert levels for response (Multiple Ministries of Korea, 2015).

Level	Conditions	Note
Attention (Blue)	o Accumulated precipitation in the area over the last 2 months falls below 80% compared with the average year, and meets at least one of the conditions stated below: - Domestic water: 10–30% reduction in water supply. - Agricultural water: Either reservoir water level is 71–80% compared to the average year or soil moisture content of the field is 61–80%. o Reservoir: Reduced to meet the actual water demand.	Conduct monitoring for signs of drought disaster
Watch (Yellow)	o Accumulated precipitation in the area over the last 2 months falls below 70% compared with the average year, and meets at least one of the conditions stated below: - Domestic water: 30–50% reduction in water supply. - Agricultural water: Either reservoir water level is 61–70% compared to that of the average year or soil moisture content of the field is 41–60%. o Reservoir: Additional reduction in water supply for in-stream flow.	Activate multi-ministry cooperation system Check preparedness
Warning (Orange)	o Accumulated precipitation in the area over the last 2 months falls below 60% compared with the average year, and meets at least one of the conditions stated below: - Domestic water: More than 50% reduction in water supply. - Agricultural water: Drought damage occurs when either reservoir water level is below 60% compared to that of the average year or soil moisture content of the field is below 40%. o Reservoir: Additional reduction of agricultural, domestic, and industrial water.	Prepare to respond
Emergency (Red)	o Drought damage has occurred, or is likely to occur, at a nationwide scale.	Reinforce response

aforementioned five government entities and aims to strengthen the comprehensive National Drought Management System. The National Drought Management Policy developed in accordance with the research findings will help establish a more systemized framework for national drought response.

4.3. Governance of the drought management framework

Cooperation across different water sectors is necessary to efficiently respond to and minimize damage caused by multiyear droughts. Thus, a drought governance structure is proposed based on lessons from the 2014–2015 drought. The purpose of this drought governance structure is to ensure cooperation among relevant line ministries on themes including drought watch and forecast, drought impact assessment, and preemptive drought response. A mere understanding of the natural event will not be adequate in properly addressing the topic of drought disaster.

Figure 7 shows the governance structure of the Drought Management Framework for Korea, which involves multi-ministry and water related agencies. The KMA, MAFRA, and MLIT are in charge of meteorological, agricultural, and hydrological drought, respectively, and all collaborate to produce comprehensive drought information and allow integrated drought monitoring and prediction. The MOE is in charge of developing a monitoring scheme for environmental impacts of drought. The MPSS is in

Table 8. Research and development conducted by the line ministries involved with drought preparedness (Korea Institute of Civil Engineering and Building Technology (KICT), 2016).

Drought Watch & Forecast	1.1. Improve accuracy of meteorological drought monitoring technologies 1.2. Improve accuracy of drought forecasting technologies 1.3. Develop long-term drought forecasting technologies reflecting climate change	Develop advanced technologies for producing drought information	KMA
	1.4. Develop technologies for agricultural drought monitoring and forecasting 1.5. Develop enhanced technologies for hydrological forecasts using weather forecast information 1.6. Develop monitoring technologies for drought-prone local water resources	Develop ICT-based technologies for integrated management of agricultural drought	MAFRA
Drought Impact Assessment & Drought Management	2.1. Develop technologies to allow pre-adaptation to and optimal management of agricultural drought 2.2. Establish a response and management system for extreme drought	Develop technologies to install drought early warning and response systems	MPSS
	2.3. Establish a drought forecasting and warning system 2.4. Develop technologies to assess hydrological drought impact	Develop technologies for long-term hydrological information forecasting and enhanced water supply systems	MLIT
Preemptive Drought Response	3.1. Develop technologies to improve water efficiency in agricultural water use		
	3.2. Develop technologies for an advanced water supply system in response to drought 3.3. Construct a portable desalination plant 3.4. Establish multi-purpose underground water plants	Develop technologies to monitor local water resource availability and secure additional water resources	MOE

charge of developing a drought early warning system, and lastly, the MAFRA focuses on coping strategies for agricultural drought.

More specifically, the MAFRA aims to renew the agricultural water supply system to ensure that the agricultural sector is capable of making early responses to drought. The MOE is establishing a seawater desalination system and groundwater dams as part of a solution to secure additional emergency water resources during times of drought. Lastly, the MLIT strengthens the capacity of the domestic and industrial water supply systems used during a drought period. This division of roles and responsibilities for drought management will allow an integrated drought management framework to function smoothly. This will not only maximize the Korean government's ability to respond to droughts at a national level, but is also expected to minimize the economic, social, and environmental impacts of droughts that have increased significantly nationwide.

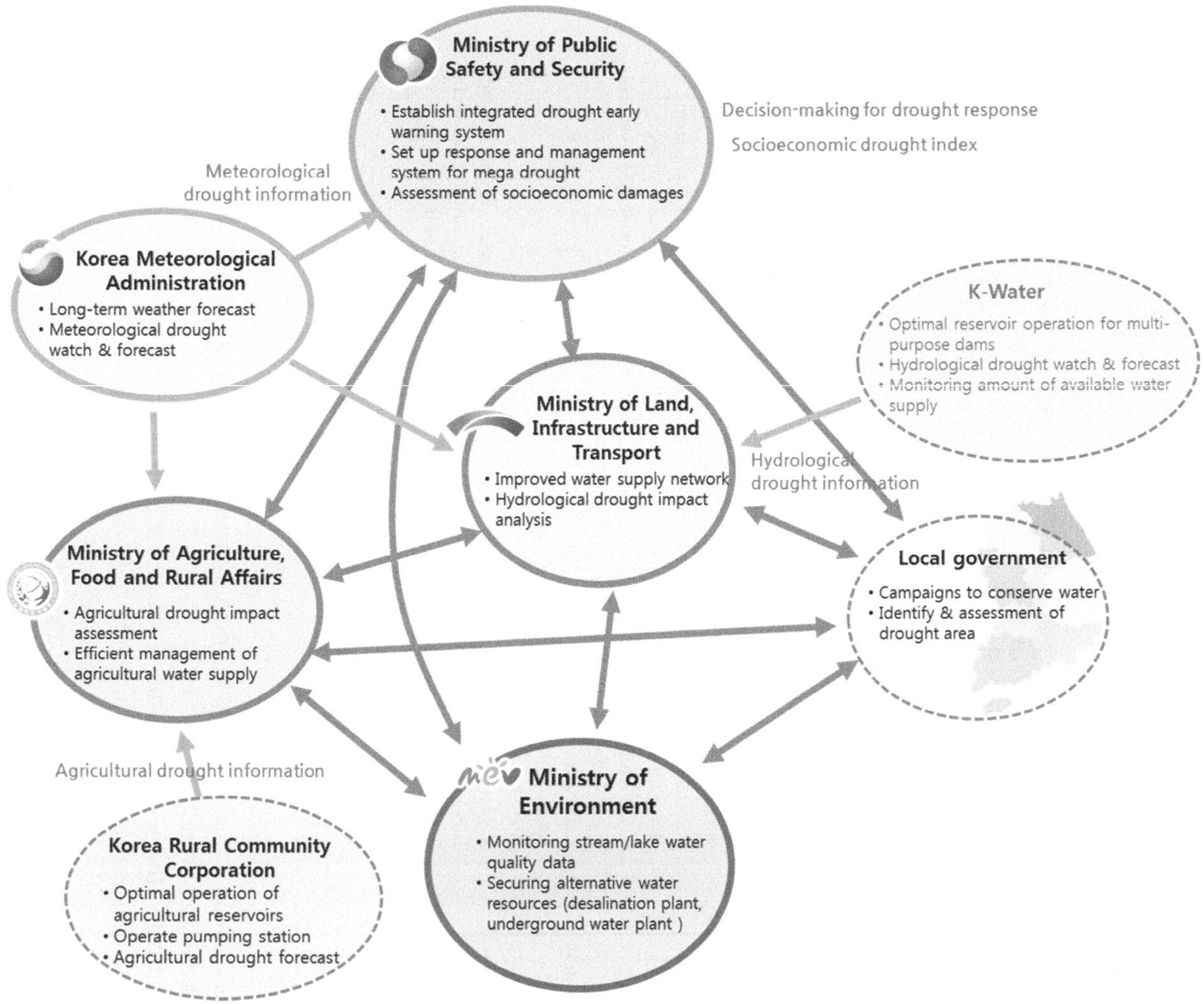

Fig. 7. The Multi-Ministry Drought Management Framework of Korea applied in response to extreme drought conditions.

5. Conclusions

Drought, like flooding, begins as part of an abnormal weather cycle, but can develop into a natural disaster depending on its duration and the scale of its negative socioeconomic effects. To establish a more precise drought management framework, Korea must establish a link between a real-time drought monitoring and prediction scheme and a drought impact assessment system.

Too much emphasis is often placed on improving the accuracy of drought monitoring and forecasting systems from an academic standpoint. This emphasis often leaves little support for the development and/or refinement of national drought management policy. It is important to strike a balance between integrating academic research findings and setting national drought management policies and measures. At the same time, academic research is not adequately reflected in government policies on drought management. This is because the majority of the studies conducted on drought are focused on a scientific-technological approach, while there is a clear need for studies related to policy measures. Such measures include establishing countermeasures to drought, establishing drought policies, and providing virtual drought response training and exercises.

Drought management measures put in place by the government require an appropriately balanced distribution of tasks between the local and central governments. Since the majority of the countermeasures to drought have thus far been established by the central government, the central government and academic research institutes must both provide support in tailoring drought management measures for administration by the local governments. Such measures should reflect water resource management policies in concert with the stance of local governments on water resources to achieve progress in the future.

From monitoring to response and recovery, drought affects human lives in myriad ways through environmental, social, and economic impacts. Thus, mechanisms to facilitate cooperation among relevant line ministries, integrate research findings from related academic fields, and administrate government policies must be strategically combined to introduce and embed systematic and effective national drought management policy in the country. Korea's drought from 2014–2015 was by far the worst incidence of drought in the past 50 years. The government's early response, appropriate drought management policy, and well-defined risk management manual helped minimize adverse drought impacts. For successful mitigation of drought in the future, it is important to secure funding for continued research and make further improvements to the drought management system.

Acknowledgements

We would like to show our gratitude to the Ministry of Land, Infrastructure and Transportation for the strong support of this work. This work was partially supported by the Korea Meteorological Administration Research and Development Program under Grant KMIPA 2015–2070, and by the National Research Foundation of Korea (NRF) Grant funded by the Korean Government (NRF-2013R1A1A2057517).

References

Kim, H. J. (2001). Drought Records from the Annals of the Joseon Dynasty. (2001). Korea Water Resources Association project research paper, Korea Water Resources Association.

Korea Institute of Civil Engineering and Building Technology (KICT) (2016). *Joint study report on establishing a preemptive drought management system to minimize damage.*

Lee, J. H., Seo, J. W. & Kim, C. J. (2012). Analysis of the tendency, regularity, and repetition rate of drought in the Korean Peninsula utilizing the drought index. *Journal of KWRA, Korea Water Resources Association 45*(1), 75–89.

Ministry of Agriculture, Food, and Rural Affairs (MAFRA) (2001). *White book for drought preparedness.*

Ministry of Public Safety and Security (2015). *National Safety Management Plan (2015–2019).*

Multiple Ministries of Korea (MPSS, MAFRA, MOE, and MLIT) (2015). *'Drought Disaster' Risk Management Manual.*

Trenberth, K. E., Overpeck, J. T. & Solomon, S. (2004). Exploring drought and its implications for the future. *Eos, Transactions American Geophysical Union 85*(3), 27.

Wilhite, D. A. (2000). Drought as a natural hazard: concepts and definitions. In: *Drought: A Global Assessment.* Wilhite, D. A. (ed.). Routledge Publishers, London, pp. 3–18.

Drought policy in Mexico: a long, slow march toward an integrated and preventive management model

Ismael Aguilar-Barajas[a,*], Nicholas P. Sisto[b], Víctor Magaña-Rueda[c], Aldo I. Ramírez[a] and Jürgen Mahlknecht[a]

[a]*Tecnológico de Monterrey, Av. Eugenio Garza Sada Sur 2501, Col. Tecnológico C.P. 64849, Monterrey, Nuevo León, Mexico*
**Corresponding author. E-mail: iaguilar@itesm.mx*
[b]*CISE (Centro de Investigaciones Socioeconómicas), Universidad Autónoma de Coahuila, Unidad Campo Redondo Edificio 'S', 3er Piso, Apartado Postal No. 680 C.P. 25280, Saltillo, Coahuila, Mexico*
[c]*Instituto de Geografía, Universidad Nacional Autónoma de México, Ciudad Universitaria, Coyoacán, Ciudad de México 04510, Mexico*

Abstract

Mexico is exposed to droughts and vulnerable to their impacts. Despite a history of serious economic and social consequences from drought, until recently national water policy did not acknowledge this type of natural disaster as a priority issue. This paper analyses how drought policy in Mexico has evolved since the creation of the National Water Commission in 1989 up to 2013, when the most specific initiative in that regard was introduced: the National Drought Program (Programa Nacional Contra la Sequía, PRONACOSE). Over that period of time drought has slowly but steadily gained importance in national water planning and management. Significant initiatives have been implemented, including a national drought early warning and monitoring system. Nevertheless, efforts still focus on compensating economic losses after the fact and little has been achieved in terms of reducing the country's vulnerability to drought. There is a great need to increase public awareness and boost the scientific knowledge necessary for the design of appropriate preventive measures. A long way remains to establish an effective, forward-looking, integrated and preventive drought management model.

Keywords: Drought policy; Mexico; National and regional water programs; National Drought Program (PRONACOSE)

doi: 10.2166/wp.2016.116

1. Introduction

1.1. Background

Mexico is the world's 11th most populated country (125.4 million inhabitants), 15th largest national economy in terms of gross domestic product (GDP) and with over 1.96 million km^2 of territory, ranks 14th in size (World Bank, 2016). Mexico's climate is relatively dry, with a mean annual precipitation (760 mm) well below the world's average (1,127 mm). Large tracts of the country normally receive much less precipitation than the national average, below 500 mm in most of the north and central regions (Figure 1). Consequently, about 54% of Mexico's territory falls under the Very Dry (BW), Dry (BS_0) or Semiarid (BS_1) categories of the Koppen-García climate classification system (García, 2004). In five northern and central states (Aguascalientes, Baja California, Baja California Sur, Coahuila, Sonora), that figure exceeds 90%.

Mexico is prone to droughts, i.e. episodes of significantly lower-than-average precipitation – see Mishra & Singh (2010) for a detailed exposition of drought concepts and categories. Exceptionally intense and prolonged droughts affecting large swathes of the country occur about every 50 years (Méndez & Magaña, 2010) but significant drought events strike almost every decade – for a historical chronicle of droughts since the fifteenth century, see Florescano Mayet *et al.* (1980). The most recent 50-year drought to affect the country occurred from late 2010 to mid-2013. At the drought's peak, up to

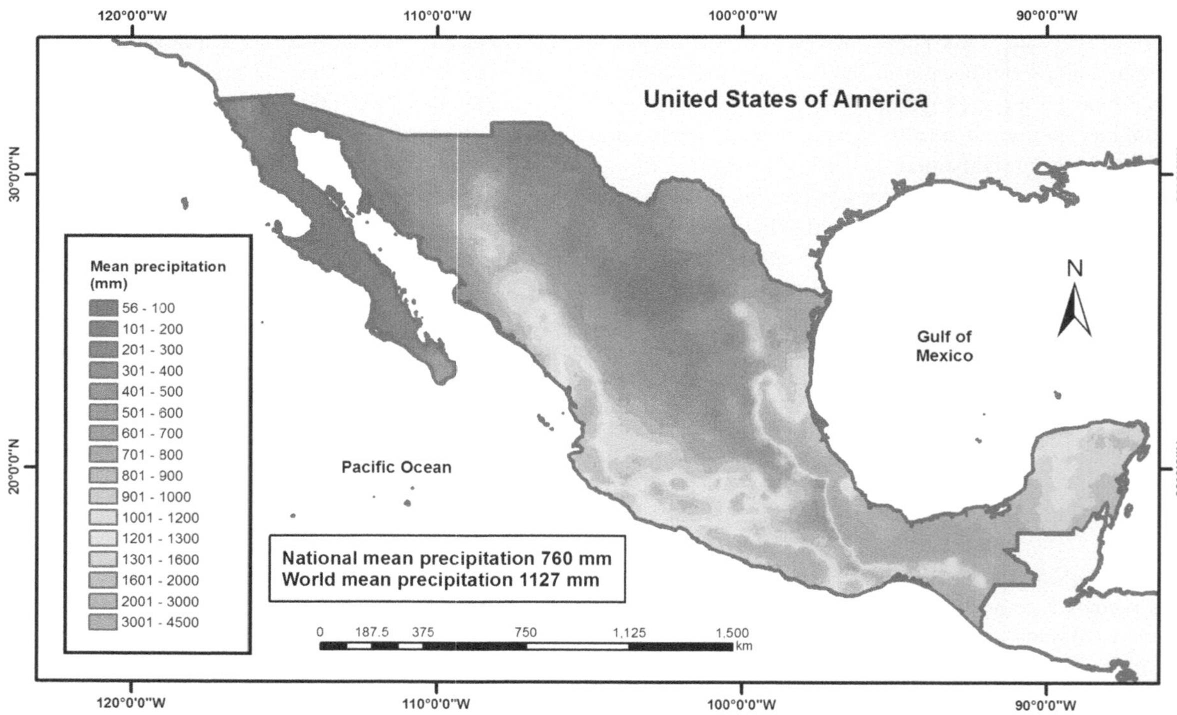

Fig. 1. Mean annual precipitation in millimetres (mm), Mexico. *Source:* CONAGUA (National Water Commission) (2015).

three quarters of the country registered abnormally dry conditions according to the standardized precipitation index (SPI) – a standard drought index first proposed by McKee *et al.* (1993). Figure 2 reports that index for 2011 and 2012; note that the relatively arid northern and central regions were the most affected. Ortega-Gaucín & Velasco (2013) estimate that this drought caused in 2011 alone, over USD1,300 million in agricultural losses (mostly in the northern and central states), reduced national GDP by 10% and affected 48 million people.

Mexico's northern, arid and drought-prone region harbors a significant share of its population and economic activities. Table 1 illustrates this with the four border states of Tamaulipas, Nuevo León, Coahuila and Chihuahua (Northeastern Mexico, NEM) that together comprise most of the Río Bravo Basin (RBB). This region encompasses more than a quarter of the nation's territory and 42% of its arid lands. Its 14 million inhabitants produce over 16% of the country's GDP, almost a quarter of its

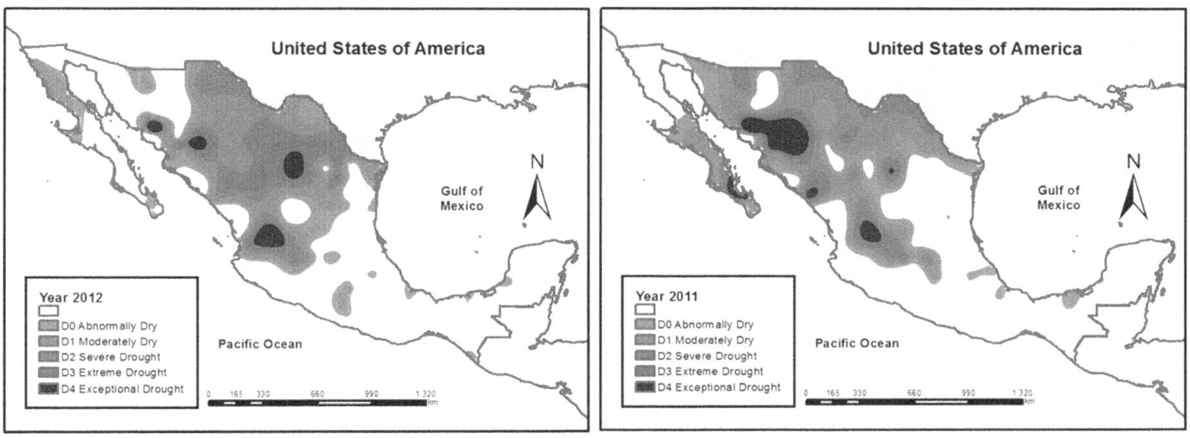

Fig. 2. SPI, Mexico, 2011 and 2012. *Source:* Authors.

Table 1. Territory, population and economy, Mexico and NEM, 2010.

	Territory (10^3 km^2)		Population (10^6 inhab.)	GDP (10^6 \$MXP)			GPD p.c. (\$MXP)
	Total	Arid[a]	Total	Total	Manuf.	Agric.	
México	1,964	1,077	112.3	11,966	2,017	390.9	106,519
Tamaulipas	77.2	55.5	3.3	381.2	63.8	14.1	116,640
Nuevo León	64.2	54.7	4.7	855.0	205.9	6.1	183,740
Coahuila	151.7	151.7	2.7	380.9	137.0	9.8	138,584
Chihuahua	250.3	195.5	3.4	326.7	67.7	21.8	95,894
NEM	543.4	457.5	14.1	1,943.8	474.5	51.8	138,086
(national %)	(27.7)	(42.5)	(12.5)	(16.2)	(23.5)	(13.3)	

[a]Surface classified as BW, BS_0 or BS_1 (Koppen-García climate classification system).
Source: Authors, with data from the Mexican federal statistical agency (INEGI).

manufacturing output and thanks to an extensive system of irrigation infrastructure, 13% of its agricultural production. The region's GDP per capita is 23% above the national figure.

The body of published literature on the social and economic impacts of droughts in Mexico, as well as the country's policy responses to the phenomenon, is limited in extent – this, perhaps surprising, state of affairs motivates this paper. A small number of contributions address impacts on agricultural output and employment (Ortega Gaucín *et al.*, 2008), spill-over effects through forward and backward economic linkages (Sisto *et al.*, 2011) and threats to urban water supply (Sisto *et al.*, 2016). Feng *et al.* (2010) corroborate quantitatively the long-held perception of a link between migration and drought, mentioned for example in Velasco *et al.* (2005); Neri & Magaña (2016) argue that the severity of drought impacts depends on the vulnerability of the region affected, including the type of water sources exploited – in Mexico aquifers provide about two thirds of water supplied to urban areas. The narrow literature on national drought policy suggests responses have been patchy and mostly reactive (Ortega Gaucin *et al.*, 2014; Arreguín Cortés *et al.*, 2015).

1.2. Objectives

This paper evaluates Mexico's institutional and policy responses to drought over the past quarter of a century. Specifically, it presents a detailed and critical analysis of how drought policy has evolved, focusing on national and regional water programs produced during that period of time. The paper is organized as follows. Section 1 offers background information on drought risks and impacts in Mexico. Section 2 shows how and to what extent the five national water programs presented since 1989 have addressed the issue of drought. This section also deals with two relevant initiatives introduced during that period: the National Fund for Natural Disasters (FONDEN) and the National Prevention Fund (FOPREDEN). Section 3 focuses on the National Program against Drought (PRONACOSE), a recent and key program intended to deal explicitly with droughts. Section 4 concludes.

2. Drought policy in Mexico's national water programs, 1989–2014

2.1. Mexico's national water management framework: a brief overview

Mexico's national water management framework is highly fragmented. The water sector's main actor, the national water commission (CONAGUA, created in 1989 by presidential decree) forms part of the federal ministry of the environment and natural resources (SEMARNAT). The federal ministry of agriculture (SAGARPA) also possesses wide authority in terms of water management. In total, 12 federal-level ministries exert influence on the matter. Moreover, 13 River Basin Organizations and 26 River Basin Councils, with a wide array of technical, administrative, legal and other functions, report directly to CONAGUA. The interior ministry (SEGOB) administers both the FONDEN and FOPREDEN programs. The former, established in 1996, provides funds to state and municipal governments facing emergencies following natural disasters (including hurricanes and droughts), while the latter, created in 2003, finances disaster prevention programs. Multiple actors, diffuse responsibilities and a lack of coordination have introduced a degree of dysfunctionality into Mexico's water management institutional structure (Organisation for Economic Co-operation and Development (OECD), 2013a).

2.2. Drought and its impacts: perceptions and knowledge

Policies should be grounded on sound information. We thus begin our analysis of Mexico's water programs with a review of what these documents reveal about how drought is perceived in relation to other water challenges, and how much is known about the impacts caused by drought events.

2.2.1. Drought as an object of concern. National water programs have systematically recognized the overall aridity that characterizes the country, especially in the north. Another recurring theme is the insufficiency of hydrological monitoring, especially in the case of groundwater. The 2001–2006 Program expressed concern over the growing pressure on groundwater, to the point of overexploitation in many areas of high population and economic growth (CNA (National Water Commission), 2001). It also emphasized that aquifers represent an essential resource for national development, as 75 million people, 1.6 million hectares of irrigated agricultural and half of the country's industrial water requirements depend on this source. The program furthermore suggested that climate change could occasion more severe, frequent and prolonged droughts, and thus reduce the availability of both surface and groundwater.

A series of regional, long-term water programs (Programas Hidráulicos Regionales de Gran Visión 2001–2025) made a substantial contribution to regional water planning as well as the understanding of river-basin scale water and policy issues. The RBB Program exemplifies this. It presented a clear assessment of that basin's main challenges: a limited natural availability of water in the face of increasing demands for both domestic and industrial uses, fueled by population and economic growth. Moreover, that program recognized the basin's vulnerability to drought as well as the relief provided by occasional hurricanes. It also argued that the existing institutional framework was too fragmented and uncoordinated to meet the basin's challenges effectively, and made the case for a more proactive governance model (CNA, 2003, pp. 22–23).

The 2007–2012 Program stressed the strategic value of water and noted the unfortunate fact that most of the country's growth occurs where water is least available. It also pointed out that per capita water availability was declining (CONAGUA, 2008). The 2014–2018 Program (CONAGUA, 2014) spoke briefly of the country's high vulnerability to drought (especially in the northern states) and emphasized the importance of adaptation to climate change, in particular with respect to a possible decline in surface flows by 2030 (using 2013 as baseline year). The program also expressed the need to deal with drought explicitly through a dedicated institutional program – this need would eventually be addressed by PRONACOSE.

2.2.2. Impacts of droughts. National water programs mostly report generic observations on the impacts of droughts, such as: reduced crop and cattle production; decreased hydropower generation; damage to flora, fauna and ecosystems; risk to urban water supplies through declining water availability; and, overexploitation of aquifers (CNA, 2000; CONAGUA, 2008). The 2002–2006 Regional Program for the RBB (CNA, 2003) went further toward describing impacts on the agricultural sector, mentioning increased activity in agricultural land markets (sales and leases) and large-scale movements of cattle herds away from drought-affected areas. It also delved into the usually ignored realm of social impacts, including: unemployment, migration (in particular, of agricultural workers), abandonment of crop lands, conflicts among water users, and even family breakups and suicides.

Relatively little hard data on specific drought events can be found in water programs. Beyond the chronology of the most severe and geographically extended droughts since the middle of the last century (1948–1954; 1960–1964; 1993–2002; and 2010–2013), few detailed economic estimates appear, with some notable exceptions. Economic impacts of drought between 1980 and 1998 are estimated to have cost up to USD500 million (CNA, 2000, p. 17); the latter year is deemed particularly difficult, with most of the country affected (23 of 32 states, particularly in the north and central regions) and damage estimated at USD310 million (CNA, 2000, p. 18). The RBB was exceptionally hard hit. Reduced surface flows in that region impeded Mexico's ability to honor its water commitments to the United States (as laid out in the 1944 Treaty), triggering a serious conflict between the two nations.

The focus of that debate centered on whether the on-going drought qualified as 'extraordinary drought' as contemplated in the Treaty – under that clause Mexico would have no obligation to deliver water to the USA from the Río Bravo tributaries considered in the Treaty. The conflict escalated to such a level that there were voices in the USA calling for restrictions on (or even suspension of) deliveries of Colorado River waters to Mexico. For many months in 2001, this issue was at the very top of the US–Mexico agenda. For more detail on these binational arrangements and their connection to droughts, see Mumme & Aguilar-Barajas (2003).

2.3. Policy responses to drought

2.3.1. The importance of drought in water programs' objectives and strategies. Objectives stated in water programs and the strategies and projects proposed to achieve those reveal a great deal about national priorities and their evolution through time. Table 2 reports the objectives presented in all national water programs since 1989; for objectives that deal with drought, Table 3 presents the proposed strategies and projects.

The 1989–1994 Program did not mention drought at all in its objectives, strategies or projects. The 1995–2000 Program did not either (at least explicitly), despite the fact that the serious aforementioned conflict with the United States over binational waters was already brewing. That program did however introduce a multi-sectoral and wide-ranging strategy relevant for drought policy: the formation of River Basin Councils. The justification behind this was to strengthen cooperation between authorities (federal, state and municipal) and water users, in order to improve water planning and management. In all subsequent programs, River Basin Councils were to form a fundamental part of national water planning and management. The 1995–2000 Program also suggested the necessity of 'controlling adverse impacts' (CNA, 1994, p. 34) from extreme weather events such as floods and droughts (a clearly reactive stance), and of addressing the problem of groundwater overexploitation.

The 2001–2006 Program for the first time introduced drought explicitly in its objectives, albeit jointly with floods and as the last of six objectives. Strategies, projects and financing however focused entirely on floods. The 2007–2012 Program's sixth objective (out of eight objectives in total) implicitly included drought as part of 'extreme weather events'. Strategies and projects mostly focused on floods, but one drought-specific strategy did appear: the development of prevention plans. The program recommended these plans include rules on water allocations between competing uses under drought conditions as well as mechanisms to verify compliance. It also proposed that River Basin Councils 'lend support' to this endeavor. The latter proposal illustrates a larger problem, common to all national water programs: the lack of definition of responsibilities among multiple actors. Programs customarily present ample lists of institutions and organizations to be involved in meeting objectives, with a brief and general description

Table 2. National water programs: objectives, Mexico, 1989–2014.

Period	Objectives
89–94	1. To develop the infrastructure necessary to improve access to water and sewer services.
	2. To promote the efficiency of water use.
	3. To improve water quality in river basins and aquifers most affected by pollution.
95–00	1. To improve access to water and sewer services.
	2. To increase wastewater treatment efforts.
	3. To grant legal certainty to holders of water use rights.
	4. To promote the transition toward sustainable development.
	5. To extend public participation in water planning and management.
	6. To increase the efficiency of water management.
	7. To promote efficiency in the allocation of water among distinct types of uses.
01–06	1. To promote efficiency in agricultural water use.
	2. To improve access to and quality of water, sewer and sanitation services.
	3. To achieve integrated and sustainable water management in river basins and aquifers.
	4. To foster the technical, administrative and financial development of the water sector.
	5. To strengthen public participation in water management and foment a culture of responsible use.
	6. To manage the risks and impacts of floods and droughts.
07–12	1. To improve efficiency in agricultural water use.
	2. To improve access to and quality of water, sewer and sanitation services.
	3. To foster integrated and sustainable water management in river basins and aquifers.
	4. To improve the technical, administrative and financial development of the water sector.
	5. To strengthen public participation in water management and foment a culture of responsible use.
	6. To manage the risks and impacts of water related disasters.
	7. To evaluate the effects of climate change on the hydrological cycle.
	8. To create a culture of compliance with the administrative dispositions of the National Water Law.
14–18	1. To strengthen integrated and sustainable water management.
	2. To increase water security in the face of droughts and floods.
	3. To bolster water supply and access to water, sewer and sanitation services.
	4. To increase the technical, scientific and technological capacities of the water sector.
	5. To ensure water for crop irrigation, energy, industry, tourism and other activities in a sustainable way.
	6. To strengthen Mexico's participation in the international water scene.

Source: Authors, with original national water programs. For 1989–1994: (CNA, 1993, 1994); for 1995–2000 (CNA, 1995); for 2001–2006 (CNA, 2001); for 2007–2012 (CONAGUA, 2008); for 2014–2018 (CONAGUA, 2014).

of their intended participation. Strategies however usually do not refer to these actors, and when they do, the language is too vague ('to lend support', 'to facilitate', etc.…) to establish clear responsibilities, let alone assess *ex post* performance.

The 2014–2018 Program elevated drought as the second most important objective (out of six) but again in conjunction with floods. It did however present a specific and innovative strategy: a drought prevention program (PRONACOSE). Given the relevance of this program, this paper's third section focuses on it.

Regional water programs on occasion have offered more concrete drought policy proposals than national programs. For example, the 2002–2006 Regional Program for the RBB (CNA, 2003) suggested: investing in new hydraulic infrastructure (dams, wells and water conveyance works); emergency distribution by water tank trucks to domestic users and ranchers; revolving suspension of water services in cities; and, specifically for the agricultural sector, investments in more efficient water

Table 3. National water programs, objectives, strategies and projects to deal with drought, 1989–2014.

Objective (period)	Strategies	Projects
N/A (95–00) & (89–94)	These programs did not present objectives, strategies or projects to deal with drought.	
6. To manage the risks and impacts of floods and droughts. (01–06)	6.2 To implement rational water use policies to better deal with periods of drought.	These programs did not present explicit projects to deal with drought.
6. To manage the risks and impacts of water related disasters. (07–12)	9. To formulate preventive plans to better deal with drought and to back their implementation.	6.9.1 To formulate a contingency drought plan for each Basin Organization.
2. To increase water security in the face of droughts and floods. (14–18)	2.1 To safeguard and increase the resilience of communities and producers in areas exposed to flood and/or drought risk.	2.1.2 To implement the National Drought Program (PRONACOSE).
	2.2 To reduce vulnerability to impacts from climate change or variability.	2.2.1 To strengthen the participation and co-responsibility of states and municipalities in adaptation measures for climate change or variability.
		2.2.2 To create or strengthen funds for climate change adaptation and for the maintenance and restoration of hydraulic infrastructure.

Source: Authors, with original national water programs. For 1989–1994: (CNA, 1993, 1994); for 1995–2000 (CNA, 1995); for 2001–2006 (CNA, 2001); for 2007–2012 (CONAGUA, 2008); for 2014–2018 (CONAGUA, 2014).

application systems for irrigated crops as well as temporary work programs. The program however clearly recognized that these measures would only serve to mitigate the impacts of drought.

2.3.2. Linking objectives with strategies and projects. The links between objectives, strategies and projects in national water programs reveal the slow evolution of drought policy during the past 25 years. Table 3 summarizes that long journey, from an initial point (the 1989–1994 and 1995–2000 periods) where no objectives, strategies or projects dealing with drought were stated, to the current period (2014–2018) where all three concepts are made explicit. The 2014–2018 Program however shows weaknesses in terms of articulating and linking objectives with strategies and projects. The program's second objective is the only one to refer directly to drought but most (if not all) of the remaining five objectives are presumably connected to drought – these connections however are not made explicit or discussed. Moreover, amalgamating drought and floods arguably muddles matters, as both types of disasters differ in nature and require different types of interventions.

Two strategies derive from the Program's second objective. The first proposes to increase resilience to floods and droughts in areas exposed to those risks however the only project presented (PRONACOSE) deals exclusively with drought. The second strategy is about reducing vulnerability to climate change and climate variability. This should encompass drought, but the relationship is not made clear. The two projects proposed: increased participation and co-responsibility for state and municipal governments, and the creation and strengthening of funds for adaptation projects, do not spell out many specifics with respect to drought. Moreover, the Program is rather vague about where the money required to fund these projects would come from.

The 2014–2018 Program establishes for its objectives a set of performance indicators, with base values for 2012 and goals for 2018, clearly a good idea for evaluation purposes. Unfortunately, it is rather difficult to find the relationship between the six objectives and eight indicators, as no coding scheme is offered to establish that correspondence.

2.4. *The wider policy-making context*

Drought policy does not emerge in a vacuum; many factors influence what possibly can be achieved and what actually is achieved. Here we focus on four factors that have influenced the development of Mexico's drought policy: the process behind the formulation of programs; the emphasis on floods; financing; and, the predominance of a reactive approach.

2.4.1. The formulation process. National water programs should ideally be formulated in concordance with other national programs, especially those that deal with national development, the environment, natural resources and hydraulic infrastructure. In reality, this does not happen. Water programs do make passing references to other relevant national programs but clearly are not developed in close parallel collaboration. This lack of coordination is plainly revealed when considering the problem posed by drought in the northern states. This problem cannot be productively addressed without considering the multiple and complex set of rights and obligations specified in the 1944 Treaty between Mexico and the United States on sharing RBB and Colorado River Basin waters – an apparently obvious point that usually goes unnoticed in national water programs.

Another layer of complexity arises from stakeholders' involvement. National public forums and meetings with experts on an array of water management issues are typically carried out during the process of drawing up national water programs. This also happens at the regional level where forums attract politicians, public servants, academics and water users' representatives and other non-governmental organizations. Experience however shows that public consultations have little impact on the design and implementation of policies in general, including drought policy. For authorities, public consultations constitute an obligation that must be met in order to fulfill legal requirements on the matter – they do not reflect a deliberate effort to incorporate a bottom-up approach in policy making. As a result, public consultations appear to represent for authorities a goal, in and of itself.

The National Water Commission – the country's maximum authority in matters of water planning and management – typically bows to more powerful and well-funded federal ministries. More than 25 years after the commission's creation, the country still suffers from a fragmented institutional framework that lacks coordination. The push toward promoting public participation through River Basin Councils has had no measurable impact on water policy in general or drought policy in particular. The Councils lack the formal and legal structure, as well as the funding, to contribute effectively to river basin-level water planning and management.

2.4.2. The emphasis on floods. In terms of water-related natural disasters, national water programs have systematically put much more emphasis on floods than drought. There are obvious reasons for that. Floods (and the hurricanes or large tropical storms that cause them) are dramatic events that produce immediate and highly visible damage to life and property. As such, they attract more attention from the mass media, the public in general and policymakers than drought – a slowly evolving, diffuse phenomenon whose impacts are not nearly as spectacular and may even go unnoticed by large sectors of the population, especially in urban centers. The National Water Commission to an extent has recognized this problem (CONAGUA, 2006, p. 235) but has stopped well short of offering any kind of a solution.

2.4.3. Financing. How much Mexico spends on drought-related activities, from policy design to actual projects, is hard to ascertain. National and regional water programs offer limited information on budgetary commitments and usually this information is too aggregated to separate out what is specific to drought, as programs and projects usually amalgamate floods, drought as well as other types of natural disasters and risks. Mexico's fragmented water sector institutional framework also poses difficulties in that respect. According to the OECD (2013b), the cost-effectiveness of the country's spending on natural disasters preventive and mitigation measures is difficult to determine as the multiplicity of actors and programs introduces a serious double-counting problem.

What information is available suggests a low order of priority for spending on drought-related activities. For example, the 2002–2006 Regional Program for the RBB (CNA, 2003) – one of the few programs to present budgetary commitments – spoke of a grand total of MXP7016 million to tackle all of its objectives. The amount dedicated to the sixth objective (on managing risks and impacts from floods and drought) amounted to MXP243 million – about 3.5% of the total. How much of this was spent specifically on drought-related activities was left unspecified, but certainly represented a small fraction.

2.4.4. Predominance of the reactive approach. Beginning with the 2001–2006 Program, national water programs have referred to risk management and prevention as key elements for water-related natural disasters policy. The 2007–2012 Program recognized the need to 'give more importance to preventive measures' in the specific case of droughts (CONAGUA, 2006, p. 235). In reality however the approach taken to deal with drought and other water-related disasters has been overwhelmingly reactive – in spite of the international consensus on the cost-effectiveness of prevention relative to disaster relief.

Monies allocated to the National Fund for Natural Disasters (FONDEN, the focus of which is to provide funds for reconstruction and disaster relief) and the National Prevention Fund (FOPREDEN, dedicated to funding preventive measures), yield some information about the federal government's actual priorities. According to the OECD (2013a), over the 2000–2010 period the former fund received on average about MXP4627 million per year and the latter, a mere 6.5% of this, i.e. MXP300 million. The same source also points out that FOPREDEN funds tend to be distributed to states with the capacity to present proposals and not necessarily on the basis of objective needs.

3. The National Drought Program

3.1. The origin of the program: circumstance

In July 2011, amid clear and ample evidence that many parts of the country were in the grip of a severe drought, the federal government produced a document ('Strategies to Deal with Drought in Impacted States') that purported to ease the difficult conditions generated by the event. On the whole the solutions presented seemed like a knee-jerk reaction. Worse, some of the solutions were at odds with well-grounded and long-standing policies – for example, the proposal to allow extraction from aquifers previously banned from exploitation because of severe overexploitation.

A few months later, the federal Congress approved a 2012 budget that did not earmark any funds to deal with the crisis (Arreguín Cortés *et al.*, 2015). This triggered a heated political conflict, with some legislators pressuring the executive branch to come up with funds to deal with the emergency. Of all potential sources of funding spread out over the spectrum of government agencies, the FONDEN was the most likely candidate. This fund however is regulated by rigid rules that prevent the quick disbursement of monies.

On January 25 2012, the federal government published another document which laid down how various existing federal programs were to be re-focused on protecting productive activities in the face of the on-going emergency, as well as providing humanitarian aid to affected communities (Arreguín Cortés *et al.*, 2015). Later in the same year, on November 22, the federal government emitted yet another document, this time providing guidelines on how to formulate mitigation and preventive measures to deal with droughts. None of the aforementioned initiatives were foreseen in the then current 2007–2012 National Water Program. Over the 2011–2012 period, authorities did react to circumstances and managed to provide a measure of relief (albeit insufficient) through temporary work programs, water and food distribution to the worst affected rural communities and help with crop insurance for farmers (Arreguín Cortés *et al.*, 2015). Nonetheless authorities' responses proved piecemeal, case by case, and revealed the absence of a coherent national policy.

In January 2013 the still on-going drought made headlines in the national press. Recognizing the ineffectiveness of the approach adopted so far, CONAGUA proposed a radical shift in strategy, from a reactive approach to a preventive one that hinged on a National Drought Program (PRONACOSE). The federal executive designated CONAGUA as general coordinator and administrator of the program and on April 5 2013 an Inter-Secretarial Commission (IC) was formally established. The IC gathered representatives from 12 federal ministries, the national electricity utility (CFE) and CONAGUA – the latter was put in charge of the technical sub-committee. A group of experts from various universities and research institutions were invited to form a special sub-committee tasked with contributing to the program's design and evaluation.

3.2. The program's design

3.2.1. Goals and features. The program's core goal is to reduce the country's vulnerability to drought through preventive and mitigation actions. For this the program contemplates three main sets of actions. The first set consists of planning; this includes a national system to detect nascent drought events, emit early warnings and monitor their evolution, as well as detailed river basin-level preventive and mitigation programs. The second set deals with providing relief to affected communities and safeguarding water supply for domestic uses. The third set is about coordination, including protocols to simplify and accelerate access to FONDEN monies.

The main implementation tool at the program's disposal comes in the form of 26 specific river basin plans (Programas de Medidas Preventivas y de Mitigación a la Sequía, PMPMS). These were adopted by the River Basin Councils between November 2013 and May 2014. The program also contemplates public information and education campaigns to raise awareness and understanding of drought and its impacts.

3.2.2. Achievements. The program has successfully implemented a national drought monitoring and early warning system. Since 2013 Mexico's National Weather Service (SMN) has provided monthly information in that respect for all river basins, state and municipalities and this information is freely available on the monitoring system's website. However, in order to become an effective management tool, the system will need to evolve and produce more information on on-going drought events, for example water consumption by type of use as well as the current state of underground and surface water sources.

Another significant achievement of the program is the formation of the IC. The ample participation of diverse federal ministries as well as national and regional water authorities makes explicit that water management problems are multi-sectoral by nature and cannot be solved by water authorities alone. This type of integration should ease the coordination problems observed so many times in the past – as the IC found out there are 114 federal programs that in some ways are relevant for drought policy (Arreguín Cortés *et al.*, 2015). Although the sub-committee of experts at the time of writing had not yet produced a first evaluation of the program, it has put together a document outlining research priorities on drought from the point of view of several disciplines.

3.2.3. Shortcomings and challenges. Since the program's inception, no significant drought event has occurred. This precludes evaluating the effectiveness of the program's operational core – the aforementioned 26 river basin plans. Nonetheless, their basic design raises doubts. These plans establish lists of actions to be carried out in case of drought, contingent on the degree of the event's severity. Proposed actions however focus on adaption to a temporary reduction in water availability and do not address the underlying factors that make a region or water use sector vulnerable to drought. This arguably represents a fundamental flaw that will limit the program's ability to fulfill its core goal of reducing vulnerability.

The program's own sub-committee of experts has expressed concerns about the paucity of impetus put into the development of the river basin plans. The original intention was to include in each of these plans, specific plans for two cities as well as for water utilities operating in the river basin – no progress to date has been registered on those fronts. The same can be said of a risk atlas, which was to be produced in coordination with the national civil defense authority.

This lack of progress can be directly ascribed to insufficient funding, but other external factors are also at play. In September 2013, Mexico suffered the consequences of two major hurricanes (Ingrid and Manuel). These put an end to the drought of 2010–2013 but also produced a radical shift in national attention towards the problem of floods, with drought as an object of concern quickly receding from memory. Later, the drop in international oil prices significantly affected the country's public finances and led to federal budget cuts in many sectors. In particular, the joint research fund administered by CONAGUA and the National Council for Science and Technology (CONACYT), which was to finance the PRONACOSE research program, was practically eliminated – despite a solid consensus on the need to expand and strengthen drought research and education (Ortega-Gaucín *et al.*, 2014).

Looking forward, the drought program faces many challenges. For the IC to function effectively, it will have to overcome the coordination roadblocks to be expected within such a diverse workgroup. Also, there is a long way to go toward the program's integration with existing federal natural disaster programs such as FONDEN and FOPREDEN. Those programs, in light of their poor performance in dealing with drought, could be modified or even re-engineered. Finally, there is the broad issue of the program's responsibilities in terms of public information, education and participation. Up to now, the River Basin Councils (the program's key element for this task) have not proven effective in that respect. As a series of personal interviews with high officials in CONAGUA reveal, there is a widely shared view that the councils have not operated successfully on the whole, save a few exceptions.

4. Conclusions

Given Mexico's high exposure and vulnerability to drought, the limited attention given to the issue in the national water programs of the last 25 years is striking. The country's drought policy has evolved and drought as a formal national priority has gained ground, although not as much as the matter deserves. It is worth noting that of the 38 initiatives presented in the National Water Agenda 2030 (CONAGUA, 2011), not a single one addressed drought, versus nine that dealt with floods.

The process behind the elaboration of national and regional water programs is problematic and needs radical improvements, if these documents are to serve as actual roadmaps for solving the problems the country faces. The programs do show awareness of, and concern for drought and its social, economic and environmental impacts. They also present conceptually valid initiatives to deal with the issue. The programs however exhibit serious limitations in terms of concrete proposals to solve identified problems and are plagued by vagueness and inconsistencies.

The need to refocus policy away from a reactive to a preventive approach has been recognized, but in reality little headway has been achieved. A risk management approach based on reducing vulnerability is still in the works. It is ironic that the national drought program, the most concrete policy accomplishment registered so far, came about in reaction to an exceptionally severe drought. Much remains to be done on several fronts, but awareness and knowledge gaps represent a significant challenge that needs to be overcome for progress to accelerate.

First and foremost, there is a general lack of public awareness about drought. This manifests itself clearly when proposals for large-scale projects to convey surface water over long distances (from where the resource is available to where it is needed) are presented. For important northern urban and industrial centers, such projects could help reduce risks to water supplies and increase robustness to drought. Experience however shows that proposals are met by apathy from the general public and fierce opposition from environmental and other civil organizations. Legitimate concerns can be raised over the negative impacts of large-scale hydraulic projects but the debate should then center on ways to minimize those. Wholesale rejection leaves no credible alternatives: if it is not acceptable to convey water, should populations and industry be relocated? River Basin Councils could play a fundamental role in improving society's understanding of the issues at hand, given their stated mission of linking water users and authorities. But for this to happen, the councils' foundations need to be firmed up in national water law and regulations and sufficient operating budgets must be provided.

Scientific and technical knowledge about drought is also insufficient. The quantification of drought risk requires an adequate characterization of both hazard and vulnerability, but such work in Mexico is only incipient (e.g. Neri & Magaña, 2016) – this presents a serious limitation for the design of preventive measures. Groundwater management is a good example of this. Aquifers form an essential part of urban and agricultural water supply, especially in the central and northern areas of the country exposed to drought risk. Currently most groundwater is obtained from overexploited aquifers; this exerts pressure on stored water resources and compromises their long-term value as a buttress against drought. Little is known however about drought-groundwater dynamics and this precludes the possibility of effective, forward-looking management based on sound operating rules (Mahlknecht *et al.*, 2015).

In the Mexican context, drought policy has to be understood as a matter of water security, i.e. ensuring a steady, predictable and sufficient supply of water to cities, industry and agriculture. By extension, the matter speaks to national development and welfare. In this, every citizen, private firm, civil organization and public institution holds a stake and all should be fully aware of the necessity to proceed accordingly. In reality, the country is still far from reaching this state of affairs.

References

Arreguín Cortés, F., López Pérez, M. & Ibáñez, O. F. (2015). National drought policy in Mexico, a paradigm change: from reaction to prevention. *Parcerias Estratégicas 20*(41), 33–56.

CNA (1993). *Política Hidráulica 1989–1994 (Hydraulic Policy 1989–1994).* Comisión Nacional del Agua, México, D.F.

CNA (1994). *Informe 1989–1994 de la Comisión Nacional del Agua (Report 1989–1994 of the National Water Commission).* Comisión Nacional del Agua, México, D.F.

CNA (1995). *Programa Hidráulico 1995–2000 (Hydraulic Program 1995–2000).* Secretaría del Medio Ambiente, Recursos Naturales y Pesca, Comisión Nacional del Agua, México, D.F.

CNA (2000). *El Agua en México: retos y avances (Water in Mexico: challenges and progress).* Secretaría del Medio Ambiente, Recursos Naturales y Pesca, Comisión Nacional del Agua, México, D.F.

CNA (2001). *Programa Nacional Hidráulico 2001–2006 (National Hydraulic Program 2001–2006).* Comisión Nacional del Agua, México, D.F.

CNA (2003). *Programa Hidráulico Regional 2002–2006 Río Bravo (Regional Hydraulic Program 2002–2006 Río Bravo).* Comisión Nacional del Agua, México, D.F.

CONAGUA (2006). *La Gestión del Agua en México – avances y retos 2006 (Water Management in Mexico – progress and challenges 2006).* Comisión Nacional del Agua, México, D.F.

CONAGUA (2008). *Programa Nacional Hídrico 2007–2012 (National Water Program 2007–2012).* Secretaría de Medio Ambiente y Recursos Naturales, Comisión Nacional del Agua, México, D.F.

CONAGUA (2011). *La Agenda del Agua 2030 (Water Agenda 2030).* Secretaría del Medio Ambiente y Recursos Naturales, México, D.F.

CONAGUA (2014). *Programa Nacional Hídrico 2014–2018 (National Water Program 2014–2018).* Secretaría del Medio Ambiente y Recursos Naturales, Comisión Nacional del Agua, México, D.F.

CONAGUA (2015). *Análisis especial de las regiones más vulnerables ante las sequías en México (Special analysis of the most vulnerable regions to droughts in Mexico).* Comisión Nacional del Agua, México, D.F. Available at: http://www.conagua.gob.mx/CONAGUA07/Publicaciones/Publicaciones/sequiasB.pdf (accessed May 15 2016).

Feng, S., Krueger, A. B. & Oppenheimer, M. (2010). Linkages among climate change, crop yields and Mexico-US cross-border migration. *Proceedings of the National Academy of Sciences 107*(32), 14257–14262.

Florescano Mayet, E., Sancho, Y., Cervera, J. & Gavilan Arias, D. P. (1980). Las sequías en México: historia, características y efectos (Droughts in Mexico: history, characteristics and effects). *Comercio Exterior 30*(7), 747–757.

García, E. (2004). *Modificaciones al sistema de clasificación climática de Koppen (Modifications to the system of climate classification of Koppen)*, 5th edn. Instituto de Geografía-UNAM, México, D.F.

Mahlknecht, J., Hirata, R. & Ledesma-Ruiz, R. (2015). Urban groundwater supply in Latin America. In: *Water and Cities in Latin America: Challenges for Sustainable Development.* Aguilar-Barajas, I., Mahlknecht, J., Kaledin, J., Kjellén, M. & Mejía-Betancourt, A. (eds). Earthscan/Routledge, Abingdon, UK, and New York, USA, pp. 126–146.

McKee, T. B., Doesken, N. J. & Kleist, J. (1993). The relationship of drought frequency and duration to time scales. In: *Preprints 8th Conference on Applied Climatology*, 17–22 January, *Anaheim, California*, American Meteorological Society, pp. 179–184.

Méndez, M. & Magaña, V. (2010). Regional aspects of prolonged meteorological droughts over Mexico and Central America. *Journal of Climate 23*, 1175–1188. http://dx.doi.org/10.1175/2009JCLI3080.1.

Mishra, A. K. & Singh, V. P. (2010). A review of drought concepts. *Journal of Hydrology 391*, 202–216.

Mumme, S. & Aguilar-Barajas, I. (2003). Managing border water to the year 2020: the challenge for sustainable development. In: *The U.S.-Mexican Border Environment – Binational Water Management Planning.* Michel, S. (ed.). San Diego State University Press, San Diego, USA, pp. 51–93.

Neri, C. & Magaña, V. (2016). Estimation of vulnerability and risk to meteorological drought in Mexico. *Weather, Climate, and Society 8*(2), 95–110.

OECD (2013a). *OECD Reviews of Risk Management Policies: Mexico 2013: Review of the Mexican National Civil Protection System*. OECD Publishing. Available at: http://dx.doi.org/10.1787/9789264192294-en (accessed May 20 2016).

OECD (2013b). *Making Water Reform Happen in Mexico, OECD Studies on Water*. OECD Publishing. Available at: http://dx.doi.org/10.1787/9789264187894-en (accessed May 20 2016).

Ortega-Gaucín, D. & Velasco, I. (2013). Aspectos Socioeconómicos y Ambientales de las Sequías en México (Socioeconomic and environmental aspects of droughts in Mexico). *Aqua-LAC 5*(2), 78–90.

Ortega-Gaucín, D., Palacios Vélez, E., Mejía Sáenz, E. & Rendón Pimentel, L. (2008). Una alternativa metodológica para la estimación del impacto socio-económico de la sequía en los distritos de riego (A methodological alternative to the estimation of socioeconomic impacts of droughts in irrigation districts). In: *La Ingeniería Agrícola: Motor del Desarrollo de la Agricultura Mexicana (Agricultural Engineering: Development Driver of Mexican Agriculture)*. Pérez Sobrevilla, L., Romantchik, E., Hidalgo, M. & García S, S. (eds). Universidad Autónoma Chapingo, México, pp. 134–146.

Ortega-Gaucín, D., Velasco, I., Güitrón de los Reyes, A., Arreguín Cortés, F. & López Pérez, M. (2014). *La Gestión de Sequías en México: Avances y Retos (Drought Management in Mexico: Progress and Challenges)*. XXXIII Congreso Nacional de Hidráulica, Puerto Vallarta, Jalisco, México.

Sisto, N., Guajardo Quiroga, R. & Aguilar Barajas, I. (2011). Estimación de los impactos económicos de una sequía (Estimation of the economic impacts of a drought). *Tecnología y Ciencias del Agua II*(2), 111–123.

Sisto, N., Ramírez, A., Aguilar-Barajas, I. & Magaña-Rueda, V. (2016). Climate threats, water supply vulnerability and the risk of a water crisis in the Monterrey Metropolitan Area (Northeastern Mexico). *Physics and Chemistry of the Earth, Parts A/B/C 91*, 2–9. (http://dx.doi.org/10.1016/j.pce.2015.08.015).

Velasco, I., Ochoa, L. & Gutiérrez, C. (2005). Sequía, un problema de perspectiva y gestión (Drought, a problem of perspective and management). *Región y Sociedad XVII*(34), 35–71.

World Bank (2016). *World Bank Open Data*. http://data.worldbank.org/indicator/all (accessed May 23 2016).

Evolution of drought management policies and practices in the United States

Eugene Z. Stakhiv[a], William Werick[b] and Robert W. Brumbaugh[c]

[a]*Corresponding author. Johns Hopkins University, Visiting Scholar, US Army Institute for Water Resources, 7701 Telegraph Road, Alexandria, VA, USA 22315. E-mail: EugeneStakhiv@jhu.edu*

[b]*Consultant, 14508 Chesterfield Lane, Culpeper, VA, USA 22701*

[c]*US Army Institute for Water Resources, 7701 Telegraph Road, Alexandria, VA, USA 22315*

Abstract

The evolution of United States (US) national and federal drought policies is examined, and the relative effectiveness of the hierarchy of federal, state and local drought management programs and practices is evaluated within the context of changing philosophies of water management. While there is no 'national' drought management policy, there is a strong federal policy that attempts to coordinate the management responses of numerous federal agencies. Drought and flood management comprise the key components of assuring national 'water security'. Early 20th century US water resources management was very effective in providing the essential prerequisites of 'water and economic security' through a rapid expansion of water storage capacity, flood control, hydropower, irrigation and conveyance facilities as the basis for a robust national economic development program. Increasing water supply capacity has proven to be the most effective strategic drought management option. Today, federal drought management policies and programs focus mainly on demand reduction strategies, buttressed by a powerful array of environmental regulatory programs that strive to attain sustainable resource use. Climate uncertainty will again require greater focus on supply-side options that increase water supply robustness and resilience, especially in the growing urban areas of the semi-arid west.

Keywords: Drought; US national drought management policy; Federal drought management; Water security

Introduction

For the purposes of this paper, the definition of drought used is one derived from the National Study of Water Management During Drought: Report to Congress (IWR, 1995):

'Droughts are periods of time when natural or managed water systems do not provide enough water to meet established human and environmental uses because of natural shortfalls in precipitation or streamflow.'

doi: 10.2166/wp.2016.017

The National Drought Study (NDS) Report to Congress is perhaps the most comprehensive technical report on various drought management approaches, practices and lessons learned, based on an accumulation of information from numerous case studies. The study also concluded that the problems in water management during drought are manifestations of problems in water management in general (IWR, 1991). Just as recessions may reveal weaknesses in the management practices of a company that made money when business was good, droughts reveal weaknesses in water management systems which are hidden in the years when water is plentiful.

The NDS also noted that, in the United States (US), water management problems come not from limited overall supply, but from problems in regional availability, management and usage. Water is not always where people want to use it, and the ways we allocate and use water have not been entirely successful in achieving economic efficiency, equity, and environmental quality. Taken as a whole, the US always has more water than it needs. About 1,400 billion ($1{,}400 \times 10^9$) gallons of fresh water per day (bgd) is available in the conterminous 48 states. A quarter of that (330 bgd) is withdrawn for all human uses, and most of that is returned to streams. In all but a few places in the US, a year-long drought so severe that it can be expected only twice a century will still produce from one half (50%) to two-thirds (67%) of the average precipitation for the year (IWR, 1994).

The focus of this paper is on the evolution of drought management policy and practices within the US. It is well understood that the special circumstances of the US do not readily travel as 'lessons learned' to developing nations where droughts are a repetitively occurring and far more widespread destructive factor in constraining socio-economic development than in the US (UN ISDR, 2011). The US has many economic and geographic advantages that are not available to many other nations. Still, there may be equivalent circumstances in many of the southwest regions of the US that can be applied to other comparable situations in other semi-arid nations that struggle with persistent droughts.

On the other hand, hydrologic conditions are relatively similar in Europe, and the European Water Framework Directive (WFD, 2000) does represent a pan-European water resources management policy that includes floods, droughts and water quality aspects. As part of the WFD, the European Commission (EC, 2011) has been tackling the problem of droughts and water scarcity since 2007, and some lessons learned can be transferred to the US, though the policy options proposed are virtually identical to those routinely considered and implemented by US federal agencies and states, albeit in a less than integrated manner. The only difference may be in the relative implementation effectiveness of the EU versus the US in terms of the limitations of their respective legal authorities. Both systems are essentially a collection of quasi-sovereign nations and states, where the EU is closer to a 'commonwealth' or 'confederation', whereas the US is a federal system, where states have considerable authority over water resources management within their respective borders (Kenney, 1997; Abrams, 2009). Seven policy options were identified by the EC in 2007 for tackling water scarcity and drought issues:

- Putting the right price tag on water
- Allocating water and water-related funding more efficiently
- Improving drought risk management
- Considering additional water supply infrastructures
- Fostering water efficient technologies and practices
- Fostering the emergence of a water-saving culture in Europe
- Improving knowledge and data collection.

An EC (2012) update noted, however, that '(T)he overall objective of the Water Scarcity and Drought (WS&D) policy – to revert the WS&D trends – has not been achieved, even if progress has taken place in implementing the seven policy instruments identified in the Commission's Communication from 2007.'

In addition, many UN agencies are currently involved with improving drought management in the developing world. UNESCO (2014) launched its International Drought Initiative in 2011, while the World Meteorological Organization (WMO), the United Nations Convention to Combat Desertification (UNCCD), the Food and Agriculture Organization of the United Nations (FAO) and the UN-Water Decade Programme on Capacity Development (UNW-DPC) jointly established a UN-Water Capacity Development Initiative to support National Drought Management Policies in 2013. As part of the effort, a high-level international meeting on National Drought Policy was held to develop a compendium of national drought management policies and practices (Wilhite *et al.*, 2014).

The goals or tenets of a 'model' national drought policy, as stated in the documents prepared leading up to the high-level meeting on national drought policy (World Meteorological Organization, 2013), were as follows:

- Proactive mitigation and planning measures, risk management, public outreach, and resource stewardship.
- Greater collaboration to enhance the national/regional/global observation networks and information delivery systems to improve public understanding of, and preparedness for, drought.
- Incorporation of comprehensive governmental and private insurance and financial strategies into drought preparedness plans.
- Recognition of a safety net of emergency relief based on sound stewardship of natural resources and self-help at diverse levels.
- Coordination of drought programs and response efforts in an effective, efficient and customer-oriented manner.

The central issue is not so much about what the core policies should be, for there is wide agreement on most of the fundamental prescriptive aspects, but whether they can be readily and uniformly implemented in a conglomeration of sovereign European nations, or equivalently in 50 relatively sovereign states of the US in order to achieve sustainable development and water security. In fact, it can be argued that the US, despite its relatively fragmented 'federal system' of governance, is well ahead of implementing the seven European policy options, mainly because of the strong advocacy of the Federal Government and wide array of financial incentives and regulatory instruments at its disposal to ensure compliance.

Despite an accumulation of countless useful polices and laws that have translated hundreds of sensible ideas related to modern sustainable water management into practical operating procedures, the US does not have a comprehensive and coherent national drought policy, in the sense of a single formal statement of principles and guiding authorities against which all actions related to drought are planned, implemented and measured. Nor does a generic national water policy exist wherein drought management, as a subset of water resources management, is embedded. Instead, drought management in the US is composed of elements drawn from a plethora of guiding principles and stratagems at all levels of government, sometimes conflicting because they adapt to the realities of multiple interpretations of 50 state water laws, investment strategies and the control over their own resources.

Most water problems are local or regional, rather than national, but the Federal Government, through a combination of regulatory authorities, budgetary resources and drought assistance programs, has progressively had a significant impact on indirectly influencing not only drought management policies at all levels, but also water resources management in general. A distinction needs to be made between policies and laws. Policies are simply a collection of principles and guidelines that are used to coordinate, consolidate and direct certain actions that are consistent with the existing body of laws. Often, there is flexibility within laws to accommodate new planning and evaluation procedures as, for example, floodplain management or drought management.

Recent severe drought events in California and the Atlanta, Georgia, metropolitan area have exposed some of the inflexibilities in existing federal laws governing the management of federal water resources systems. These laws cannot be overcome by simply issuing new 'policies'. Drought management policies are further constrained by law, and directed by the Federal Government, through a suite of agricultural, environmental and water quality laws and regulations (National Drought Policy Commission, 2000). The reinforcement of many federal environmental regulations has had a profound impact on water management, especially in the western states.

The core issue that is routinely glossed over in focused flood and drought management policies, however, is that flood and drought management are the two bookends of integrated water resources management (IWRM) whose main purpose is to provide 'water security' for its populace. Water security has two basic components: reducing the risks and damages of extreme hydrologic hazard events before they occur (floods and droughts), and reliably providing a wide array of water-based services to the various sectors, which include energy, agriculture, public water supply and ecosystems. Much of what the EU, US and UN drought management policies address is preparing for droughts and mitigating their most damaging effects, along with post-drought relief efforts. The more important objectives that are typically neglected in contemporary drought management policies are strategic measures that prevent the most damaging drought impacts, such as increased storage, water conveyance and interconnections between vulnerable public water systems.

Most contemporary drought management policy statements and plans de-emphasize the notion that 'water security' depends very much on preventing the worst damages of drought, especially in developing nations that are highly dependent on irrigated agriculture, rather than simply focusing on mitigating post-drought losses. The primary purpose of water supply infrastructure systems is as a defense against the damaging and long-lasting impacts of drought, and turning those natural hazards into a beneficial instrument of economic development (Stakhiv, 2011). Most non-structural drought management solutions that are routinely touted as drought mitigation measures, such as water conservation, pricing, water banking, forecasting, etc. inherently depend on an existing robust, resilient and reliable water storage and conveyance system as the core of a practical and comprehensive drought management strategy.

Federalism and water policy

Water is not mentioned by name in the US Constitution as a distinct federal responsibility, but a 'federal interest' was created, mainly based on an 1824 Supreme Court decision that established the federal right to regulate matters of interstate commerce, including navigation. This was the era of canal building and waterborne transportation. The US federal role has grown incrementally over the decades, with interstate commerce improvements as the constitutionally accepted vehicle for federal interventions in state water management responsibilities. Section 8 of the Constitution reads as follows:

'The Congress shall have the power ...To regulate Commerce with foreign Nations, and among the several States, and with the Indian Tribes...'

Since the 1824 Supreme Court decision, the evolving federal policy pressed against the presumption that the primary responsibility for water management, and drought management in particular, begins in each of the 50 states (Reuss, 2004). The most effective federal intervention related to water resources management came about in the 1970s, with the introduction of a series of powerful legislated federal laws regulating public health and safety that addressed poor water quality and ecological degradation of streams, wetlands and watersheds. There is no current aspect of local, state, or federal water resources management actions that does not fall under the provisions of some federal regulatory authority, which does not possess a *de facto* veto threat over any planned water-related management activity in the 'waters of the United States'.

Rogers (1993) makes a useful distinction between the 'federal interest' and the 'national interest' when discussing water policies. The 'national interest' must account for the diverse interests of 50 relatively sovereign states. Furthermore, he contends that there is no need for a 'national' water policy since the US is also diverse in its natural water resources – with a very dry western half and a relatively humid and water-rich eastern half. Rogers also maintains that since water is more of a regional and local resource, with many different permutations of customs, legislation, availability and demand, water policy should reflect that diversity, rather than attempting to achieve an artificial uniformity through a 'national policy'. The report of the US Senate Select Committee (1961) on National Water Resources essentially makes the same case:

'It is unlikely that the Nation will have a uniform policy covering the details of all water resources activities in all parts of the country. Conditions vary from place to place, and from year to year. Many water resources problems are basically local problems. In some places disparate conditions are found within a single State or within a single river basin.'

Nevertheless, there exists a *de facto*, loosely connected *federal* policy, which directs the policies and procedures of the vast bureaucracy of federal water management, science and regulatory agencies that have greatly expanded their influence on overall water management practices in the US since 1970. Hence, federal water management and drought management policies, buttressed by a plethora of environmental regulatory authorities, have essentially formed the core of a loosely connected, *de facto* national policy. However, expecting consistency in applying the various laws, rules, regulations and Executive Orders (EOs) among those many agencies and state governments is a monumental achievement in itself, given the fact that most EOs extend only for the duration of a presidential administration, and are revised or repealed with each new Administration. The nature of the US political system itself introduces many of the policy vagaries that constantly require heightened coordination and collaboration.

The major problem is that Congress has created an awkward system of water management, that can be termed 'water federalism' (Abrams, 2009), wherein state sovereignty over water management is acknowledged and promoted, while the federal 'water management footprint' has increased dramatically since 1970, through its regulatory powers to license a wide range of water uses. This includes the Federal Energy Regulatory Commission, the Nuclear Regulatory Commission, Federal Emergency Management Agency and many other environmental regulatory agencies, with an ever-increasing array of environmental protection laws. This increasing power of federal regulatory agencies, coupled with Congress's legislative authorities, has created a powerful 'federalist' position that is increasingly difficult for states to counter in their pursuit of controlling intra-state water management decisions.

Extending an ever-growing proliferation and constant revision of federal water management initiatives, objectives and regulations to 50 states simply amplifies the coordination difficulties among federal agencies, which remains one of the most frequently criticized barriers to effective implementation of existing drought policies (PPIC, 2016). Hence, as part of revisions to federal policy on water management, a great deal of emphasis is always placed on better coordination, collaboration and partnerships amongst the numerous agencies, as the basis for more effective implementation of drought management actions. Unfortunately, when responsibility is shared among an ever-increasing number of entities, no one entity is holistically managing drought or flood risk and ensuring that actions of the many stakeholders involved are effectively combined to reduce risk.

For most of the 20th century, US national water policy formulation advanced progressively with two principal goals in mind: using water resources development to promote economic development, while protecting against natural hazards, with the ever-evolving need for environmental sustainability and an equitable distribution of benefits and costs associated with federal water projects. More than 25 national commissions established by Congress or the President since 1900 (CSIS, 2014) have confronted that ever-changing balancing of societal demands, while adapting to the evolving notions of sustainability. Water resources management for economic development, with its complexity of management objectives and water uses, was always the main topic of these commissions. It is only recently, within the last two decades, that drought management itself has become a specialized topic of concern, mostly concentrating on how to deliver federal drought relief more effectively to the states, farmers and local municipal water supply systems. Drought prevention and flood control have always comprised the core issues of water resources development. Drought management was thought of more in terms of how to prepare for and alleviate and mitigate drought losses after a drought event, given the water resources management capacity and conditions *in situ*.

The most recent Congressionally-sanctioned Presidential Commission was the National Drought Policy Commission (NDPC, 2000). The NDPC was preceded by the Western Water Policy Review Advisory Commission Report (WWPRAC, 1998), which explicitly dealt with the unique water management problems of the 17 western states, where droughts have especially pernicious adverse impacts. The WWPRAC Report updated many of the key recommendations of the 1973 Water Policy Commission, but had little to say about drought management. Instead, it focused on the unique relationship between the Federal Government and the western states, because the Federal Government is the major landowner in the west and thereby claims significant water rights (Fort, 1997). The WWPRAC attempted to implement many of the emerging policy recommendations of the President's Council on Sustainable Development (PCSD, 1997) within the context of the unique needs of western water management. The WWPRAC report created a significant backlash from the Western Governors Association.

In a series of actions, the Western Governors' Association (WGA, 2015) responded to the serious drought episodes of the 1970s to 1990s by preparing a Drought Action Plan of 1996, which became the framework for a number of specific drought actions. The plan stated '*(T)he western governors believed that a comprehensive, integrated response to drought emergencies is critical... It is important to work together and cooperatively with other affected entities to plan for and implement measures that will provide relief from the current drought and prepare for future drought emergencies.*' The WGA and a separate federal drought initiative led to the formation of a partnership called the Western Drought

Coordination Council (WDCC, 1999). Its mission was to develop and implement model drought policies and management/mitigation measures that reduce impacts associated with droughts and that promote economic and environmental sustainability in the western states (Motha, 2011).

According to Eisel & Aiken (1997), in a report written for the WWPRAC, the major force presently driving many water resources policies, management and development decisions in the 17 western states is the Endangered Species Act (ESA). This act affects the nature and direction of municipal water supply planning and projects in the front range of Colorado; development of new storage projects in Wyoming; permitting of existing and proposed diversion and storage structures on federal lands, and the operation and management of existing water management systems in the river basins of the west. The ESA has triggered a wide number of studies in major river basins (Columbia, Missouri, Florida Everglades, etc.) regarding their management principles with respect to aquatic restoration, especially for endangered species. This Act, and other environmental legislation and regulatory decisions has spawned many studies by the National Research Council of the National Academy of Sciences (NRC, 2002, 2004a, 2004b, 2007, 2011) on improving water management for aquatic ecosystems.

If anything, the emerging federal water management policies that attempt to achieve 'sustainable development', reinforced by the President's Council on Sustainable Development (PCSD, 1999), are most clearly reflected in the manner in which federal agencies have been adapting these new policies in western water management. In reality, it is the relative success of the federal agencies in implementing the new philosophy of sustainable development, coupled with the principles of IWRM that has exacerbated water conflicts with and among the states, especially in the west.

In response, the Western Governors Association developed a strongly worded comprehensive water policy statement (WGA, 2015) that reiterated states' authority over water management within their borders:

> '***State Primacy in Water Management**: As the preeminent authority on water management within their boundaries, states have the right to develop, use, control and distribute the surface water and groundwater located within their boundaries, subject to international treaties and interstate agreements and judicial decrees.*'

On the other hand, both the WGA and the National Governors' Association (NGA, undated) believe federal investments should assist states in implementing state water plans designed to provide water for municipal, rural, agricultural, industrial and habitat needs, and should provide financial and technical support for development of watershed and river basin water management plans when requested by states. Integrated water management planning should also account for flood control, water quality protection, and regional water supply systems. Water resource planning must occur within a framework that preserves states' authority to manage water through policies which recognize state law and the financial, environmental and social values of the water resource to citizens of the western states today and in the future.

Many templates for water management have been proposed, tested, rejected and transformed since the progressive era of Theodore Roosevelt and the Reclamation Law of 1902. Only once has the US Congress established a multipurpose governmental institution with comprehensive jurisdiction over a major river basin – namely the Tennessee Valley Authority (TVA) in 1933. This decision was influenced by the Great Depression and Midwestern 'Dust Bowl' drought during that period. However, growing state and public opposition to such federal institutions soon prevented the creation of other such institutions (Rogers, 1993).

The federal role in water supply storage and drought management

The core of 'water security' and, by extension, drought management, has a supply component as well as a demand management component, particularly during times of scarcity. Over the past century, the US has developed a formidable system of water storage and conveyance projects. At the Federal level, the Corps of Engineers alone built 465 reservoirs, which now comprise about half of the total constructed reservoir storage space in the US (USACE, 2016). The Corps is the largest producer of hydropower in the US. Municipal water supply is an authorized purpose for all Corps of Engineers reservoirs, but the allocation of space for that purpose is approved on a case by case basis and a local sponsor, typically a city or a water utility, must pay the costs for developing and maintaining the storage.

The Bureau of Reclamation is the nation's largest wholesale water supplier, operating 337 reservoirs. The Bureau provides irrigation water for 10 million acres of farmland, accounting for 20% of all western farms. That land produces 60% of the nation's vegetables and one quarter of its fresh fruit and nut crops. The Bureau is the second largest producer of hydropower in the US and operates 53 hydroelectric power plants that have annually produced an average 40 billion kilowatt-hours for the last 10 years, and these reservoirs deliver 10 trillion gallons of water to more than 31 million people each year (Bureau of Reclamation, 2016).

It is important to note that the major federal agencies responsible for managing the nation's water resources – the Corps of Engineers, Bureau of Reclamation, TVA and the Natural Resources Conservation Service – have planned, designed and managed most of the federal water storage and conveyance projects within the US. As such, they have been incrementally adapting to the ever-changing philosophies, policies and objectives of federal water management, focusing on reducing floods and drought hazards, while significantly improving the ecological viability of aquatic ecosystems. In addition, the Federal Government's land management agencies – such as the Bureau of Land Management (which controls over 1 million km^2), the Forest Service (which manages 0.8 million km^2) and the Department of Agriculture (which influences the nation's agricultural practices on over 4 million km^2) – all have a considerable influence on the pace and direction of the Federal Government's sustainable development policies, as well as drought management itself.

Reducing vulnerability to droughts and preventing the worst impacts through water resources development, as the most reliable form of 'water security', began early in the nation's history, mainly through private investments. This was followed by the 1902 Reclamation Act, and expanded fairly quickly through a series of Congressional initiatives, which led to a rapidly expanding federal role during the Depression era, and after WWII. Over the course of the last century, a sense of a loosely connected and marginally coordinated national drought management policy came to be defined. While the Federal Government – through Congressional authorization and appropriations power, and executive branch implementation authorities – began to exercise a greater role in national water resources development, the states continued to exercise their rights regarding water management allocation and distribution of water supplies, especially that of groundwater resources.

Were it not for the fairly extensive water storage and conveyance infrastructure built up since the Reclamation Act of 1902, much of the west would not have been developed to the extent that it is today, serving as an economic and agricultural powerhouse. The adverse social, economic and ecological effects of droughts in the western US would be far more severe without the system that was largely planned during the 'Dust Bowl' era, and built up during and after World War II. From a water management standpoint, we are living off the foresight and capital investments of the previous two generations.

The agricultural sector is, by far, the largest user of water in the US, accounting for 115 billion gallons per day (bgd) in 2010, and representing the lowest levels since before 1965. Irrigation withdrawals, all freshwater, accounted for 38% of total freshwater withdrawals for all uses, or 61% of total freshwater withdrawals for all uses excluding thermoelectric power. Surface water withdrawals (65.9 bgd) accounted for 57% of the total irrigation withdrawals, or about 12% less than in 2005. Groundwater withdrawals were 49.5 bgd in 2010, about 6% less than in 2005. About 62.4 million acres were irrigated in 2010, an increase from 2005 of about 950,000 acres (1.5%). The number of acres irrigated using sprinkler and micro-irrigation systems continued to increase and accounted for 58% of the total irrigated lands in 2010 (Maupin *et al.*, 2014).

Federal drought policies are dominated by responses to the large agricultural sector, and largely focus on financial and tax incentive programs that persuade farmers to use water more efficiently. According to Schaible & Aillery (2012), irrigated farms accounted for roughly 40% (S118.5 billion) of the value of US agricultural production; nationwide, the average value of production for an irrigated farm was more than three times the average value for a dryland farm. Roughly three quarters of US irrigated agriculture occurred in the 17 western states, although irrigation has been expanding in the more humid eastern states.

There is also a wide array of drought relief and crop insurance programs that subsidize losses that farmers may incur as a result of drought. Municipal water systems, which account for only about 10% of national water use, have progressively been most responsive to water conservation initiatives, mainly because their water is metered, and pricing is used as an effective water conservation measure. But municipal systems have their own drought vulnerability problems, key among them being fragmented into thousands of small water supply systems. The US Environmental Protection Agency (US EPA, 2013) reports that there are over 155,000 public water systems across that US that they regulate for water quality. This fragmentation makes it difficult to deal with as part of any comprehensive drought preparedness and management effort. Fortunately, only about 4,200 systems serve 250 million people, or 75% of the total population of the US, which makes it a bit more manageable. One of the strategic drought management options is to interconnect the smaller public systems of towns and villages, increasing reliability of supplies in the event of shortages.

The key aspect of the EPA's regulatory authorities regarding public water supply systems, via the Clean Water Act (CWA) (1972) and the Safe Drinking Water Act (SDWA) (1996), is the fact that the EPA has used these provisions to control new sources of water supplies for growing municipal systems. Although water supply planning development for municipal water supply systems falls largely to publicly owned or private utilities within state boundaries, and states have primacy over water management within their boundaries, the Federal Government has a great deal of control over these decisions because of provisions within the SDWA, CWA and Endangered Species Act (ESA, 1973), among others. For example, the 'Siting Requirement' under the SDWA has been extended to require that States adapt water supply expansion criteria that include drought conditions:

> '*Before a person may enter into a financial commitment for or initiate construction of a new public water system or increase the capacity of an existing public water system, he shall notify the State and, to the extent practicable, avoid locating part or all of the new or expanded facility at a site which: (a) Is subject to a significant risk from earth-quakes, floods, fires* or other disasters which could cause a breakdown of the public water system or a portion thereof;'

The EPA has used this provision to require that any municipality, seeking to expand its water supply capacity, undertake and demonstrate a series of demand management actions before capacity expansion can even be considered as an option. An example is the recent 'Partnering Agreement' (2016) that was signed between the Texas Water Development Board, the EPA and the Corps of Engineers to 'improve coordination on large water supply projects' that are subject to the regulatory requirements of the CWA and SDWA, and which may require an Environmental Impact Statement. In Texas, as in many similar situations around the US, the EPA is constraining capacity expansion as a principal strategic instrument of increasing overall water security.

Similar federal constraints on state and municipal water supply expansion initiatives as part of emergency drought conditions in California have led to legislative initiatives in Congress to allow greater flexibility in federal agency regulations. For example, legislation was introduced in the House of Representatives (HR 1709, 2015) to 'amend the Safe Drinking Water Act to provide for the assessment and management of the risks of drought to drinking water (provided by public water systems), and for other purposes'. Part of the rationale for this legislation was to prompt the EPA to give greater consideration to relaxing regulatory constraints during droughts.

The recent 5-year California drought exposed many of the same problems related to stringent federal regulations that constrained flexible management solutions and adaptation to drought conditions. Hanak *et al.* (2015) laid out 'ten ways the feds can help ease drought in the west'. Among them were relaxing requirements in the SDWA for small communities to abstract emergency sources of water supply, and to allow the Secretary of Interior to facilitate water transfers between federal and state water projects, and to facilitate water trading. It should be understood that the changes require Congressional approval, and are not within the authority of the federal agencies or the Executive Branch to change unilaterally.

Federalism and drought management

Although water resources management is ostensibly carried out in the context of a Federal system where all powers not ascribed to the Federal Government in the national Constitution remain with sovereign states (Kenney, 1997), the Federal Government has greatly expanded its role during the past 40 years, as can be seen from previous examples. This has created a growing tension between federal agencies and states since, constitutionally, states retain sovereignty over water within their respective boundaries. Yet the Federal Government is by far the largest landowner in the US, controls most of the large dams and water supply systems, and has the largest budgetary resources, taxing authority and regulatory controls to achieve its desired goals.

While the federal role in water resources development evolved rather slowly and incrementally over the course of the first 200 years of US history for specific reasons in response to specific needs (Holmes, 1972), it accelerated from the 1970s, with the beginning of the environmental era. Although it may appear that the federal role is dominant and pervasive because many of the federal principles advocated by the various federal agencies are being implemented on a wide range of federally-sponsored projects and programs, the states still wield considerable authority over their internal resources, and, as a consequence, often resist federal policy implementation via EOs, and frequently challenge these provisions in court. This creates a practical implementation dilemma for state and federal agency planners, and adds complexity to a national water management framework, fragmented by 50 state management perspectives.

In addition, a strong tradition of property rights, individualism, markets and private ownership exists in the US. There is also general resistance to federally imposed land use planning, which makes water related disaster risk reduction management difficult since many viable solutions that contribute to infrastructure resilience must integrate current and future land uses (CSIS, 2014).

Congress routinely requires that the federal agencies defer to or comply with state law in the planning, construction and operation of federal facilities pertaining to allocation, control, or distribution of water. At the same time, as owner of hundreds of thousands of acres of public domain land, much of it in the semi-arid southwestern US, the Federal Government is the 'owner of the right to use the waters pertaining to the public domain lands, the right to use of which has not passed into private ownership under authority of the US or an earlier sovereign' (National Water Commission, 1973). The Federal Government also holds reserved water rights – although in many cases unquantified – for reservations of federal lands withdrawn from the public domain (e.g., national forests, national park lands, tribal lands and wilderness areas).

Nevertheless, despite all these constitutional barriers, US drought (and flood) management policies have evolved rapidly over the past three decades, driven mainly by the Federal Government, to encompass two interrelated resource management philosophies: sustainable development and IWRM, with its complementary objectives of efficient resource use and environmental protection (Delli-Priscoli & Stakhiv, 2015). Federal Government agencies are the implementers of Congressional legislation, judicial rulings and EOs, that are buttressed by the veto power of federal regulatory agencies, which oversee both public and private sector development. The federal establishment has steadily steered fundamental water resources development and management policy formulation and implementation towards programs and policies that increasingly advocate and rely on non-structural measures and water conservation. This was accomplished despite the tradition of 'federalism', wherein the Constitution gives the states broad authorities over the management of its resources.

A key critical function of federal agencies is flood, drought and related disaster risk reduction management (Dyson, 1988). It underpins all other federal initiatives and, as such, has become the most powerful entry point for federal agencies to indirectly overcome or usurp state authority in water resources management. The most recent example of such federal actions, in response to the 5-year drought in California (2011–2015), most clearly explains the functioning of this dichotomous system of US 'federalism', with the issuance of a Presidential Memorandum (March 21, 2016) on 'Building National Capabilities for Long-Term Drought Resilience' (White House, 2016a, 2016b):

> *'Sec. 2. Policy. It is the policy of the Federal Government to coordinate and use applicable Federal investments, assets, and expertise to promote drought resilience and complement drought preparedness, planning, and implementation efforts of State, regional, tribal, and local institutions. In addition, where appropriate, the Federal Government shall seek partnerships with such institutions and the private sector in order to increase and diversify our Nation's water resources through the development and deployment of new technologies and improved access to alternative water supplies.'*

During the same time that the Presidential Memorandum was being drafted and vetted, a legislative bill was introduced (S.1894, 114th Congress, 2015), entitled: '*A BILL To provide short-term water supplies to drought-stricken California*'. This legislation was intended to provide federal technical and financial assistance, reallocate water deliveries from federal projects and provide regulatory relief from a myriad of federal environmental regulations and water allocation laws that constrained the diversion of emergency water supplies from federal projects to the state of California.

The legislation demonstrates the dividing line between federally managed projects by the Corps of Engineers and the Bureau of Reclamation, and those built and managed by California, and the respective authorities of the President, those of Congress and the Governor of California. The President can marshal the existing resources of various federal agencies, as long as they are consistent with laws passed by Congress. The Presidential Memorandum is an example of those limited federal authorities, in that it highlights and encourages greater coordination among federal agencies, data-sharing, better planning, communication and technological development.

To go beyond that, i.e. to change water allocation in federal reservoirs or to ease strict regulatory water withdrawal criteria, Congress must pass new legislation (e.g. S. 1894), as it has done periodically, after major flood and drought events. Thus, each flood and drought incrementally adds new legislative principles and management precedents, some of which contradict other existing laws, thereby expanding the federal role and influence in water resources management over the past century, while inadvertently creating new conflicts and confusion about responsibilities and legal authority.

Whether it be Colorado river basin management by the Bureau of Reclamation, or the Corps' dominant management role in the Appalachicola-Chattahoochee-Flint water supply system, and controversy over Atlanta's water supply, states only possess 'nominal' rights to manage their intra-state water systems, because a wide array of 'superior federal rights' exist, and are routinely exercised by the Federal Government through a vast network of federal regulatory agencies. The system will not change much until the states come to an agreement amongst themselves about basic water allocation decisions in the form of an interstate compact (Abrams, 2009). State water allocation priorities, policies and interests do not account for much, as existing federal laws that specify the conditions under which federal agencies must exercise their respective authorities and operate, maintain and manage their respective water resources systems.

Congress has not enacted any comprehensive or overarching change in federal water resources management or national water policy since enactment of the 1965 Water Resources Planning Act (P.L. 89-80; 42 USC. §1962). Although an assessment of the nation's water resource *conditions* was last conducted in 1975 (WRC, 1978) and several entities have studied selected aspects of water policy and management, the last systematic and comprehensive review of nationwide federal water *policy* was conducted by the National Water Commission (1973). This was followed soon after by President Jimmy Carter's 1978 Water Policy Task Force Report (ASCE, 1979).

In reality, the major problems surrounding the most recent drought situations in California would not be solved by a 'national' drought management policy, for it is not policies and principles that are required, but a new approach to drought management that is authorized by Congress. For, according to many scholars (Abrams, 2009; Hanak *et al.*, 2015), most of the problems that have emerged in many recent drought management instances are because of the relative inflexibility of federal laws, many of which were written in the late 1960s and early 1970s. These federal laws did not foresee the unique circumstances of population growth and shifts in water uses over the decades, nor the dominance of environmental protection laws that trump many other legislated water management actions.

This sort of incremental federal policy formulation, dedicated to solving specific problems, inevitably increases decision making fragmentation even as there are greater calls for more coordination. Greater coordination and collaboration among agencies and interest groups does not necessarily lead to better decisions about resource management. Criticism of the fragmented nature of federal water policy has been a recurrent theme for decades. Historically, countless commissions, councils, and studies have called for new directions in water policy and better planning, evaluation, and coordination of federal actions. Options used in the past have included formal and informal coordination entities within the executive branch, nongovernmental commissions tasked with reviewing past policies and laws, and a legislative branch committee made up of key committee leaders (Copeland & Carter, 2009).

Any attempt to untangle the complexities of current water policy involves many constituencies with differing interests, which becomes politically difficult to sustain. Instead of comprehensive or overarching legislation, Congress has enacted numerous incremental changes, agency by agency, statute by statute. The recent proposed legislation to alleviate California's severe drought, in the form of Senate Bill 1895 (S.1894, 2015) is an example of such incrementalism. Nevertheless, S.1894 stands as a blueprint for the types of changes that a national law should encompass regarding federal–state management roles, and increasing the flexibility of federal environmental protection laws and operations of federal water projects to alleviate some of the difficulties California was having in coordinating the reallocation and delivery of water supplies between federal and state water projects (Hanak *et al.*, 2011; Hanak *et al.*, 2015; PPIC, 2016).

Whenever focused coordination of federal activities has occurred, it has been driven largely by pending crises, such as potential threatened or endangered species listings, droughts, floods, and hurricanes; or by local or regional initiatives. Concerns about water supply and its development routinely bolsters interest by various members of Congress to either establish a national water commission to assess future water demands, study current management programs, or develop recommendations for a comprehensive national water management strategy (Copeland & Carter, 2009). However, Congress has not been effective in acting on any of the Commission recommendations, the most recent example being their inaction on the National Drought Policy Commission (2000) recommendations, though President Obama incorporated most of them in his recent Presidential Memorandum (White House, 2016a, 2016b) on building long-term drought resilience.

Performance of US demand management policies and strategies

Water demand management, rather than capacity expansion, currently forms the core of most federal water resources management strategies, and comprises the essential component of drought management as a contributor to national and regional 'water security'. This strategy is possible only because of the foresight of previous generations to build adequate storage capacity in order to serve the needs of future generations. Many academicians and water policy experts, including international development lending banks and multilateral lending institutions, lament the absence of a national water policy as a prerequisite for efficient and sustainable water use, which inherently relies on demand management.

Current water management ideology, in the form of IWRM, requires that sustainable water management consist of a set of enabling institutional factors as prerequisites: a national water policy, a national water management plan, a comprehensive water code and river basin management agencies, as part of a comprehensive approach to IWRM and prerequisite for lending for water infrastructure projects (GWP, 2004; World Bank, 2007).

Ironically, the US does not fulfil many of these prerequisites (Stakhiv, 2003), even as its foreign aid arm, the US Agency for International Development routinely advocates all of these prerequisites as a condition for its own assistance programs in developing nations (Lautze *et al.*, 2011; Giordano & Shah, 2014). Yet it can readily be argued that a US national water policy exists, at least as a collection of disparate and incrementally developed and implemented federal, state and local policies that are relatively consistent in purpose, if not in implementation. They have been quite effective in achieving the sort of outcomes generally desired by the EU Water Framework Directive, as well as stated IWRM and sustainable development goals and objectives. The principal reason is the extraordinary role and effectiveness of a broad array of federal incentives that underpin those goals, as well as a strong federal regulatory structure, with an accountable enforcement arm to ensure that the goals are met.

How can it be demonstrated that such an accumulation of eclectic policies has been effective, and that US water security has improved over time? What is the best way to measure the overall performance and effectiveness of an accumulation of diverse policies in terms of such broad goals as sustainable development or drought management? A wide array of performance metrics has been advocated to explain and measure the relative reliability of water management or control systems, and their effectiveness in reducing community risks and regional vulnerabilities with respect to floods and droughts.

The notion of a 'national water security' performance metric has generally been addressed in terms of simple per capita water availability measures (e.g. Falkenmark, 1986). The concept of water stress is relatively simple: it applies to situations where there is not enough water for all uses, whether agricultural, industrial or domestic. Defining thresholds for stress in terms of available water per capita is more complex, however, entailing assumptions about water use and its efficiency. Nevertheless, it has been proposed that when annual per capita renewable freshwater availability is less than 1,700 m^3, countries begin to experience periodic or regular water stress. Below 1,000 m^3, water scarcity begins to hamper economic development and human health and well being. For comparison, the total annual renewable water resources for the continental 48 US states is 7,400 m^3/capita/yr (UNESCO WWDR, 2003), compared to average US water use of 1,600 m^3/capita/yr, or 20% of available supply (OECD, 2005). The average per capita water use for OECD countries, which includes the US, is 900 m^3/capita/yr. In addition, the US has approximately 950 km^3 of storage. With a population of 325 million in 2015, this storage translates to nearly 3,000 m^3/capita/yr, or three times the minimum Falkenmark threshold for water security.

The problem with most indices is that they do not effectively capture the major missing component of institutional deficiencies that lead to poor water management (Fekete & Stakhiv, 2015). Saleth & Dinar (2004), in their study of institutional performance in the water sector, note that '...water scarcity whether quantitative, qualitative, or both, originates more from inefficient use and poor management than from any real physical limits on supply augmentation'. According to their analysis, the crisis in the water sector is mainly a function of limitations of contemporary institutions, which allocate and manage water, and they advocate a series of institutional reforms that are key for successful implementation of IWRM.

Governance reforms, however, are very difficult to implement, and it is more difficult still to evaluate their relative effectiveness on improvements in water use efficiency, economic productivity, poverty reduction or improvements in environmental sustainability. Saleth & Dinar (2004) tried to develop a mechanism for evaluating these important institutional changes. The quantitative results were quite mixed and in most cases inconclusive. The reason was that effective water governance is predicated on a series of prerequisites, beginning with a well-defined system of water rights or water law and enforcement. Other related water policies, administrative changes and water extraction regulations, including those for privatization, cost recovery and water transfers, were all intertwined with the fundamental requirement to define and enforce water rights.

IWRM, of course, is the fundamental comprehensive management platform for attaining what is termed 'water security' and sustainable development (GWP, 2016). The most widely acceptable definition of water security would read as 'the availability of an acceptable quantity and quality of water for health, livelihoods, ecosystems and production, coupled with an acceptable level of water-related risks to people, environments and economies' (Grey & Sadoff, 2007). The definition is firmly embedded in the concept of sustainable development, with its aim to ensure a triple bottom line of social, environmental, and economic development outcomes.

The water engineering profession measures system performance somewhat differently, more accurately reflecting the nature of system delivery performance under climate variability. While economists value water for its abstract economic production values, engineers measure performance against the specific objectives for which a project or system was designed, which includes economic performance. A number of indices are proposed in literature, which measure the performance of water resources systems in terms of reliability, resilience and vulnerability of water resources subjected to climate extremes (e.g., Hashimoto *et al.*, 1982). Reliability is the probability that monthly precipitation (or discharge) is larger than its long-term monthly mean value. Vulnerability, in this context, refers to the likely magnitude of a failure (e.g. maximum drought intensity) if one occurs. Resilience may be interpreted as a measure of how quickly a system is likely to recover from failure once failure has occurred. Vulnerability and resilience are effectively complementary notions. Inherently, the ensemble of drought management measures (or flood control measures) that are implemented by any locality or region aims to increase system reliability, resilience and robustness, while decreasing vulnerability. Over the past 30 years, the largest contributor to reduced drought vulnerability has been water conservation measures that were a by-product of the water quality regulations of the Clean Water Act (1972).

Notwithstanding all the performance metrics available for evaluating the cumulative effectiveness of all federal and state water management policies consisting of governance adjustments (enabling legislation, regulatory regime, judicial interventions, conservation measures, etc.), as well as technological innovation and engineering adaptation, there exists a very important and useful performance measure (not indicator) of the relative success of US water policies over the last 40 years: the reduction in average annual demand and freshwater withdrawals in the US (Rogers, 1993). Figure 1 shows that since the peak use in 1980, water use has declined to the levels of 1970, despite a 33% population increase during that same period.

Overall, the US was able to achieve this remarkable reduction in water use, not through the conventional top-down IWRM management and governance ideology, but mainly through a bottom-up emphasis on an enabling institutional framework that focused on regulating, monitoring and enforcing a suite of water quality and environmental laws passed in the 1970s (Stakhiv, 2003). Among them was the Clean Water Act (1972), National Environmental Policy Act (1970), Endangered Species Act (1973) and the Safe Drinking Water Act (1974). These mechanisms were much more effective (if not altogether economically efficient), in controlling municipal, industrial and agricultural water use through stringent water quality standards. As seen in Figure 1, water withdrawals began to drop after 1980, and have continued declining to the present day.

The decrease is even more dramatic when juxtaposed with the projections of leading expert commissions of that period. Figure 2 shows actual water use superimposed over projections of six nationally respected forecasts, all starting in the same year, 1970. Only the US Water Resources Council (US WRC, 1978), offering a single projection of water use, projected an 18% decline in water use, from 400 bgd in 1975 to 306 bgd in 2000. This came only 2 years after the National Water Commission (NWC, 1973) report, which projected an increase of water use to 880 bgd by 2000. The 1975 WRC projection was bold, and based on their understanding of how the Clean Water Act of 1972 would dramatically increase industrial water recycling, thereby decreasing withdrawals, as well as a reduction in irrigation withdrawals.

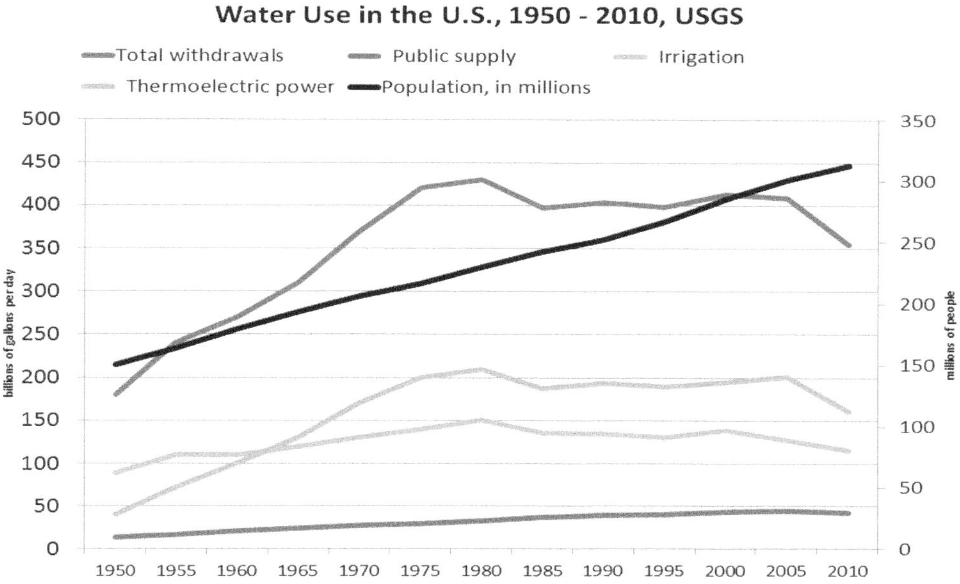

Fig. 1. Water use in the US. *Source:* US Geological Survey.

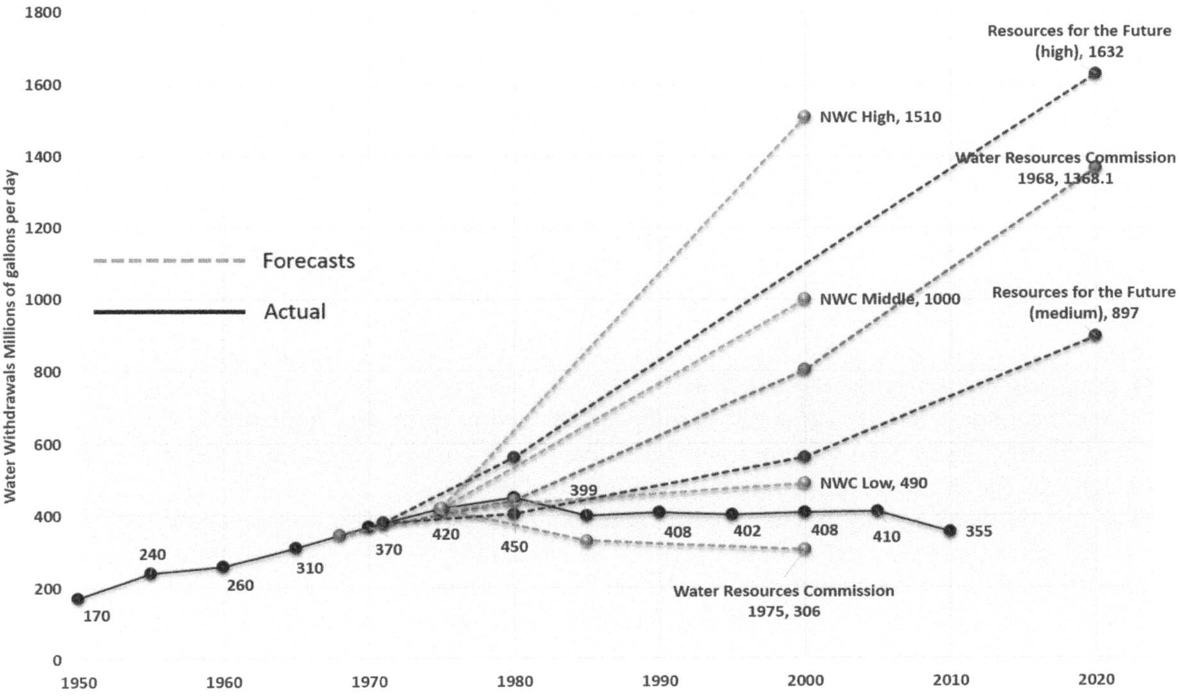

Fig. 2. Actual water withdrawals vs projected water withdrawals.

Drought damages

Another imperfect measure of the relative effectiveness of US drought management policies is the ability to contain the growth of drought-related economic damages, and mitigate the economic impacts through a wide array of federally-sponsored insurance and compensation programs. Despite the absence of a national water policy and, more specifically, a uniformly applied drought management policy, the US does not incur substantial economic damages caused by extreme droughts, in comparison to many developing nations, where floods and droughts routinely depress annual GDP by 2–5% during and after each frequently occurring catastrophic event (Hallegatte *et al.*, 2016). No nation can develop its socio-economic structures and grow progressively with such periodically occurring devastating natural hazard events, which sap the nation's resources during each recovery phase. This is what is fundamentally meant by 'water security': to protect the citizens, economy and social institutions from such catastrophic events, and reduce and mitigate the damages to an 'acceptable' or 'tolerable' level of risk, so that it does not routinely threaten the lives and livelihoods of its populace and prevent sustainable development (Grey & Sadoff, 2007).

Droughts in the US caused approximately $9.5 billion damages annually during the period from 1980–2016, mainly in the agricultural sector (NOAA, 2016). As a consequence of the continuous adaptation by the various economic sectors associated with agriculture, as well as the growing suite of drought mitigation and compensation policies and programs of various governmental institutions, this is a miniscule percentage of the $18.6 trillion (GDP) US economy, and approximately 5% of the total $180 billion farm agricultural sector production of the US in 2015 (Schnepf, 2016). Even during the height of the most recent California drought, the direct and indirect economic costs attributed to the drought in 2015 amounted to $2.17 billion, or 4% of the $54 billion total agricultural production and 0.1% of California's $2.5 trillion GDP. In contrast, floods, severe storms and tropical cyclones cause about 70% ($25 billion/yr) of all economic damages annually in the US (Howitt *et al.*, 2015).

There are a number of reasons why the US is in a reasonably good position with respect to effective water and drought management, even under a broad range of climate change scenarios:

1. The US has a relatively abundant natural fresh water supply of 1,400 bgd, on average, while using only 350 bgd (Maupin *et al.*, 2014).
2. The US has a large water storage capacity (950 km^3) that is able to effectively augment natural supply deficiencies in the western US.
3. The US economy and its various water-dependent sectors (primarily agriculture, the power sector and industry) have been adapting very effectively to increased competition for water resources, water scarcity and evermore stringent environmental regulations.
4. The natural water supply is enhanced by a robust national water storage and distribution system which stores, approximately, more than an average annual supply of runoff, and in some basins, four times the average annual runoff (Missouri and Colorado River basins).
5. Since the 1970s, an array of federal and state water policies and incentive programs, supplemented by regulatory controls and enforcement measures, has resulted in a reduction of fresh water withdrawals to 350 bgd in 2010, from a peak of 450 bgd in 1980 – a decrease of 20% despite a 50% increase in population from 1970–2010.

6. Since 1975, the federal agencies and state agencies have been more effective in planning for droughts, and engaging in preparatory drought management exercises, especially in the semi-arid southwestern US.
7. The US Federal Government and state agencies have a broad array of insurance programs and disaster assistance programs that mitigate most economic losses incurred by droughts, especially in the agricultural sector.
8. The US per capita income is among the highest in the world, allowing it to devise and implement more costly technological solutions that save water and mitigate the worst impacts of drought in a timely manner.
9. Industry, the agricultural sector and municipalities have all been implementing cost-effective conservation measures and technologies that have contributed to a substantial reduction in per capita water use over the past 40 years.
10. Unlike floods and hurricanes, droughts are slowly evolving phenomena, providing ample time to respond and adjust, in a staged and structured manner. Droughts allow time for triggering a series of planned contingency measures, including the implementation of emergency structural measures (e.g. tapping conservation storage in reservoirs, deeper groundwater sources with treatment, pipeline interconnections of groundwater and surface water conjunctive systems, etc.) as the drought severity and extent increases.

Drought trends and climate uncertainty

Despite global data that show a unidirectional warming trend, the observational data on a global scale as well as subcontinental US scale do not show any discernible increase in drought frequency or severity. Hao *et al.* (2014), analyzing global data for the past 30 years (1982–2012), demonstrated that there was a peak in drought activity (severe droughts) in the 1990s that covered 10–20% of the globe. In any given year, about 25% of the globe is in somewhat dry conditions. Since 2000, however, drought frequency and severity has diminished considerably.

Andreadis & Lettenmaier (2006) found a similar pattern of a decrease in drought severity and frequency in the US, using a different analytical approach. Sheffield & Wood (2011) discussed the 'skill' of climate models in representing 20th century droughts. They noted that 'climate models do not provide a perfect representation of the climate system' and that there is a '…general overestimation by the models of long-term drought frequencies and mean drought duration…' (p. 177). This assessment of substantial climate model uncertainties conforms with that of Kundzewicz & Stakhiv (2010) and of Koutsoyiannis *et al.* (2008).

Irrespective of the uncertainties in climate forecasting, water management has historically dealt with such extreme climate uncertainties, and has developed solutions that overcome most of the problems associated with climate variability. One of the pioneering studies related to climate change related droughts and agricultural adaptation strategies, was the MINK Study (Rosenberg, 1993) to examine the impacts of a recurring 'Dust Bowl' era event under current climate conditions and the projected climate change conditions of 2030. Overall, the authors found that there would be minimal impacts, even in 2030, in the agricultural sector, because of a continuously adapting economy that employs new markets, agricultural techniques and technologies and adjusts to scarcity in many different ways. The MINK study did not even account for an extensive network of new water storage projects that were built since the great Dust Bowl catastrophe in the region, which provides considerable quantities of irrigation water, and has improved overall water security of the region, i.e. increased reliability, resilience and robustness.

The key point is that proactive drought management and risk reduction actions depend, to a large degree, on strategically justified infrastructure projects that rely on more reliable quantitative estimates of risks based on the probability and duration of drought and flood events. Recent work and thinking on this conundrum has been undertaken by the World Bank (Ray & Brown, 2015) and UNESCO (Mendoza *et al.*, 2016) by developing a decision-analytic framework termed 'decision-scaling'. What this framework does is to identify and focus on current vulnerabilities, applying probabilistic estimates of failure mode and time to failure, and then expanding the evaluation to include an increasingly uncertain range of plausible climate scenarios, and evaluating the performance of the system with the incremental addition of various risk-reduction management options.

Regardless of disparate drought trends and climate uncertainty, droughts have gripped the western US repeatedly, especially in the Colorado River basin and California. Hence, the recent drought management responses of both California and the respective states, as well as the federal agencies whose responsibility is to deal with drought management, are of interest. Of particular interest is whether the strong imprint of federal actions and support, as part of the ensemble of solutions, is deemed effective.

Notable US droughts

Like earthquakes and floods, droughts are natural events that become problematic for mankind because of the development decisions people make, which also bring benefits. Tree ring records suggest there was a great drought in the Southwest between 1276 and 1299 AD that was partially responsible for the disappearance, through migration, of the Anasazi or 'Pueblo' people (Benson *et al.*, 2007; Woodhouse *et al.*, 2010). Droughts of that magnitude and severity can be associated with the coincidence of negative Pacific Decadal Oscillation conditions and a positive Atlantic Multidecadal Oscillation.

Droughts in the late 1500s and early 1600s may have doomed two European settlements, one in Roanoke, North Carolina, and the other in Jamestown, Virginia (Stahle *et al.*, 1998). Of course, these settlements were entirely at the mercy of weather and climate, without the ability to harness the extremes of nature. Nineteenth century droughts in the US were similar narratives – nature damaging man – but water supply engineering offered the possibility that drought impacts could be mitigated.

In the semi-arid west, the Appropriation Doctrine underlying state laws about the right to use water established a transparent process for allocating water during a drought: the first to use water would be the last to lose the right to withdraw from dwindling supplies during a drought. During George Washington's first term (1789–93) the largest US cities in the 13 eastern states all had relatively abundant water supplies (Philadelphia, New York, Boston and Charleston) for their populations. Starting with Philadelphia in 1801, these municipalities began to develop water supply systems for their growing populations, primarily to reduce groundwater contamination and other waterborne illnesses.

Later, even eastern cities had to consider drought impacts in the design of their water systems. Epidemics of yellow fever and cholera were common in the New York metropolitan area in the 1830s because of dependence on groundwater supplies and unsanitary wastewater disposal conditions. Frequently recurring droughts exacerbated these health conditions, and contributed to the 'Great Fire of 1835', when there was inadequate groundwater to put out the raging fires in lower Manhattan. The great loss of life and accumulation of disastrous events led to the design and construction of the

Croton watershed water supply system north of the city, in Westchester county. Water was stored and conveyed through an aqueduct to a reservoir in what is today Central Park, completed in 1842 (Hinchey, 1987).

A severe drought in the 1960s led to an interstate conflict, because New York City's withdrawals from the Delaware River threatened the Philadelphia water system. This drought stimulated the development of new federal water planning principles, a series of comprehensive water supply studies and recommendations for securing the water supplies of major metropolitan areas in the North Atlantic Region (Boston, New York, Washington DC) (Stakhiv, 1976; US Army Corps of Engineers, 1976; Major & Schwarz, 1990). Any substantive climate-related changes could make even the 1960s north-eastern US droughts pale by comparison (Pedersen *et al.*, 2013) and increase the vulnerability of the New York metropolitan area water supply.

However, it was the great mid-western 'Dust Bowl' drought of the 1930s, in conjunction with the great economic Depression (1930–40) that were the dominant historical events that led to the creation of US Federal drought aid programs and an acceleration of water resources development programs to assure water supplies to farmers, industry and municipalities. By 1934, the droughts covered 75% of the US, severely affecting 27 states. By 1936, some 21% of all rural families in the Great Plains received Federal emergency relief; in some counties, it was as high as 90%. In 1937, the Works Progress Administration reported that drought was the main reason for relief in the Dust Bowl region. More than two thirds receiving assistance were farmers. Total assistance was estimated at $1 billion (in 1930s dollars). The report found that losses in the Dust Bowl affected the entire national economy (National Drought Mitigation Center, undated).

Overview of US drought management perspectives and practices

Until the 1970s, drought management had traditionally been carried out following a reactive approach, i.e. by implementing emergency measures after a severe drought had already been recognized as a natural disaster. A shift toward a proactive (or risk-management) approach with measures planned in advance has been progressively advocated by a variety of congressionally mandated commissions such as the National Water Commission (1973), the Western Water Policy Review Advisory Commission (1998) and the National Drought Policy Commission (2000).

'True' drought management initiatives that are strategic in nature, and which implement pro-active, anticipatory risk-reduction measures, address actions that should be taken *in anticipation of* a drought, as well as actions that can be undertaken to mitigate the adverse consequences of water shortages *during* a drought, by reallocating existing storage and supplies for emergency use purposes, escalating emergency water use measures, increasing conservation measures, etc. (IWR, 1993, 1994). They are true drought management actions, for they deal with the *existing system capacity* and available resources once a drought is proclaimed, triggering a series of contingency measures. There is usually no ability to significantly expand the capacity of water availability during a drought – you work with the dwindling resources and attempt to ameliorate their worst adverse consequences – either as financial compensation or by reallocating resources to their highest priorities and human needs.

A more fundamental, and strategic form of drought management deals with expanding resource availability, and ensuring greater reliability of services over a wider range of unanticipated drought scenarios. Usually, this means expanding water supply capacity by reallocating existing storage for municipal or irrigation purposes from other existing lower priority purposes, adding emergency supply by building new reservoirs, emergency water intakes, pumping stations, or interconnecting existing systems for more flexibility, diverting water through inter-basin transfers, or technologically increasing supplies via desalting or conjunctive use of groundwater, surface water and/or desalting. These are all considered pre-planned strategic drought management measures that ensure adequacy of supplies and increase system reliability.

There is a recent trend in natural hazards management towards quantifying the risks of various hazards, their consequences, and in evaluating risk reduction options, especially in the context of climate change scenarios. Risk, typically expressed as a function of hazard and vulnerability (UNISDR, 2011), serves as a measure of anticipated consequences and can indicate an improvement (or not) after the implementation of a mitigation option. In addition, traditional cost–benefit analysis could be used to support the selection of the appropriate measures (Baumann *et al.*, 1998). Much of this work is done under the auspices of 'vulnerability assessments', which do not provide the same degree of analytical rigor in evaluating cost-effective risk-reduction options, as for example, the traditional water resources project justification procedures of the World Bank, Corps of Engineers or Bureau of Reclamation. The problem is that the relation between vulnerability and risk is not commutative: reduced vulnerability always means reduced outcome risk, but reducing the outcome risk does not always reduce vulnerability. (Sarewitz *et al.*, 2003).

A good part of this asymmetry can be attributed to the manner in which the quantification of risk is used in normative economic analysis as part of conventional project justification. Stakhiv (2011) delineates a 'quadruple discount dilemma' in the manner in which extreme hydrologic risks are 'discounted' in a typical analysis of water resources project alternatives. Contemporary water project justification criteria are generally incompatible with the search for 'no regrets' climate adaptation solutions that reach out far into the future under very uncertain, non-stationary climate scenarios. The socio-economic effects of uncertain, low probability–high consequence events, such as extreme floods and droughts, are sequentially discounted by the choices of probability distributions that diminish the importance of low probability events (Haimes, 2004; Botterill & Cockfield, 2013). The economic consequences of such low probability events are further discounted when the economic discount (interest) rate that is typically used for benefit-cost analysis is too high for dealing with adaptation mechanisms for intergenerational phenomena such as climate change, and associated increases in drought duration, intensity and variability.

Analyzing vulnerabilities and evaluating drought management options from a risk-analytic perspective is quite difficult, because of the ill-defined and slowly evolving characteristics of drought and its many intertwined meteorological, hydrological, agricultural, economic and social facets. Hence, there are many drought policy incentives and measures that focus on preventive measures that contribute to vulnerability reduction, but not necessarily 'risk reduction', i.e. reducing physical consequences and mitigating the socio-economic impacts. For example, 'drought triggers' (Palmer *et al.*, 2002) are threshold values for a series of escalating actions usually based on water quantities (reservoir or river levels, streamflow, rainfall deficits, etc.). They are not based on quantitative risk analysis, using probabilities, since drought intensity, duration and extent are very difficult to quantify in probabilistic terms. Hence, both preparatory planning actions or a collection of non-structural drought mitigation measures are difficult to evaluate within a traditional benefit–cost framework that is typically used to justify structural projects. In this case then, more rigorous assessment and evaluation protocols need to be developed that quantify vulnerability reduction in quasi-risk-cost terms (Fontaine & Steinemann, 2009).

In general, though, strategic drought management is about reducing future drought vulnerabilities and risks, employing the principles and tools of risk-benefit-cost analysis (Baumann *et al.*, 1998). Tactical drought management focuses on reducing local vulnerabilities once a drought has been recognized, while emergency drought management focuses on reducing and mitigating the worst effects of drought on vulnerable segments of the population and priority uses during a drought (Stakhiv, 1976; IWR, 1993).

The hierarchy of drought management practices

Drought management practices reflect the governance structures within which they are deployed. Not every sensible idea can be employed with uniform effectiveness in every state of the US. Each state is a laboratory for implementing drought management measures that are compatible with its laws and management priorities. As Rogers (1993) noted, in any complex system of governance, there are many discrete levels of management options that exist – a vertical hierarchy, consisting of top-down federal and state policy-setting and regulatory controls to the lowest units of users, such as irrigation districts, industry and self-supplied users, as well as a 'horizontal system' of integration, that coordinates uses, actions and drought management measures at the local or sub-state levels, in a given area.

This governance complexity is why Rogers (1993) questioned whether a 'national water policy' could ever be effective in a complex 50-state, federal governance system. Each vertical level of governance, has a corresponding horizontal system of coordination, matched with a somewhat different array of tools, policies, authorities and responsibilities, so it is very difficult to generalize about the effectiveness of drought management policies and programs at any of these levels, especially in the context of a very dense and dendritic network of federal regulations and authorities exercised by a wide array of federal agencies. A 'national policy' would do little to correct the many inconsistencies in federal law that prevent more flexible responses to drought management by the states and federal agencies. That requires a comprehensive 'organic' legislative act.

It is important to understand that there exists a hierarchy of drought management measures, which overlap, to a certain degree, as in a Venn diagram (Knutson *et al.*, 1998). Strategic measures, i.e. those that expand water availability or capacity, are proven to increase the reliability, flexibility and robustness of a system, and form the fundamental core of drought management. They are the foundational components of 'water security', by reducing vulnerability to unforeseen periods of drought. Tactical measures are those that increase an existing system's effectiveness in delivering water supplies, and improving efficiencies in use, and comprise those measures that contribute to system resiliency in withstanding a drought, and hastening the recovery from a drought episode. Emergency measures are inherently those associated with unforeseen drought conditions, or those that exceed the designed capacity of a given system.

Strategic water (drought) management is forward looking, anticipating supplies and demands 30–50 years from now, and includes both supply and demand aspects – developing or augmenting supplies and implementing long-term demand management measures. Thus, water supply planning and state laws governing the right to use water are key parts of strategic drought management, such as conjunctive use of surface water supplies and storage with groundwater. Strategic drought management should also include both tactical and emergency actions (US Army Corps of Engineers, 1976; IWR, 1995). However, tactical response measures generally are associated with modifying the rules and regulations associated with an existing water storage and delivery system, either as reallocation of water storage in reservoirs or changing the operating rules of a system during a predefined sequence of emergency measures.

Tactical measures are inherently associated with altering the short-term or permanent operation of an existing supply system, or temporarily changing the rules and requirements for drought aid, or modifying environmental regulations or water withdrawal criteria associated with a particular river, groundwater aquifer, region or situation, in order to enhance water supplies and ease delivery to drought stricken areas. Emergency drought management generally deals with an orderly curtailment of demands, and/or a reallocation of existing supplies towards higher priority needs, as determined by the political system (IWR, 1995; US Army Corps of Engineers, 2011).

Drought contingency plans are part of a suite of measures that comprise a preparatory subset of responses usually based on a set of pre-determined hydro-meteorological or capacity 'triggers' for initiating a series of actions depending on the circumstances of particular drought but according to a strategy for dealing with all future droughts (IWR, 1995; Palmer *et al.*, 2002). When the drought exceeds the capacity of the existing water supply system and planned measures are inadequate, emergency measures must be taken, such as graduated demand curtailments or temporary water transfers. The interplay of these three levels of management is driven in part by forecasts of future supply and demand (IWR, 1995).

Federal water management agencies have been under a great deal of external scrutiny these past 30 years, as well as Congressional oversight and Executive Branch implementation pressure to adapt rapidly to emerging and ever-changing constructs of sustainability, ecological restoration, risk management and climate adaptation. As a consequence, most Corps reservoirs, as well as those of the Bureau, have expanded their formal system of adjusting ever-changing policy requirements, within a framework defined as the 'reservoir master manual' (US Army Corps of Engineers, 2016). There are many opportunities for updating the drought and flood contingency plans based on new information that could improve the overall resiliency, robustness and reliability of an existing water delivery system.

At the federal level, all federal agencies coordinate their drought management actions so that they are applied in a reasonably consistent manner and reinforce one another. In that respect, there exists a collection of 'federal' water policies that are applied consistently, if not entirely uniformly. Hence, the recent Obama Executive Memorandum (White House, 2016a, 2016b) on 'Building National Capabilities for Long-Term Drought Resilience', referred to the implementation authorities of federal agencies to initiate and promote several aspects of drought management that fall within the authorities of federal agencies. This Executive Memorandum was, in effect, a policy to enhance the planning and preparatory aspects for subsequent drought management actions by the agencies of the Federal Government, in support of state and local authorities that are charged with drought management. This was a federal policy that largely dealt with tactical rather than strategic drought management measures, i.e. the available tools and mechanisms that could be marshalled to improve planning by the states and local agencies in preparation for, and monitoring and responding to, the next drought, relying on existing legislative authorities and federal programs.

However, while the Federal Government has been proactively advocating anticipatory drought risk management approaches, and implementing them within their own respective programs, the states have been somewhat slower in adapting these policy initiatives. According to a recent study by Fu *et al.* (2013), which examined 44 state drought plans, nearly all states have developed drought plans, suggesting that these plans would be effective in reducing drought losses, to some extent. However, most of these plans were developed to facilitate drought responses, during and after a drought event, and such a crisis management approach toward droughts still lacks the necessary proactive risk-management strategic dimensions.

A similar recent appraisal focusing on the drought preparedness programs of the 19 western states (Fontaine *et al.*, 2014), found the state approaches were comparable to those reviewed by Fu *et al.* (2013). Few states conducted post-drought assessments to prioritize the most vulnerable areas, and fewer still undertook strategic studies to reduce the risks and mitigate the impacts of future droughts. Most of the drought preparedness protocols were geared towards drought formal declarations that were required by the Secretary of Agriculture for federal disaster designation and assistance. To qualify for a federal drought disaster designation, states need to assess drought impacts at the county level, within each state.

Although many of the state drought preparedness plans include mitigation (i.e. strategic) measures that require new capital intensive water supply projects and/or conveyance infrastructure, as well as new legislation that would ease regulatory restrictions during a drought, neither Fu *et al.* (2013), nor Fontaine *et al.* (2014) found much evidence that there was any substantive follow-up on assessing post-drought performance as the basis for planning mitigation measures in anticipation of the next drought.

A somewhat different, more strategic approach was exhibited by the federal system in 2000, in mustering the authorities of all the federal agencies to ameliorate and mitigate the impacts of that recent Midwest drought of 1998. In July 1998, the 105th Congress enacted Public Law 105-199, the National Drought Policy Act. This law established 'an advisory commission to provide advice and recommendations on the creation of an integrated, coordinated Federal policy designed to prepare for and respond to serious drought emergencies'. The law directed the Commission to 'conduct a thorough study and submit a report on national drought policy'.

The National Drought Policy Commission (NDPC, 2000) Report, 'Preparing for Drought in the 21st Century' was issued in 2000. Improved interagency coordination is invariably one of the most predictable of all report recommendations, as is information sharing, which was the focus of the NDPC. Nevertheless, the NDPC Report stands as one of the best sources of contemporary policy advice, noting that '…drought preparedness and mitigation actions formed the cornerstone of the Commission recommendations'. The report, however did not mention increasing water supplies through storage capacity expansion, interconnecting existing supplies, or building new conveyance facilities for inter-basin transfers, because the Commission, consisting largely of Federal agency heads, focused on the inherent existing and legislatively constrained authorities and programmatic capabilities of federal agencies, and did not propose new legislation. The following Policy Statement was prominent in the Executive Summary of the NDPC:

> *'The Commission believes that national drought policy should use the resources of the Federal Government to support but not supplant nor interfere with state, tribal, regional, local, and individual efforts to reduce drought impacts. The guiding principles of national drought policy should be:*
> 1. *Favor preparedness over insurance, insurance over relief, and incentives over regulation.*
> 2. *Set research priorities based on the potential of the research results to reduce drought impacts.*
> 3. *Coordinate the delivery of federal services through cooperation and collaboration with nonfederal entities.*
>
> ***This policy requires a shift from the current emphasis on drought relief.*** *It means we must adopt a forward-looking stance to reduce this nation's vulnerability to the impacts of drought. Preparedness – especially drought planning, plan implementation, and proactive mitigation – must become the cornerstone of national drought policy.'*

While the role of the states has grown in drought preparedness, as their technical capabilities have increased over the past 30 years, partly in response to citizen demands on their resources, the states have been relatively weak in grappling with strategic drought management needs. Preparedness inherently assumes a status quo availability of water during a drought – it merely develops a set of options and priorities for allocating scarce water. Structural measures, such as reservoirs, pumping stations, pipelines for interconnections and river diversions are costly, and federal benefit–cost procedures have been substantially modified to limit such projects, and federal cost-sharing for such projects has declined markedly during the past two decades. Finally, concern for the environment, and the voices of many powerful environmental interest groups, have stymied such structural measures repeatedly, as can be seen in California, for example (Hanak *et al.*, 2011; PPIC, 2016).

Nevertheless, it is clear that the Federal Government plays a substantial role in drought management, and states often look to a more prominent federal role in order to overcome their own internal political difficulties in implementing reforms (IWR, 1995). It is an uneasy partnership that tries to protect a state's right and sovereignty, while demanding ever more federal resources to improve each state's water security.

The components of a prospective National/Federal drought policy

For reasons specified earlier, we have to differentiate between a 'national' policy that would encompass both the roles of the states and the Federal Government, from a conceptually simpler 'federal' policy, which would only address the roles and authorities of federal agencies in providing the enabling institutional, operational and financial support to the respective states for drought preparedness activities. The recent 5-year California drought, along with the 15-year Colorado river drought and other comparable instances in the southeastern US, centering on the city of Atlanta's struggles with downstream states in allocating water for its growing municipal needs (Abrams, 2009), exposed the real problem of a wide array of rigid federal laws that cannot readily adjust to changing socio-economic circumstances and drought management needs.

If a federal drought policy for the future were to be developed, what would be its key features and components? Much of what exists in the US is essentially about tactical drought management measures: preparing for a drought with an existing water availability infrastructure that requires curtailments in water use and deliveries to satisfy priority needs (Knutson *et al.*, 1998). The existing federal and state drought preparedness policies and programs are adequate for coordinating actions during and after a drought. A federal drought policy that addresses future national water needs under increased climate variability would have to consider drought management within the broader context of enhancing national 'water security' for a growing population that is increasingly concentrating in urban centers.

This prospective or hypothetical federal policy, presumably developed by a future Presidential 'Drought Management Commission', would promote a re-evaluation of existing, higher-level water allocation agreements at the river basin scale that have been previously legislated by Congress and adjudicated in courts. It must deal with the very real issues of state sovereignty in water management, and the profound differences between western appropriative water law and eastern riparian water laws. It should foster and provide incentives for new drought management elements and technological advances, such as water banking, and including desalinization, conjunctive groundwater use and broader use of insurance programs. It should foster state water management compacts for shared interstate resources, as for example the Great Lakes–St Lawrence River Water Resources Compact, enacted by Congress in 2008, and signed into law by President Bush.

A prospective Drought Management Commission, however, would have to tackle the plethora of existing conflicting laws and barriers to effective federal–state management of existing water resources, as highlighted by the recent problems in California. Unlike the NDPC (2000), which confined itself to improving the delivery of federal drought management services during and after a drought, a prospective

Commission must deal with untangling the myriad of existing laws that hamper the rapid delivery of services and are inflexible in responding to drought emergencies, such as those in California or the Colorado basin.

Certainly, a model national drought management policy could be drafted, but it would be very difficult to implement at the 50-state national level. It would be easier to start at the coordinated federal agency level, via an EO. The WMO & GWP (2014) drafted guidelines for a national drought management policy development process. In fact, the guidelines essentially addressed how to organize a commission or committee to develop such guidelines. However, the underlying objectives for such guidelines were mostly the consideration of tactical measures, not a comprehensive water security approach.

Of course, each US federal resources management agency has developed and possesses its own drought management strategies for the resources it is responsible for managing. Thus the US Forest Service (2016) has developed its own strategy, as has the Natural Resources Conservation Service (NRCS, 2015), and the Bureau of Land Management (BLM, 2013), among the many US federal agencies. While this is a considerable degree of overlap between the drought management policies of water resources agencies that are responsible for water supply and deliveries to the municipalities, farms and irrigation districts, clearly there is a very wide array of land-based drought management practices that cannot be covered here that would encompass farming practices and rangeland management, as well as forestry practices.

Conclusions

Drought is a pervasive condition in many arid and semi-arid nations, and is an occasional problem in the US. In general, the US does not have serious drought management problems, as yet. But it could get worse, as the uncertainties of climate change begin to manifest themselves, especially in the semi-arid western US, disproportionately affecting already stressed urban metropolitan areas. California's response to such changes may serve as guides for other states and nations.

Current federal drought policies are sufficient to deal with contemporary drought management problems, and improved delivery of federal aid and services to drought-stricken areas, and states are increasingly better prepared in dealing with droughts. The major discord between the federal agencies and states lies in the relative inflexibility of existing federal laws governing the operation of federal water projects and federal regulatory standards, to rapidly and effectively assist the states in meeting unforeseen circumstances during a drought. A 'national policy' will do little to correct the many barriers and inconsistencies in current federal law that prevent more flexible responses to drought management by the states and federal agencies. In that respect, the creation of a national drought management policy is essentially superfluous, because what is really needed is a Presidential Commission that is able to recommend changes to the numerous existing overlapping laws that hamper federal–state cooperation during drought episodes, and are too rigid to offer any substantive relief during emergency conditions.

The humid areas of the eastern US have a different type of drought management problem from the Midwest or southwestern US. Concentrated population centers require water deliveries from distant watersheds. The problem of eastern municipal areas is one of adequate water storage, and delivery reliability. The 17 western states, mostly semi-arid, are closer in drought management practices to those of developing nations, and can be used as models for effective responses to water security, and inherently to drought management.

From the standpoint of searching for practical management alternatives, there are about ten feasible and cost-effective strategic drought management measures that are likely to have the greatest overall benefit in the near future in enhancing state and regional 'water security', beyond what is called for as part of improved federal drought management coordination and data sharing (IWR, 1995). Most fall within the responsibilities of local communities and states, and do not require federal assistance:

1. Regionalize small private and public water utilities via interconnections to increase flexibility, reliability and lower costs (economies of scale).
2. Price water to reflect costs associated with operation, maintenance and long-term capital recovery costs and environmental costs.
3. Foster improvements and cost effectiveness for desalting technologies and portable systems for small towns as well as conjunctive use of groundwater and surface water supplies.
4. Continue improving real-time, satellite forecasting techniques to provide more reliable information for rational and measured drought contingency responses.
5. Adapting to climate change and its uncertainties demands more storage for ultimate water security.
6. Expand water trading and marketing, and water banking.
7. Expand cost-effective water use efficient technologies, and wastewater recycling for outdoor uses, lawn watering, etc.
8. Update all reservoir management master plans and associated drought contingency plans to include climate scenarios and recalculate 'safe yields' for water supply and other purposes.
9. Large municipal water supply utilities must continuously update their water supply drought contingency plans under a variety of climate change scenarios, as a preparatory activity.
10. Continue to focus on increasing efficiencies in irrigation water use, as this is the largest user of water in the western US.

References

Abrams, R. H. (2009). Water federalism and the Army Corps of Engineers' role in eastern states water allocation. *U. Ark. Little Rock L. Rev. 31*, 395.

American Society of Civil Engineers (ASCE) (1979). *Water Policy Initiatives: Positions of the National Water Policy Committee of the ASCE on the President's June 1978 Statements*. New York, NY, p. 272.

Andreadis, K. M. & Lettenmaier, D. P. (2006). Trends in 20th century drought over the continental United States, Geophys. *Res. Lett. 33*, L10403.

Baumann, D. D., Boland, J. J. & Hanemann, W. M. (1998). *Urban Water Demand Management and Planning*. McGraw-Hill, Inc., New York, p. 349.

Benson, L., Petersen, K. & Stein, J. (2007). Anasazi (Pre-Columbian Native-American) Migrations during the Middle-12th and Late-13th Centuries – Were they drought induced? USGS Staff Pub. Research Paper 750. See: http://digitalcommons.unl.edu/usgsstaffpub/750.

Botterill, L. C. & Cockfield, G. (eds) (2013). *Drought, Risk Management, and Policy: Decision-Making Under Uncertainty.* CRC Press, Taylor Francis, Boca Raton, London, New York, p. 226.

Bureau of Land Management (2013). *Drought Management.* Instruction Memorandum No. 2013-094. US Dept of Interior, Washington, DC.

Bureau of Reclamation (2016). *Budget Justifications and Performance Information for FY 2017.* US Dept of Interior. See: https://www.doi.gov/sites/doi.gov/files/uploads/FY2017_BIB_BH035.pdf.

Center for Strategic and International Studies (2014). Achieving Disaster Resilience in US Communities: Executive Branch, Congressional, and Private-Sector Efforts, by S. S. Kostro and G. Riba.

Clean Water Act (1972). Federal Water Pollution Control Act Amendments of 1972. Public Law 92-500, 33 U.S.C., Chapter 26, Sections 1251–1387.

Copeland, C. & Carter, N. (2009). *Water Infrastructure Funding in the American Recovery and Reinvestment Act of 2009*. Congressional Research Service. 7-5700, www.crs.gov, R40216.

Delli-Priscoli, J. & Stakhiv, E. (2015). Water-related disaster risk reduction (DRR) management in the United States: floods and storm surges. *Water Policy 17*(2015), 58–88.

Dyson, L. K. (1988). *History of Federal Drought Relief Programs*. Economic Research Service, US Dept of Agriculture. ERS Staff Report No. AGES880914. Washington, DC.

Eisel, L. & Aiken, J. D. (1997). *Platte River Basin Study*. Report to the Western Water Policy Review Advisory Commission.

Endangered Species Act (ESA) (1973). Endangered Species Act of 1973, Public Law 93, 205, 16 USC §1531 et seq.

EU Water Framework Directive (WFD) (2000). Directive 2000/60/EC of the European Parliament and of the Council establishing a framework for the Community action in the field of water policy. Official Journal (OJ L 327), 22 December 2000.

European Commission (2011). *Water scarcity and droughts in the European Union*. Report from the Commission to the European Parliament and the Council.

European Commission (2012). *Report on the Review of the European Water Scarcity and Droughts Policy*. Communication from the Commission to the European Parliament, the Council, the European Economic and Social Committee and the Committee of the Regions. See: http://eur-lex.europa.eu/legal-content/EN/TXT/?uri = CELEX:52012DC0672.

Falkenmark, M. (1986). Fresh water: time for a modified approach. *Ambio 15*, 192–200.

Fekete, B. & Stakhiv, E. (2015). Performance indicators in the water resources management sector, chapter 2. In: *The Global Water System in the Anthropocene*. Bhaduri, A., Bogardi, J., Leentvaar, J. & Marx, S. (eds). Springer Water. Springer Cham Heidelberg, New York, Dordrecht, London.

Fontaine, M. & Steinemann, A. (2009). Assessing vulnerability to natural hazards: impact-based method and application to drought in Washington State. *Natural Hazards Review 10*(1), 11–18.

Fontaine, M., Steinemann, A. & Hayes, M. (2014). State drought programs and plans: survey of the Western United States. *Nat. Hazards Rev. 15*, 95–99.

Fort, D. (1997). The western water policy review advisory commission: another look at western water. *Natural Resources J. 37*, 909–922.

Fu, X., Svoboda, M., Tang, Z., Dai, Z. & Wu, J. (2013). An overview of US state drought plans: crisis or risk management? *Nat. Hazard. 69*, 1607.

Giordano, M. & Shah, T. (2014). From IWRM back to integrated water resources management. *International Journal of Water Resources Development 30*(3), 364–376.

Global Water Partnership (2004). *Catalyzing Change: Handbook for Developing IWRM and Water Efficiency Strategies, Technical Committee, No. 1*. Stockholm, Sweden.

Global Water Partnership (GWP) (2016). *Increasing Water Security: the Key to Implementing the Sustainable Development Goals, Technical Background Paper No. 22*. Stockholm, Sweden.

Grey, D. & Sadoff, C. W. (2007). Sink or swim? Water security for growth and development. *Water Policy 9*, 545–571.

Haimes, Y. Y. (2004). *Risk Modeling, Assessment and Management*. Wiley & Sons, Inc., New Jersey, p. 837.

Hallegatte, S., Bangalore, M., Bonzanigo, L., Fay, M., Kane, T., Narloch, U., Rozenberg, J., Treguer, D. & Vogt-Schilb, A. (2016). *Shock Waves: Managing the Impacts of Climate Change on Poverty*. Climate Change and Development Series. World Bank, Washington, DC.

Hanak, E., Lund, J., Dinar, A., Gray, B., Howitt, R., Mount, J., Moyle, P. & Thompson, B. (2011). *Managing California's Water: From Conflict to Reconciliation*. Public Policy Institute of California, San Francisco, CA, p. 482.

Hanak, E., Lund, J. & Mount, J. (2015). *Drought: 10 Ways the Federal Government can Help. Viewpoints*. Public Policy Institute of California San Francisco, CA.

Hao, Z., Agha Kouchak, A., Nakhjiri, N. & Farahmand, A. (2014). Global integrated drought monitoring and prediction system. *Scientific Data 1*, Article number 140001, Nature Research Journal.

Hashimoto, T., Stedinger, J. R. & Loucks, D. P. (1982). Reliability, resiliency, and vulnerability criteria for water resource system performance evaluation. *Water Resources Research 18*(1), 14–20.

Hinchey, M. (1987). The New York city water supply system – an assessment. *Pace Environ. Law Rev. 5*, 149.

Holmes, B. (1972). *A History of Federal Water Resources Program 1800–1960.* US Department of Agriculture, Economic Research Service, Washington, DC.

Howitt, R., MacEwan, D., Medellín-Azuara, J., Lund, J. & Sumner, D. (2015). *Economic Analysis of the 2015 Drought for California Agriculture.* UC Davis Center for Watershed Sciences ERA Economics. UC Agricultural Issues Center, p. 31.

Institute for Water Resources (IWR) (1991). *National Study of Water Management During Drought.* 91-NDS-1. US Army Corps of Engineers, Alexandria, VA.

Institute for Water Resources (IWR) (1993). *Lessons Learned from the California Drought.* 93-NDS-5. US Army Corps of Engineers, Alexandria, VA.

Institute for Water Resources (IWR) (1994). *Drought Atlas.* 94-NDS-4. US Army Corps of Engineers, Alexandria, VA.

Institute for Water Resources (IWR) (1995). *National Study of Water Management During Drought: Report to Congress.* 94-NDS-12. US Army Corps of Engineers, Alexandria, VA.

Kenney, D. (1997). *Resource Management at the Watershed Level: An Assessment of the Changing Federal Role in the Emerging Era of Community-Based Watershed Management.* Report to the Western Water Policy Advisory Commission.

Knutson, C., Hayes, M. & Phillips, T. (1998). *How to Reduce Drought Risk.* Western Drought Coordination Council with National Drought Mitigation Center, Lincoln, NB, p. 43.

Koutsoyiannis, D., Efstratiadis, A., M, N. & C, A. (2008). On the credibility of climate predictions. *Hydrol. Sci. J. 53*(4), 671–684.

Kundzewicz, Z. W. & Stakhiv, E. Z. (2010). Are climate models "ready for prime time" in water resources management applications, or is more research needed? *Hydrological Sciences Journal 55*(7), 1085–1089.

Lautze, J., da Silva, S., Giordano, M. & Sanford, L. (2011). Putting the cart before the horse: water governance and IWRM. *Natural Resources Forum 35*(1), 1–8.

Major, D. C. & Schwarz, H. (1990). *Large-Scale Regional Water Resources Planning: The North Atlantic Regional Study.* Kluwer Academic Publ. p. 174.

Maupin, M. A., Kenny, J. F., Hutson, S. S., Lovelace, J. K., Barber, N. L. & Linsey, K. S. (2014) *Estimated use of Water in the United States in 2010.* US Geological Survey Circular 1405, p. 56, Reston, VA.

Mendoza, G., Matthews, J. & Jueken, A. (2016). *Water Resources Planning & Design for Future Uncertainties: Collaborative Risk-Informed Decision Analysis [CRIDA].* UNESCO International Center for Integrated Water Resources Management, Alexandria, VA.

Motha, R. (2011). *Elements of National Drought Policy in USA.* Drought Mitigation Center Faculty Publications. Paper 57. National Drought Mitigation Center, U. Nebraska. See: http://digitalcommons.unl.edu/droughtfacpub/57.

National Drought Mitigation Center (undated). *Economics of the Dust Bowl.* See: http://drought.unl.edu/DroughtBasics/DustBowl/EconomicsoftheDustBowl.aspx.

National Drought Policy Commission (NDPC) (2000). *Preparing for drought in the 21st Century, National Drought Policy Commission Final Report.* US Department of Agriculture Office of Communications, Washington, DC. May 2000.

National Environmental Policy Act (NEPA) (1970). National Environmental Policy Act of 1969. Public Law 91-190, 42 U.S.C. § 4321 et seq.

National Governors' Association, NR-3 (undated). *Water Resource Management Policy.* See: http://lobby.la.psu.edu/_107th/116_WI_SRF/Organizational_Statements/NGA/NGA_NR3_Water_Resource_Mgmt_Policy.htm.

National Oceanic and Atmospheric Administration (2016). *Billion-Dollar Weather and Climate Disasters: Overview.* National centers for Environmental Information. See: https://www.ncdc.noaa.gov/billions/overview.

National Research Council (2002). *The Missouri River Ecosystem: Exploring the Prospects for Recovery.* National Academy Press, Washington, DC, p. 175.

National Research Council (2004a). *Managing the Columbia River: Instream Flows, Water Withdrawals, and Salmon Survival.* National Academy Press, Washington, DC, p. 246.

National Research Council (2004b). *US Army Corps of Engineers Water Resources Planning: A New Opportunity for Service.* National Academy Press, Washington, DC, p. 108.

National Research Council (2007). *Colorado River Basin Water Management: Evaluating and Adjusting to Hydroclimatic Variability.* National Academy Press, Washington, DC, p. 210.

National Research Council (2011). *Missouri River Planning: Recognizing and Incorporating Sediment Management.* National Academy Press, Washington, DC, p. 152.

National Water Commission (1973). *Water Policies for the Future*. Final Report to the President and to the Congress of the United States. US Government Printing Office, Washington, DC, p. 579.

Natural Resources Conservation Service (2015). *NRCS and Drought: Investing in Water Conservation and Resilience, 2012–2015*. Drought Fact Sheet 5-14-15. US Dept of Agriculture.

OECD (2005). Water Consumption, OECD Factbook 2005. Environment • Air, Water and Land Environment • Air, Water and Land ISBN 92-64-01869-7 ©OECD 2005.

Palmer, R. N., Kutzing, S. L. & Steinemann, A. C. (2002). Developing drought triggers and drought responses: an application. *Georgia Journal of Water Resources Planning and Management 132*, 164–174.

Pedersen, N., Bell, A. R., Cook, E. R., Lall, U., Devineni, N., Seager, R., Eggleston, K. & Vranes, K. P. (2013). Is an epic pluvial masking the water insecurity of the Greater New York City region? *Journal of Climate 26*, 1339–1354.

President's Council on Sustainable Development (PCSD) (1997). *Building on Consensus: A Progress Report on Sustainable America*. White House, Washington, DC.

President's Council on Sustainable Development (PCSD) (1999). *Towards a Sustainable America: Advancing Prosperity, Opportunity, and a Healthy Environment for the 21st Century*. White House, Washington, DC.

Public Policy Institute of California (2016). *Accounting for California's Water*. San Francisco, CA.

Ray, P. A. & Brown, C. M. (2015). *Confronting Climate Change Uncertainty in Water Resources Planning and Project Design: The Decision Tree Framework*. The International Bank for Reconstruction and Development/The World Bank, Washington, DC, p. 125.

Reuss, M. (2004). The development of American water resources: planners, politicians, and constitutional interpretation. In: *Managing Water Resources, Past and Present*. Trottier, J. & Slack, P. (eds). Oxford U. Press, Oxford, UK, pp. 51–72.

Rogers, P. (1993). *America's Water: Federal Roles and Responsibilities*. Twentieth Century Fund. MIT Press, Cambridge, MA, p. 285.

Rosenberg, N. (Ed.) (1993). *Towards an Integrated Impact Assessment of Climate Change: The MINK Study*. Kluwer Academic Publishing, Dordrecht, Netherland.

Safe Drinking Water Act (SDWA) (1996). Safe Drinking Water Act Amendments of 1996. Public Law 104–182 6 August 1996. 42 U.S.C. 300f-300j.

Saleth, R. M. & Dinar, A. (2004). *The Institutional Economics of Water: A Cross-Country Analysis of Institutions and Performance*. World Bank, Dordrecht, The Netherlands and Edward Elgar, Cheltenham, UK, p. 398.

Sarewitz, D., Pielke Jr, R. A. & Keykyah, M. (2003). Vulnerability and risk: some thoughts from a political and policy perspective. *Risk Anal. 23*(4), 805–810.

Schaible, G. & Aillery, M. (2012). *Water Conservation in Irrigated Agriculture: Trends and Challenges in the Face of Emerging Demands*. Economic Information Bulletin No. (EIB-99). Economic Research Service, US Dept of Agriculture, p. 67.

Schnepf, R. (2016). US Farm Income Outlook for 2016. US Congressional Research Service Report R40152, Washington, DC, USA. https://www.fas.org/sgp/crs/misc/R40152.pdf.

Sheffield, J. & Wood, E. F. (2011). *Droughts: Past Problems and Future Scenarios*. Earthscan Ltd, London, p. 210.

Stahle, D. W., Cleaveland, M. K., Blanton, D. B., Therrell, M. D. & Gay, D. A. (1998). The lost colony and Jamestown droughts. *Science* New Series, *280*(5363), 564–567.

Stakhiv, E. Z. (1976). Washington metropolitan area water supply study: people, places, programs. In: *Proc. 96th Annual Conf., New Orleans*, Amer Water Works Assn.

Stakhiv, E. Z. (2003). Disintegrated water resources management in the US: Union of Sisyphus and Pandora. *J. Water Resources Planning and Management* ASCE / May/June 2003 / 1, 151–154.

Stakhiv, E. Z. (2011). Pragmatic approaches for water management under climate uncertainty. *J. Amer Water Resources Assoc. 47*(6), 1183–1196.

UNESCO (2003). *WWDR1: Water for People, Water for Life*. United Nations World Water Development Report, Paris, France.

UNESCO (2014). *International Drought Initiative*. United Nations Educational, Cultural and Scientific Organization. International Hydrological Programme. Paris, France, p. 8.

United Nations International Strategy for Disaster Reduction (UNISDR) (2011). *Global Assessment Report on Disaster Risk Reduction*. Geneva, Switzerland. See: http://www.preventionweb.net/english/hyogo/gar/2011/en/home/download.html.

United States Senate Select Committee (1961). *Report of the Select Committee on National Water Resources*. 87th Congress, 1st Session, Report No. 29. US Government Printing Office, Washington, DC.

United States Water Resources Council (1978). *The Nation's Water Resources, 1975–2000: Second National Water Assessment*. US Government Printing Office, Washington, DC.

US Army Corps of Engineers (1976). *Water Conservation Measures for the New York Metropolitan Area*. Northeast US Water Supply Study. Special Studies Branch. January 1976.

US Army Corps of Engineers (2011). *Drought Contingency Plan for the Apalachicola- Chattahoochee-Flint River Basin*. South Atlantic Division. Mobile District. March 2011.

US Army Corps of Engineers (2016). *Status and Challenges for USACE Reservoirs*. National portfolio Assessment for water Supply Reallocation. IWR Report 2016-RS-1. Institute for Water Resources, 7701 Telegraph Road, Alexandria, VA 22315.

US Environmental Protection Agency (EPA) (2013). *Providing Safe Drinking Water in America*. 2013 National public water systems compliance report. Office of Enforcement and Compliance Assurance. Washington, DC EPA document 305R15001.

US Forest Service (2016). *Effects of Drought on Forests and Rangelands in the United States: A Comprehensive Science Synthesis*. Gen. Tech. Report WO-93b. US Dept of Agriculture.

US House of Representatives (HR 1709) (2015). To amend the Safe Drinking Water Act to provide for the assessment and management of the risks of drought to drinking water, and for other purposes.

Western Drought Coordination Council (WDCC) (1999). *The Western Drought Experience, The Western Drought Coordination Council's Report to the National Drought Policy Commission*. Western Governors Association Publication, Denver, Colorado. May 1999.

Western Governors' Association (2015). *Water Resource Management in the West*. Western Governors' Association Policy Resolution 2015–08. See: http://www.westgov.org/policies/301-water/989-water-resource-management-in-the-west.

Western Water Policy Review Advisory Commission (1998). Water in the West: Challenge for the Next Century. Report to Congress. Washington, DC.

White House (2016a). *Long-Term Drought Resilience Federal Action Plan of the National Drought Resilience Partnership* (Mar, 2016). See: https://www.whitehouse.gov/sites/default/files/docs/drought_resilience_action_plan_2016_final.pdf.

White House (2016b). *Presidential Memorandum: Building National Capabilities for Long-Term Drought Resilience (Mar 21, 2016) Memorandum for the Heads of Executive Departments and Agencies*. See: https://www.whitehouse.gov/the-press-office/2016/03/21/presidential-memorandum-building-national-capabilities-long-term-drought.

Wilhite, D. A., Sivakumar, M. V. K. & Pulwarty, R. (2014). Managing drought risk in a changing climate: the role of national drought policy. *Weather and Climate Extremes 3*, 4–13.

Woodhouse, C. A., Meko, D. M., MacDonald, G. M., Stahle, D. W. & Cooke, E. R. (2010). A 1,200-year perspective of 21st century drought in southwestern North America. *Proc US National Academy of Sciences 107*(50), 21283–21288.

World Bank (2007). *Strategic Environmental Assessment and Integrated Water Resources Management and Development*. Economic and Sector Work Environment Department, Washington, DC, p. 150.

World Meteorological Organization (WMO) (2013). *High Level Meeting on National Drought Policy*. WMO, Geneva. Switzerland. See: http://www.hmndp.org.

World Meteorological Organization (WMO) & Global Water Partnership (GWP) (2014). *National Drought Management Policy Guidelines: A Template for Action (D.A. Wilhite)*. Integrated Drought Management Programme (IDMP) Tools and Guidelines Series 1. WMO, Geneva, Switzerland & GWP, Stockholm, Sweden, p. 48.

Drought management policies in Spain and the European Union: from traditional emergency actions to Drought Management Plans

Teodoro Estrela[a,b,c] and Tomás A. Sancho[d,e]

[a]*Corresponding author. Standing Committee on Water, World Council of Civil Engineers, Colegio de Ingenieros de Caminos, Canales y Puertos, C/Almagro 42, Madrid 28010, Spain. E-mail: testrela@chj.es*

[b]*Júcar River Basin Authority, Spanish Ministry of Agriculture, Food and Environment, Av/Blasco Ibañez 48, Valencia 46010, Spain*

[c]*Research Institute of Water and Environmental Engineering (IIAMA), Technical University of Valencia, Camino de Vera s/n, Valencia 46071, Spain*

[d]*SERSC/Alustante 1, Madrid 28002, Spain*

[e]*World Council of Civil Engineers, Colegio de Ingenieros de Caminos, Canales y Puertos, C/Almagro 42, Madrid 28010, Spain*

Abstract

Droughts are current challenges to the economic, social and environmental development of a region, as reflected in European Union (EU) water policies. In addition, the impacts of droughts can be aggravated by climate change effects, adding pressure to already water stressed areas in the EU. This paper presents a general overview of drought management policies in the EU and in Spain. The paper studies the specific case of Spain, where drought management policies have evolved from traditional emergency actions to Drought Management Plans (DMPs). Historical droughts and the main drought episodes occurring over the last 30 years in Spain, together with the actions taken and the effects produced, are analysed. The National Indicator System for Droughts has played a significant role in drought management since 2005, serving to declare drought situations, and also to apply measures and actions to be carried out by the Spanish Administration. In addition, DMPs for all River Basin Organisations (RBOs) were approved in 2007, and they have represented strategic tools with positive results in drought impact mitigation and have stressed the importance of public participation in the decision-making process for drought events. These plans must be updated in 2017 and are now being revised, taking into account the experience acquired since their approval.

Keywords: Drought; Drought indicator system; Drought Management Plan; European Union; Spain; Water policies

doi: 10.2166/wp.2016.018

1. Introduction

Water is a strategic resource for the economic, social and environmental development of a region. However, droughts are current challenges to this development. Droughts are a relevant temporary decrease of the average water availability, refer to important deviations from the average levels of natural water availability and are considered natural phenomena (European Commission, 2007a). Although by itself a drought is not a disaster, whether it becomes one depends on its impacts on society and the environment (Wilhite & Buchanan-Smith, 2005).

The impacts produced by droughts are numerous. Water supply to populations may be affected, irrigated crops can have severe restrictions and river ecosystems may suffer the consequences of low river flows, among other problems. The duration and related impacts of droughts can greatly vary in different countries. While in those countries lacking water storage infrastructures, where water supply is directly dependant on rainfall, a decrease in rainfall during some months can become a drought, in other countries with enough storage infrastructures, the greatest impacts occur when there are consecutive water deficits for longer periods of several years. The effects of droughts can be aggravated when occurring in regions already presenting low water resources levels, with imbalances between available resources and water demands (European Environment Agency, 2001).

In addition, it is expected that climate change will produce negative direct impacts on the available water resources in the most vulnerable European Union (EU) regions (Intergovernmental Panel on Climate Change, 2007). Furthermore, some extreme weather events have increased, with more frequent heat waves, forest fires and droughts in southern and central Europe (European Commission, 2013). However, although the expected impacts of climate change on water resources are negative, the Intergovernmental Panel on Climate Change (2014) concluded that 'there is low confidence in observed global-scale trends in droughts, due to lack of direct observations, dependencies of inferred trends on the choice of the definition for drought, and due to geographical inconsistencies in drought trends'.

Traditionally, most countries have responded to drought situations with efforts mainly directed at providing water and food and to mitigating economic and environmental impacts. In the last few years, these policies have been changing all over the world. Following droughts impacts in the past few years, some EU Member States have moved from a crisis management approach to drought risk management and the associated measures often result in comprehensive drought risk management plans with water stress area mapping, alert levels, warning systems, etc. Stringer *et al.* (2009) concluded that further efforts to integrate local adaptation strategies within adaptation policies to climate change, climate variability, drought and desertification could increase local resilience to environmental change in different countries in southern Africa. The National Drought Programme PRONACOSE (Federman *et al.*, 2014) in Mexico has focused on reducing vulnerability through the implementation of planned preventive actions under a comprehensive and participative approach. Kampragou *et al.* (2011) referred to the importance of engaging risk assessment and management practices and identified policy gaps and requirements for further improvement of a drought management policy framework at all levels of governance – at EU, national and river basin levels. Wilhite *et al.* (2014) stated that 'the time for adopting drought risk reduction approaches is now, given the spiralling impacts of droughts in an ever-increasing number of sectors and the current and projected trends for the increased frequency, severity and duration of drought events in association with a changing climate'.

In the EU, relevant policy tools have been developed in the last years which include, among their objectives, fighting the negative effects of droughts. Some examples of these tools are the Water

Framework Directive (WFD) 2000/60/EC, the Communication 'Addressing the challenge of water scarcity and droughts in the European Union' (European Commission, 2007b), the 'Blueprint to Safeguard Europe's Water Resources' (European Commission, 2012) and the Communication 'An EU strategy on adaptation to climate change' (European Commission, 2013).

Spain is an EU country which has a high irregularity in the temporal and spatial distribution of water resources, and many areas are affected by water scarcity and frequent droughts (Ministerio de Medio Ambiente, 2000). The implementation of policies and actions such as the development of an indicator system on a national scale, the implementation of Drought Management Plans (DMPs) in River Basin Organisations (RBOs), and the implication of users and other interested parties, are minimising socio-economic and environmental drought impacts (Ministerio de Medio Ambiente, 2008). Most of these policies have their origin in the severe drought which occurred in Spain between 1991 and 1995, described below. Its high economic, social and environmental impacts served to raise awareness about the need for which administrations considered drought as a normal and recurring fact of climate in Spain, shifting drought management policies from risk/management actions to a planned approach.

This paper examines first the status of implementation of the recent policies and actions on drought management carried out in the EU and in Spain. Then, drought events and main impacts in Spain are described, including historical drought events and the most recent droughts corresponding to the periods 1990–1995 and 2004–2008. Next, the paper describes the objectives and current status of the DMPs elaborated by the Spanish RBOs as well as the National Drought Indicator System. The last issue raised in the paper concerns the main measures applied in Spain to deal with droughts, within the framework of the DMPs.

2. Drought management policies in the EU

Drought episodes have occurred repetitively in Europe over the past centuries. More recently, drought events have been analysed. The data provided by EU Member States have made it possible to estimate the direct economic impact of drought events over the past thirty years at a minimum of €100 billion (1,000 million) and even this figure is an underestimate of the overall impact on the economy. The annual economic impact due to droughts is estimated to have doubled between 1976–1990 and 1991–2006, reaching an annual average of €6.2 billion from 2001 to 2006 (European Commission, 2007b).

EU water policy developments are essentially linked to the implementation of the EU WFD, which requests EU Member States to develop a robust integrated water resources management system in Europe, built upon the principles of river basin planning (Quevauviller, 2014). The directive is very demanding in terms of the fulfilment of environmental objectives. However, a temporary deterioration of the status of water bodies will not be considered a breach of the WFD requirements if it is the result of natural circumstances or 'force majeure', such as a prolonged drought, when some conditions have been met (Estrela *et al.*, 2006). Nevertheless, it should be noted that it is not always easy to separate natural circumstances from others. This is the case, for instance, in areas that use groundwater resources from overexploited aquifers.

Unlike floods, for which a specific Directive was adopted in 2007, droughts are not yet the subject of EU law. However, increasing drought trends in multiple EU regions over the last decades have led the European Commission, Member States, scientists and stakeholders to reflect upon specific measures that should be taken to better manage drought risks and impacts. This is particularly reflected by the 'Water

Scarcity and Drought' Communication (European Commission, 2007b) and by the 'Blueprint to Safeguard Europe's Water Resources' (European Commission, 2012).

The above-mentioned 'Water Scarcity and Drought' Communication underlines that water conservation must become the priority, and that all possibilities to improve water efficiency must be explored prior to increasing supply. In addition, it states that policymaking should be based on a clear water use hierarchy established through participative approaches. The communication lists possible measures to cope with water scarcity and droughts and recommends the development of DMPs, shifting from a risk/emergency to a planned drought management approach (Estrela & Vargas, 2012). This shift has become evident in other areas such as the United States (Wilhite *et al.*, 2000). The communication also supports establishing a European drought strategy, considers using European funds when suffering prolonged droughts, and proposes establishing a European drought observatory (EDO).

Within the Common Implementation Strategy of the WFD, in 2007 the European Commission elaborated a technical report, the 'Drought Management Plan Report, including agricultural, drought indicators, and climate change aspects' (European Commission, 2007a), the main objective of which was to be a useful tool to elaborate DMPs in EU Member countries providing general criteria, structures, and recommended measures.

The third Follow up Report (European Commission, 2011) to the above-mentioned European Commission Communication gave further details on the extent of water scarcity and droughts in the EU and the measures being put in place to address both situations. According to this report, activities to integrate water scarcity and droughts into sector policies have been undertaken by several Member States, in particular efforts to reduce water consumption and adaptation to climate change. Most Member States (except the United Kingdom, Spain and Belgium) did not envisage setting up water markets to address water scarcity, and a relevant point highlighted in the report is that the prototype of the EDO had already been developed.

The EDO was established by the Joint Research Centre of the European Commission as part of ongoing efforts to integrate droughts into water policy. The main purpose of the EDO is to enhance the knowledge of droughts using efficient alert systems which are an essential dimension of risk management and improve the drought preparedness of the relevant authorities. Since 2011, the EDO has been the leading disseminator of drought-relevant information and maps of indicators derived from a range of different data sources, including precipitation measurements, satellite measurements and modelled soil moisture content. The EDO integrates relevant data and research results, drought monitoring, detection and forecasting on different spatial scales, from local and regional activities to a continental overview at the EU level. The EDO produces reports about severe drought events detailing the situation to better inform policymakers (Stein *et al.*, 2016).

The 'Blueprint to Safeguard Europe's Water Resources' (European Commission, 2012) outlines actions that concentrate on a better implementation of current water legislation, the integration of water policy objectives into other policies, and filling the gaps in particular as regards water quantity and efficiency. Its main objective was to ensure that a sufficient quantity of good quality water is available for people's needs, for the economy and the environment throughout the EU. The assessment and prevention of drought damages resulting from these new policies are a key issue for EU Member States, as they are derived from the WFD.

In the last few years, different studies have been developed that will help to bridge the important knowledge gaps on water scarcity and droughts in the EU. These studies are being incorporated into a Policy Review of the Strategy for Water Scarcity and Droughts, which is part of the 'Blueprint to safeguard European waters' and cover, among others, the following areas: resource and economic efficiency of water distribution networks in the EU; assessment of water scarcity and drought aspects in a selection of EU River Basin Management Plans; GAP analysis of water scarcity and droughts; buildings and water efficiency; leakage reduction in water distribution networks; activities to halt desertification in Europe; water footprinting and product labelling; and water use in agriculture, including water pricing and water saving. In addition to these studies, the most recent works within the WFD's Common Implementation Strategy are focusing on further developing water scarcity and drought indicators that could be commonly used by Member States, and on a policy review and assessment of EU-related strategy to be integrated into the 'Blueprint to safeguard European Waters'.

At EU level, drought policy objectives outline three issues: promoting risk management policies, promoting drought preparedness and mitigation and planning measures, and consideration of financial assistance tools (Wilhite *et al.*, 2014). Risk management approaches including the characterisation of drought episodes, risk indicators' development, identification, selection and prioritisation of measures to alleviate the effects of droughts or the analysis of the role of economic instruments for risk mitigation, are needed and are now being developed in the EU (Iglesias *et al.*, 2009). Ensuring transparent public participation processes, agreements among the interested parties, collaboration among the water administrations as well as active participation, will be essential elements to guarantee the successful application and follow-up of DMPs (Nelson *et al.*, 2008). In-depth guidance on how to undertake or evaluate participation will also be needed.

Table 1 shows how the advances that have occurred in the planning and management of droughts in the EU have been very important in recent years and have contributed to reaching these objectives, especially since the adoption of the EU WFD at the beginning of this century, in 2000.

3. Drought management policies in Spain

Spain is an EU country characterised by a high temporal and spatial variability in water resources, with important areas experiencing water scarcity and long periods of droughts. Mean annual precipitation is about 670 mm/year, varying from 2,200 mm in northern areas to 120 mm in the South-East (Figure 1). Similarly to rainfall, a high variability in runoff distribution occurs. The mean annual runoff is about 220 mm/year varying from 50 mm/year (south-eastern areas of Spain, La Mancha and Canary Islands) to more than 800 mm/year (Northern areas and some mountainous areas). Total renewable water resources in Spain are about 110,000 hm^3/year, varying from 50,000 hm^3/year in the driest years to almost 200,000 hm^3/year in the wettest years.

Spain has a population of 46.5 million inhabitants (in 2015), mainly concentrated in the largest cities (Madrid, Barcelona, Valencia, Seville…) and in the coastal areas. Coastal economic and tourist development, coupled with highly productive agriculture, have caused the highest water demands to occur in areas where available water resources are scarce. The mean annual water demand in Spain for consumptive uses is about 31,122 hm^3/year. This total demand is distributed per sector as follows: 25,049 hm^3/year (80%) for irrigation, 4,842 hm^3/year for urban uses (16%) and 1,231 hm^3/year (4%) for industrial purposes (DGA-CEDEX, 2016). This distribution shows the importance of water demand for irrigation in Spain.

Table 1. Drought relevant policies in the EU since the beginning of the 21st century (modified and updated from Kampragou *et al.* (2011)).

Policy-related document	Year	Link to drought management
WFD 2000/60/EC (European Parliament and Council of the European Union, 2000)	2000	The protection of water bodies will contribute to mitigating the effects of droughts.
Communication on 'Addressing the challenge of water scarcity and droughts in the European Union' COM (2007) 414 final (European Commission, 2007b)	2007	Underlines that water saving must become the priority, and improving water efficiency must be explored prior to increasing supply. It states that policymaking should be established through participative approaches. It lists measures to cope with water scarcity and droughts, and recommends the development of DMPs.
DMP Report, including agricultural, drought indicators and climate change aspects, Technical Report 2008-023 (European Commission, 2007a)	2007	General, non-binding guidelines for the development of DMPs elaborated jointly by EC and some leading EU countries (Spain, France and Italy).
European Commission follow-up reports to COM (2007) 414 final (European Commission, 2008; European Commission, 2010; European Commission, 2011).	2008, 2010 and 2011	Reporting on progress made by Member States, with regard to addressing the challenge of water scarcity and droughts in the EU.
EDO developed by Joint Research Centre of European Commission	2011	EDO enhances knowledge of droughts using efficient alert systems which are an essential dimension of risk management and will improve drought preparedness of relevant EU authorities.
Blueprint to Safeguard Europe's Water Resources (European Commission, 2012)	2012	Outlines actions that concentrate on the better implementation of current water legislation, integration of water policy objectives into other policies, and filling gaps, in particular regarding water quantity and efficiency.
Development of water scarcity and droughts indicators to be commonly used by Member States.	2013–2016	Will help to bridge the important knowledge gaps regarding water scarcity and droughts in the EU. These studies are being incorporated into the Policy Review of the Strategy for Water Scarcity and Droughts.

The high variability in water resources, the water scarce areas and frequent drought episodes in Spain explain, in part, the ancient building tradition of hydraulic works. The spatial and time variability of water resources has made the construction of numerous hydraulic works necessary to adequately meet water demands. Spain is the ninth country in the world in terms of number of dams, with approximately 1,200. The current capacity of its reservoirs is 56,000 hm^3, i.e. ten times the capacity of the mid-20th century (in 1950).

On the other hand, since the mid-20th century the intensive use of groundwater in the south-eastern areas has allowed the extension of traditional surface water irrigation, taking advantage of good soils and climate, and fostering great economic and social development. Groundwater mining produces benefits but these benefits are sometimes also associated with serious economic, administrative, legal and environmental problems (Custodio *et al.*, 2016). Spain is rich in groundwater and fortunately the major aquifers are located in regions where water shortages are greater, which to some extent allows drought effects to be coped with by using groundwater.

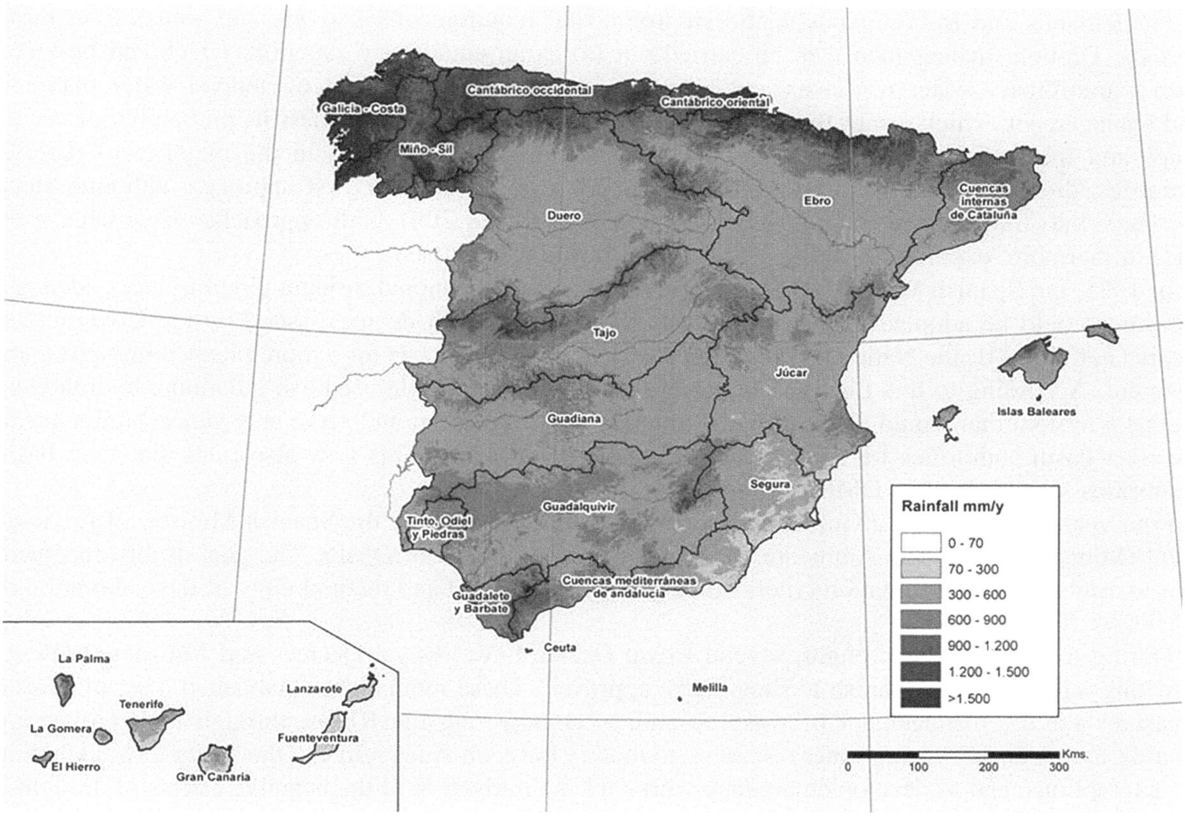

Fig. 1. Mean annual rainfall in Spanish river basin districts.

Today, most natural water resources are already regulated by large dams and groundwater is highly exploited in many areas. A significant increase of conventional water resources is therefore not foreseen in the future. In the areas most vulnerable to water scarcity, non-conventional water resources, such as treated wastewater or sea water desalination, arise as supplementary or alternative resources. Wastewater reuse increases the guarantee of a water supply in the most water stressed areas near the coast. Seawater desalination is currently helping to meet water demands along the Spanish Mediterranean coast and the Balearic and Canary Islands. At present, about 3.5 million m^3 of desalinated water are produced daily, making Spain the fifth largest producer in the world.

These water balance issues can worsen due to climate change impacts. These impacts were recently assessed by the Centre for Hydrographic Studies of CEDEX (2012) for the whole Spanish territory. The projections for the A2 scenario lead to reductions in runoff in Spain of 8% for the period 2011–2040, 16% for the period 2041–2070 and 28% for 2071–2100. The reductions for the B2 scenario are lower, at 8%, 11% and 14%, respectively. These and other similar studies show that the runoff reduction varies strongly depending on the region; the largest decreases take place in the river basins located in the southern area of Spain (Estrela *et al.*, 2012).

Droughts are considered a recurring phenomenon in Spain, which coupled with the scarcity of resources in some areas of the country and the effects of climate change, make it difficult to meet water demands and to comply with the environmental requirements in rivers and wetlands in these periods. Drought management can be carried out (a) as an emergency situation, which can be faced with extraordinary water resources and measures and (b) as an element of general water planning and management, which means that a risk analysis must be carried out to assess its probability of occurrence and measures to be applied must be planned ahead. In Spain, as in the majority of the EU countries, droughts have been traditionally managed according to the first approach, although since the entry into force of the National Hydrological Plan Act in 2001 both approaches have been used and, furthermore, experiences show a shift to the planning approach.

In 1985, the Spanish Water Law established that during exceptional drought circumstances adequate measures could be adopted to overcome these situations through decrees issued by the Government. Years later, in 2001, the National Hydrological Plan Law set the basis for a more planned drought management. According to this Law, the Ministry of Environment would establish a national hydrological indicator system that would allow drought situations to be foreseen and serve as a general reference to the river basin authorities for the formal declaration of droughts. This Law also rules that river basin authorities should develop DMPs.

Four years later, a general Guidelines Document was elaborated by the Spanish Ministry of Environment (Ministerio de Medio Ambiente, 2005) for the development of DMPs. The goal of this document was to establish the coordination criteria about the scope, content and methodology of the elaboration of DMPs by RBOs.

During the 2004–2008 drought, several Royal Decree-Laws, Royal Decrees and Ministerial Orders affecting large areas of Spanish territory were approved. These regulations established a set of special measures for the management of water resources, empowering the RBOs, through their Governing Boards, to reallocate existing water resources, to modify the economic regime of the Water Law, including fee exemptions, and to develop emergency works for the mitigation of the negative effects of droughts.

In 2005, the Spanish Drought Observatory was established on the web page of the current Ministry of Agriculture, Food and Environment, including the National Drought Indicator System. This observatory integrates information coming from all Spanish water administrations with competences in the field of water, in order to constitute a centre of knowledge, forecast, mitigation and monitoring of drought effects in the Spanish territories.

Later, while the 2004–2008 drought developed in large areas of Spain, the RBOs elaborated the DMPs, which were approved by Ministerial Order in 2007. The main objective of these plans was to minimise the environmental, economic and social impacts of droughts.

In January 2016, River Basin Management Plans were approved by Royal Decree 1/2016. This Royal Decree established that all the DMPs approved in 2007 should be reviewed before 31 December 2017.

As outlined in Table 2, the progress made in the planning and management of droughts in Spain has been very important in recent years. It is very similar to the progress made in the EU, especially since the approval of the National Hydrological Plan Act in 2001, which coincided in time with the adoption of the EU WFD.

4. Stakeholder participation in water management in a drought context

Water management in a drought context is part of the Spanish Water Governance System, whose key aspects are: (a) legal certainty; (b) River Basin Authority; (c) users' corporations; (d) participative

Table 2. Drought relevant policies in Spain since the beginning of the 21st century.

Policy-related document	Year	Link to drought management
National Hydrological Plan Act	2001	Basis for a planned drought management. The Ministry of Environment should establish a national drought indicator system that would allow the foreseeing of drought situations and serve for the formal declaration of droughts. It was regulated that River Basin Authorities should develop DMPs.
DMP Guidance Document	2005	General guidelines for the development of DMPs by RBOs in Spain.
Different Royal Law-Decrees, Royal Decrees and Ministerial Orders during the 2004–2008 drought.	2004–2008	These regulations empowered the RBOs to reallocate existing water resources, modify the economic regime of the Water Act, and establish emergency works to mitigate the negative effects of droughts, among others.
Implementation of Drought Observatory on the web page of Ministry of Agriculture, Food and Environment, including National Drought Indicator System	2005	Constitutes a centre of knowledge, forecast, mitigation and monitoring of the drought effects in the Spanish territories integrating information from different administrations.
DMPs elaborated by RBOs were approved by Ministerial Order MAM/698/2007	2007	The objective of these plans was to minimise the environmental, economic and social impacts of droughts using a drought indicator system and to establish a related programme of measures.
Royal Decrees during the 2015–2016 drought in the southeast of Spain	2015–2016	These regulations, affecting the RBOs of Segura and Júcar in the southeast of Spain, empowered these RBOs to reallocate the existing water resources and establish emergency works to mitigate the negative effects of droughts.
Approval of River Basin Management Plans by Royal Decree 1/2016	2016	Revision and updating of DMPs in 2017.

management; (e) hydrological planning; (f) technical and scientific knowledge; and (g) investment in hydraulic infrastructures.

The participative management carried out in RBOs is a key aspect of good governance to manage drought events. Figures 2 and 3 present stakeholder participation in the main water management issues as well as the real participation of users in the governing, management and planning bodies of Spanish RBOs. The figures show that the participation of users is very high in most RBO bodies, especially those concerning management, although greater participation from the economic, social and environmental sectors of society would be desirable. The criteria to appoint representatives for users and economic, social and environmental organisations in the different participation bodies within the RBOs are regulated by Royal Decrees. The Users' Assembly includes all the users in the basin and the representatives for each body are chosen amongst them.

In addition to these general bodies, stakeholders take part in commissions derived from the previous ones, as in the so-called Permanent Drought Commission, which is derived from the Governing Board of the River Basin Authority and whose main objective is drought management in water resources systems where a drought event has been declared.

Table 3. Water management: decentralisation, participation and transparency.

	1		2				3	
	Who?		Degree of Participation				Transparency on	
	Central government	RBO	Information	Consultation	Agreed consultation	Co-decision	Collective decisions	Individual rights
Legislation	■		■	■	■		■	
Administrative authority		■	■					■
National planning	■	■	■	■	■		■	
River Basin planning		■	■	■	■		■	
Large infrastructures								
a) Approval	■		■	■			■	
b) Investment and management		■					■	
Basin operation								
a) Operation rules		■	■	■	■			■
b) Water distribution agreements		■	■	■	■	■		■
c) Storage and withdrawal regime		■	■	■	■		■	
d) Aquifer exploitation		■	■	■	■			■
Conflict management		■	■	■			■	■
International agreements	■		■	■			■	
Extreme situation management (floods)		■	■	■			■	

The Permanent Drought Commissions are formed by representatives of the Administrations and users with a right to speak and vote in the committee, and by representatives from the economic, social and environmental organisations with a right to speak. The Presidents of the River Basin Authorities are authorised to adopt as many measures as necessary for the efficient compliance of the agreements approved by the Permanent Commission. The creation and composition of these commissions has been carried out through the approval of Royal Decrees by the Government.

One of the key aspects of participative water management is the appropriate integration of users. This participation takes place through participative bodies with user representation, in charge of promoting consensus with regards to objectives, challenges, strategies, policies, programmes, projects and actions between the institutional water authorities and duly accredited water users. The real representativeness of users should be highlighted at this point, since it is an important basis in order to understand the quality

Competences	Stakeholders participation
Legislation	National Water Council (Government advisory body)
Administrative Authority (licenses and sanctions)	–
Water planning	River Basin District Water Council (RBOs' planning body)
Large infrastructures (implementation, operation and maintenance)	Hydraulic Works Board (RBO´s management body)
River Basin Management	Users' Assembly + Exploitation Board + Reservoir Withdrawal Commission (RBO´s management bodies)
Conflict Management	Governing Board (RBO´s governance body)
	Users' Assembly + Exploitation Board + Reservoir Withdrawal Commission (RBO´s management bodies)

Fig. 2. Stakeholder participation in the main water management issues.

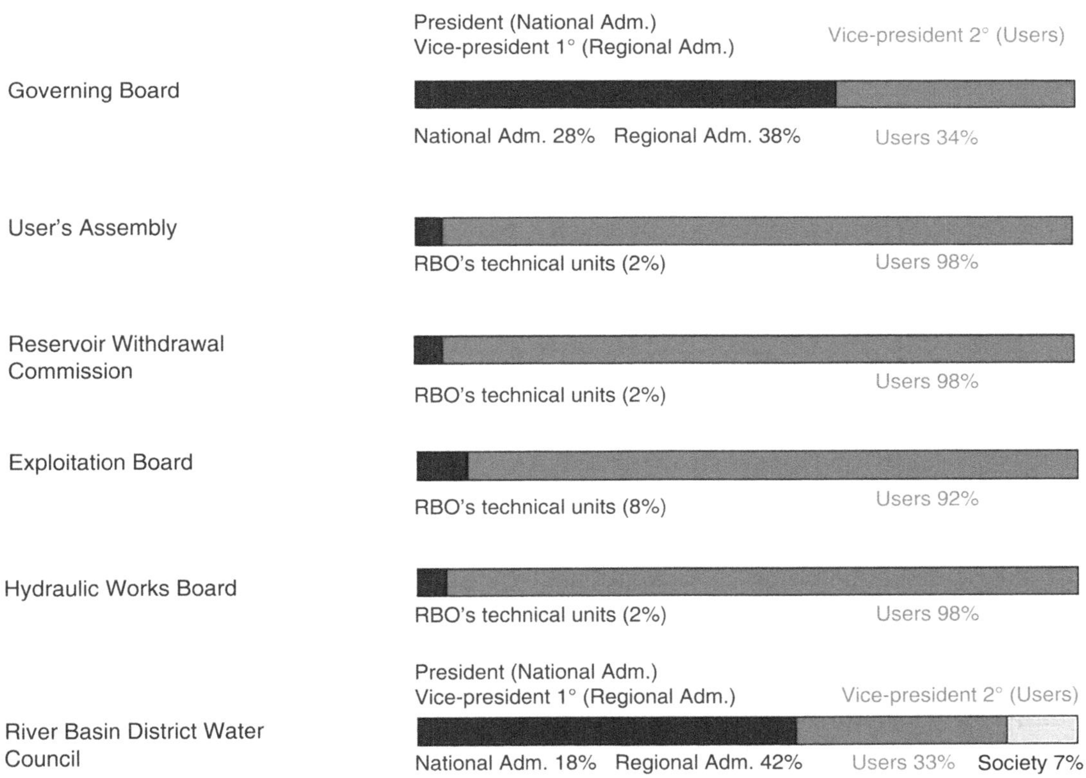

Fig. 3. Participation of users in the governing, management and planning bodies of the Spanish RBOs.

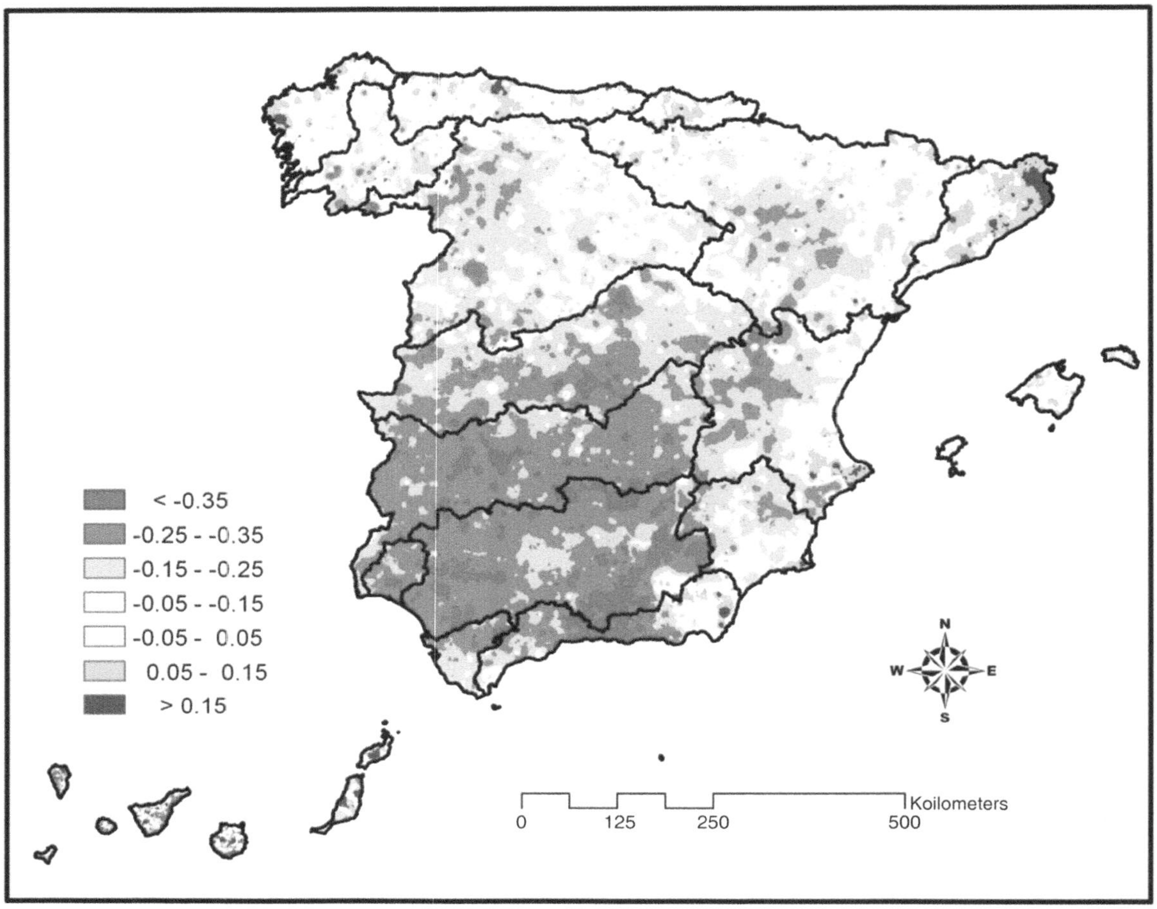

Fig. 4. Reduction in mean annual precipitation in the period 1990/91–1994/95 in respect of the mean annual precipitation in the period 1940/41–2015/2016.

of their participation in the management. In Spain, users are appointed to be part of the Users' Assembly and the management committees using only hydraulic criteria, as they are chosen based on types of use in the basins, regardless of political divisions. To some extent this guarantees their independence from political interference and the effective representation of sectoral interests. Moreover, Spanish users have a wide scope of self-management with regards to water. In fact, their users' associations have legal standing, are considered public water administrations within the Spanish Water Governance System and have water administration and management competences, even for inspection, sanction and budget management.

The participation of stakeholders other than national/regional authorities in water management does not eliminate the fundamental role of these authorities, but it does modify it: the national/regional executive power cannot act as sole manager or simply as an authority – it must lead the basin. Moreover, the government must still be able to intervene when the rest of the stakeholders cannot reach an agreement on the objectives or on how to fulfil them. This is particularly necessary in drought management, especially in water scarcity areas where there is a risk of the users acting against future generations (for instance, abstracting more water than what is ecologically sustainable). As a consequence of the shift in emphasis from command and control to negotiation and persuasion, governmental decision-makers must obtain a new set of skills. Alongside the typical management skills (planning, organising, staffing, directing, coordinating and budgeting), the skills of public decision-makers with regards to governance are activation, orchestration and modulation. These skills are essential to adequately manage drought situations.

Table 3 shows the decentralisation and participation levels in the most important actions in the basin. Column 1 indicates who must carry out the competence: the central government or the basin organisation. Column 2 portrays the degree of possible participation: information to those interested, occasional consultation of their queries, agreed consultation (i.e. with structures that meet periodically) or co-decision of those affected. Column 3 depicts whether the desirable transparency involves collective and/or individual decisions.

Although in Spain there is a long tradition where the role of the administration is to supply water for users, the experience of the Permanent Drought Commissions has shown that it really is possible to make users and environmental and social organisations sit around a table not only to distribute the available resources but also to develop actions from the demand side during drought periods.

5. Drought events in Spain

The study of historical drought events provides an understanding and explanation of current drought management in Spain. Next, a brief review of the main historical droughts that have occurred since the year 1200 is developed, focusing attention on the two most recent droughts: those corresponding to the periods 1990–1995 and 2004–2008.

5.1. Historical droughts

Although the network of meteorological stations in Spain was not sufficiently dense and systematic until the 20th century, there are numerous historical references to droughts. A compilation of historical droughts identified by different authors from the 18th century to the present day in Spain was carried out by Álvarez-Rodríguez *et al.* (2008), who concluded from records going back to the turn of the 19th century that droughts were more frequent, intense and persistent in the 1940s and from the 1980s to the present time. Recently CEDEX (2013) developed a catalogue of historical droughts in Spain, which includes droughts identified from three different types of historical information: reconstructed precipitation series, works and historical climatology papers by different authors, and, lastly, from dendrochronological analysis. The set of information covers an extended period dating back to hundreds of years B.C. in the case of the analysis of historical documentation, approximately 1,000 years in the case of the dendrochronological estimates and about 200 years when working with series of recorded rainfall. In this research, 'historical drought' refers to drought prior to the year 1940. From that year on, the volume of information available is considerably greater, which allows the identifying and characterising of droughts more accurately (CEDEX, 2013). The catalogue collects historical information for about 184 drought events. The first drought with data took place around 1059 BC, affecting the Mediterranean area, while the last drought occurred in 1938–1939.

Table 4 shows the major droughts that have affected Spain, which according to the criteria followed by the authors, are those which affected large areas, produced serious impacts and were identified in numerous bibliographical references (CEDEX, 2013).

From 1940, the most serious droughts were concentrated in four periods: 1941–1945, 1979–1983, 1990–1995 and 2004–2008. These two latest droughts are described below.

5.2. The 1990–1995 drought

During the 1990–1995 drought, 30% reduction in mean annual precipitation from the mean annual recorded in the period 1940/41–2015/2016 took place in large areas of Spain, especially in the Southern part of the country (Figure 4). Consequently, reductions in runoff occurred in most of the Spanish territory, average 40% and up to 70% in the Guadiana and Guadalquivir river basins districts.

Restrictions in supply in the cities of Granada, Jaen, Seville, Malaga, Toledo, Ciudad Real and Puertollano were particularly severe, with up to 30% in some cases and 9 and 10 hours a day water cuts and 12 million people suffered drinking water restrictions (Ministerio de Medio Ambiente, 2008). Supply problems occurred even in cities where, in principle, it seemed difficult to imagine that lack of water

Table 4. Relevant historical droughts affecting Spain (adapted from CEDEX, 2013).

Period	Description of drought
1219–1220	General drought. Hunger in Spain.
1333–1334	Widespread and especially serious drought in Galicia, North plateau and the Ebro Valley. Poor harvests in Spain and Portugal.
1472–1475	Main drought of the century. In 1472 the drought affected a large part of the territory and between September 1473 and October 1475 it mainly affected the north half of the country.
1537–1542	Intense and long drought in Spain: dry rivers, crop failures and locust plagues, scarcity of food, hunger, plague outbreaks. 1539 was a year of severe drought, which brought famine and then the plague that decimated the population of Spain.
1566–1567	Widespread drought in Spain in the years 1566 and 1567. 1567 was considered 'year of hunger' with an intense drought in the basin of the Segura. There are sunspots described in these 2 years.
1595–1598	Droughts in Galicia between 1595 and 1598. The drought of 1595 was especially severe in the North plateau. Andalucía was very affected, with hunger in Seville in 1598–99.
1626–1635	Widespread drought in Spain, which began in the South. Drought so serious that new levels of religious prayer were created. The drought caused hunger in Seville in 1626–28 and in Murcia in 1628.
1680–1683	General drought between 1680 and 1683 with numerous prayers and low crops.
1748–1755	Drought of great impact which widely affected Spain between 1749 and 1753. Intense episodes of hunger and popular uprisings in 1753. Many prayers in all of Spain.
1779–1784	The decade of 1780 was a period of great dryness in the North Plateau and on the Mediterranean side.
1803–1807	Persistent droughts during 1803–1807 in the Southeast, with frequent prayers to the maximum level, with outbreaks of plague and food price increases. 1803 was very dry in Spain, although it was in the Mediterranean side where it acquired greater virulence. The year 1803 made history as 'the year of hunger'.
1872–1880	The major drought of the 19th century in Spain. The period 1872–1879 was especially severe in its second half, particularly in the East.
1909–1915	Widespread drought in 1911 and very intense in 1913–15 in the North of Spain and in 1909–14 in the East. During the period 1909–14 the lands of South-eastern Spain were greatly affected, where the progress of agricultural activity was limited and had devastating consequences: loss of land dedicated to growing and trees, non-compliance with tax obligations, emigration to North Africa between 1908 and 1914.

would happen, such as Bilbao and Vitoria. The Andalucía regional government estimated its farmers' loss of income during this period of drought at €1,743 million. On the other hand, Martínez Cachá (2004) assessed the economic impact on irrigation at €81 million in all the provinces of Valencia, Castellón, Alicante, Murcia and Almeria, estimating the overall impact on employment at €329 million.

The Guadalquivir RBO adopted a series of measures that included a ban on irrigation from 1993 to 1995. In the eastern part of the Guadiana river basin it was not possible to meet the demands of irrigation from 1992, a situation that was aggravated in 1995 when it was not possible to supply the required water to the population. Most frequent measures, in addition to the anticipated imposition of restrictions and special procedures of exchange between users, consisted of carrying out interconnection works between basins, the use of new groundwater resources and the exploitation of non-conventional resources. The first type included the installation of a large number of pipes to transfer alternative resources, increasing the flexibility of public water supply. Examples of this type of work are the connection between the lower Guadalquivir and Arcos-Bornos, transfer tunnels to the reservoir of Conception, or the diversion of the river Alberche to supply Madrid (Ministerio de Medio Ambiente, 2008).

The search for new groundwater resources was carried out on a large scale. The city of Granada, for example, with a population of 300,000 inhabitants and a demand of about 34 hm^3 per year, was completely supplied from groundwater. In the metropolitan area of Madrid, which had a pumping capacity of about 4 m^3/s (20% of the annual demand), a greater use of groundwater resources was also carried out. In the cities of Santander, Pamplona, Burgos, Segovia, Avila, Benidorm and Alcoy emergency supply to the population with groundwater took place, as well as in many other cities. In the Ribera del Júcar, a significant capacity of groundwater pumping was developed, although it was not used to its maximum capacity (100 hm^3/year), while in Las Vegas in the Segura River Basin some 50 drills were made and users were authorised to open wells to save trees. To issue these authorisations, a specific regulation on drought wells was elaborated. In the Segura river basin alone, almost 2,000 well licences were processed and overexploitation increased by about 166 hm^3/year (Ministerio de Medio Ambiente, 2008).

Royal Decrees 531/1992 and 134/1994 established a set of special measures for the management of water resources, empowering the RBOs, through their Governing Boards, so that they constitute the so-called permanent drought commissions. These commissions, among other duties, can reduce or suspend any water exploitation and force users to install adjustment devices for modulation and measurement in the irrigation canals, and can build small works considered as emergency measures.

During this drought period, previous plans specifically designed for this type of emergency were not developed. Various actions were implemented when the drought extended and its effects were most severe. In those four years, it transpired that a drought situation should not just be managed as an emergency because that answer is not adequate and sufficiently effective. The drought in 1990–95 provided experiences that resulted in what would be the future DMPs.

5.3. The 2004–2008 drought

The hydrological year 2004–2005 was the driest in the series of rainfall in Spain since 1947, with a value of annual precipitation 40% lower than the mean annual value. Rainfall was higher during hydrological year 2005–2006 but it was still lower than the average. By the end of 2005–2006, the status of the water reserves at reservoirs was lower than the value corresponding to the same date in the previous hydrological year. Already in the following hydrological year, 2006–2007, precipitation during spring was higher than the historical average, but in certain areas of the southeast of the peninsula, reserves were not enough to meet normal water demands. In the last year of the drought, in 2007–2008, the rainfall in Spain was marked at 16% below normal values. In this year, the drought situation was limited to the territories of the Southeast and South of Spain.

Water shortage in many areas of the country during the drought of 2004–2008 made it necessary to implement some instruments in order to manage drought. Different Royal Decree-laws and Royal Decrees were approved based on article 58 of the revised text of the Water Act, which in extraordinary drought situations results in the adoption by the Government of measures for overcoming such situations. The Ministry of Environment also approved several Ministerial orders.

The Royal Decree-laws have the status of Law because they result in Law amendments. The Royal Decree-laws approved during this drought allowed urgent measures for the regulation of transactions for water use rights to be taken. Royal Decree-Law 15/2005 authorised the exceptional use of the Tagus–Segura aqueduct for water trade between users of different river basin districts. The existence of this hydraulic infrastructure made possible the water reallocations during this drought period. Additionally, these Royal Decree-laws established fee exemptions creating economic advantages for users affected by water restrictions, who were exempt from payments for water use.

Royal Decrees, with less regulatory status than the Royal Decree-laws, put in place exceptional administrative measures for the management of water resources and the correction of drought effects, or urgent improvement and consolidation of irrigation works for water conservation. Through these Royal Decrees, the Governing Boards of the River Basin Authorities were authorised to temporarily modify the conditions of use of the public hydraulic domain, and particularly to: (a) reduce water supply; (b) modify priority criteria for resources allocation to different water uses, while respecting the priority of supply; (c) implement the substitution of licensed flows for those of other origin; (d) modify the conditions set in the discharge authorisations; and (e) temporarily modify allocations and reserves foreseen in the hydrological plans. The Royal Decrees also established the creation of Permanent Drought Commissions.

Additionally, these Royal Decrees contained different measures and emergency works to guarantee public water supply, mitigate negative effects on the irrigation sector or mobilise extraordinary resources. However, Urquijo *et al.* (2015) consider that the use of these exceptional instruments is not always justified, because drought is not always an extraordinary event and because often the implementation time frame of the water works included in the approved Royal Decrees exceeds the duration of the drought episode and, consequently, they consider that these types of measures should be included in ordinary water resources planning.

Finally, Ministerial orders, with lower regulatory status than previous, were used to modify territorial areas subject to Royal Decrees and Royal Decree-laws or to approve DMPs in the different RBOs.

Innovative measures were also put into practice during this drought period, which meant considerable water savings in the agricultural sector (Ministerio de Medio Ambiente, 2008). Reductions of 50–60% for irrigation practices in the most critical areas were established (as shown in Figure 5, where the reduction of volumes supplied to the traditional irrigation area of the Júcar River from 2005 to 2008 can be seen). In addition to this, water use rights exchanges were carried out, with farmers who renounced irrigating their lands during drought periods receiving an economic compensation from the River Basin Authorities or reallocating their water use rights among themselves, which ensured economic efficiency.

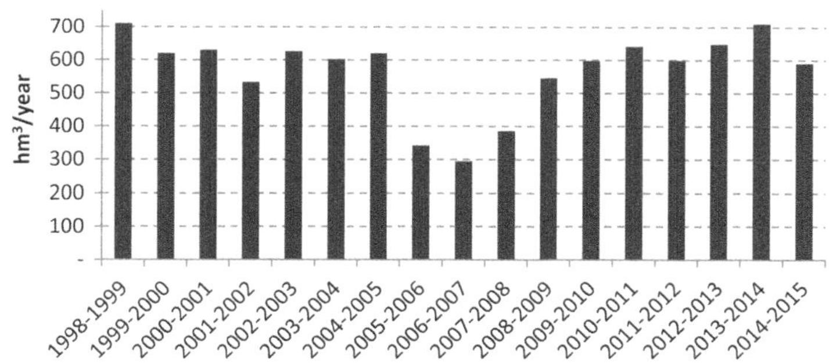

Fig. 5. Volume supplied to the traditional irrigation area of river Júcar. Period 1998/1999–2014/15.

The drought effects also had an impact on the environment. In some areas, due to the low flows, there were episodes of mortality caused by the low water level in the reservoirs. Strong occasional storms in population centres resulted in important non-treated discharge, causing dissolved oxygen decrease in rivers (Ministerio de Medio Ambiente, 2008).

During the drought period, actions to monitor and control the status of water bodies increased. It is common, in these situations, to abstract water from aquifers above their rates of renewability, which can negatively affect both aquifers and surface streams and connected wetlands. Specific monitoring and control was developed in the most vulnerable aquifers to these situations, such as the aquifers in the plain of Valencia, connected with the Albufera de Valencia Natural Park.

DMPs required by the National Hydrological Plan Act were developed during 2006 in all the Spanish RBOs. These plans were approved in March 2007 with the participation of stakeholders under the framework of a very complete participatory process. A public consultation process for the different plans was carried out. Suggestions and observations made by stakeholders were studied, answered and incorporated in the final version of the plans in each river basin. At the same time, a strategic environmental assessment was made, including the elaboration of the initial documents, the scoping and elaboration of the environmental sustainability report and its public consultation. Also, in compliance with the National Hydrological Plan Act, the National Drought Indicator System was developed.

Moreover, where mitigation of the effects of drought could not be reached only by management measures, emergency works were executed, exceeding a total of €650 million in the period 2004–2007. These were mainly focused on increasing water supply to agriculture through the modernisation of irrigation systems and improvement of domestic water supply, followed by the construction or upgrade of reservoirs and irrigation ponds, treated water reuse, desalination, wastewater treatment, environmental protection, groundwater wells and other infrastructures (Urquijo *et al.*, 2015).

The Ministry of Environment created an Experts Committee whose main objective was to evaluate the water scarcity situation and advise the Ministry on actions to develop in order to manage droughts. The Committee was formed from specialists in different subjects: water economy, water management, social communication, climatology and environment. Monthly follow-up reports on drought were drawn up and they enabled the adoption of the most appropriate measures in each situation.

This drought's management, including the measures applied and their effects is extensively documented in a report elaborated by the Spanish Ministry of Environment (Ministerio de Medio Ambiente, 2008). A comparison of the impacts of the droughts in 1991–1995 and 2004–2008, similar in intensity and duration, shows clearly the positive effects of the measures implemented during the 2004–2008 drought when compared to the 1991–1995 drought (Ministerio de Medio Ambiente, 2008). During the drought of 1991–1995 there were severe restrictions in supply in large cities, such as Granada, Jaen, Seville, Malaga, Toledo, Ciudad Real and Puertollano. During the drought period 2004–2008 there were no severe restrictions in any Spanish large city. The economic impacts on irrigation were lower during the drought of 2004–2008. The approval of DMPs in 2007 and of protocols previous to these plans in 2006 enabled the implementation measures of water saving and water restriction for irrigation. The environmental impacts were higher during the drought of 1991–1995 than during that of 2004–2008: i.e., whereas the middle section of the Júcar river, as it passes through the plain of La Mancha, dried up during the drought of 1991–1995, in 2004–2008 the exhaustive control of water use, reservoir management and the farmers' decision to reduce groundwater abstraction for irrigation in areas closer to the river (in exchange for economic compensation), prevented the river from drying up.

6. Current drought management in Spain

6.1. Drought management plans

The specific objectives of DMPs are: (a) to guarantee water availability required to sustain population life and health; (b) to avoid or minimise negative drought effects on the status of water bodies, especially on the environmental water flows, avoiding in any case, any permanent negative effects; and (c) to minimise negative effects on public water supply and on economic activities, according to the prioritisation of uses established by River Basin Management Plans. To meet these objectives, DMPs identify the most adequate mitigation measures for the different stages of a drought. During a normal phase, measures derive from regular management practices. As the drought progresses and a more critical situation takes place, types of measures go from control and information to conservation and restriction, prioritising uses.

The main contents of the DMPs are: drought diagnosis, a programme of measures and management, and a follow-up system. The drought diagnosis analyses and characterises historical droughts as well as lessons learnt, taking into account local and regional acquired knowledge and technical experiences. It also incorporates one of the most relevant elements of the plans: the indicators and drought phase definitions. A crucial and innovative aspect of DMPs in Spain is to establish an adequate link between basin drought status and actions to be taken. A DMPs programme of measures defines the different types of measures that can be applied in each area of the basin, according to the drought status. This programme consists of a catalogue of actions, ranging from enforcing demand reduction strategies to establishing a priority of users to allocate scarce water or approving emergency works (Garrote *et al.*, 2007). Their action methods and established measures must be applied once the interested parties have agreed upon them: administrations, civil society, scientific community, NGOs etc. The main mitigation measures included in the DMPs can be grouped into different categories: structural measures (new pumping wells, new pipes, greater use of desalination plants for public water supply and water reuse in agriculture ...) and non-structural measures (changing the priority of the users, water conservation and demand reductions, increase in the use of groundwater, etc.). Last, the DMPs include a management and follow-up system that allows the analysing of the implementation of measures, using corrective measures in case the established objectives are not met.

DMPs have provided the basis for a more planned drought management in Spain, establishing drought phases and describing the measures that should be progressively applied, as well as the monitoring and follow-up processes needed. Knowledge and experiences gained in drought management in Spain have allowed for the current river basin management plans approved by Royal Decree in January 2016 to include DMPs as a relevant tool. This Royal Decree established that all the DMPs approved by order MAM/698/2007 will have to be reviewed before 31 December 2017. To carry out this review in a harmonised way, the Ministry of Agriculture, Food and Environment is drafting technical instructions, in particular for the hydrological indicators allowing the separate diagnosis of drought emergencies and water scarcity situations.

One of the deficiencies of the DMPs from 2007, still in force, is that they do not appropriately contemplate drought impacts on different economic sectors. For this reason, the afore-mentioned technical instruction will include criteria to evaluate these impacts in the different river basin districts. This means that the plans to be approved in 2017 will probably contain a homogeneous and adequate estimation of the economic impacts of droughts in the whole Spanish territory.

6.2. The National Drought Indicator System in Spain

The implementation of drought indicator systems allows drought situations to be anticipated and the seriousness they present to be assessed. They are also instruments of aid to decision making in the management of water resources in drought situations. Some advantages of the use of drought indicator systems are that: they allow an objective characterisation of the drought phases, making it possible to plan in advance measures and actions in each phase; they facilitate the progressive implementation of measures associated with the threshold values of the indicators; and they facilitate public participation and transparency. This objectivity is important to involve water users in decision making to mitigate the negative effects of drought, as well as to serve as a reference for the formal declaration of droughts (Arqued *et al.*, 2013).

The speed of action to minimise drought effects is a key issue, so the system of indicators used must be agile, simple and allow an adequate assessment of the problem. This means that the information used for the calculation of the indicators should be available in a systematic way and should not require complex data collection and calculation systems. This philosophy of using indicators that are simple, objective, give quick gain/understanding, and are able to describe the real situation has been implemented in Spain within the DMPs, connecting these indicators with progressive measures to mitigate adverse socio-economic and environmental drought impacts (Arqued *et al.*, 2013).

The current National Drought Indicator System consists of selected control points distributed throughout the Spanish river basin districts which include the following information: volume stored in surface reservoirs; groundwater in aquifers; river flows discharges; reservoir inflows and precipitations in those areas where they are significant in relation to the water resources availability to supply the main water demands.

From the values of the indicators in the selected control points in each water resources system, River Basin Authorities obtain, by different processes of aggregation (taking into account the importance of the supplied water demand), an index representative of the status of each water resources system defined in its territorial scope. This index has an important advantage over other meteorological and edaphic indices in the drought management of a water resources system because it is linked to the demands to be met and the water resources that are used to meet them. The standardised values of the indicator (ranging from 0 to 1) define the water resource drought status: normal, pre-alert, alert and emergency. A normal situation is associated with a better hydrological situation than mean values. Other levels are established to differentiate situations below the average and are useful to apply the different measures to mitigate the effects of droughts established in the DMPs.

All the data produced by the RBOs are sent periodically to the Directorate General for Water of the Ministry of Agriculture, Food and Environment, where a common database is kept, and reports with maps, graphs and statistics are published monthly on the Ministry's website (www.magrama.es). Figure 6 has been obtained using data from the National Drought Indicator System on the Ministry's website.

The system allows graphs to be drawn of the temporal evolution of the indexes for each water resources system. In Figure 7 the evolution of the index representative of the hydrological status at the Júcar river basin for the period 2001–2016 is shown. This Mediterranean river basin has a drainage area of 21,600 km^2 and is located in the Eastern area of Spain where the negative effects of water scarcity and droughts are significant. The values of the drought indicator clearly define the drought which extended from 2004 to 2008.

6.3. Measures in the DMPs to cope with droughts

The measures included in the DMPs are classified into three different types: strategic, tactical and emergency. Strategic measures are long-term actions as part of the programme of measures for the

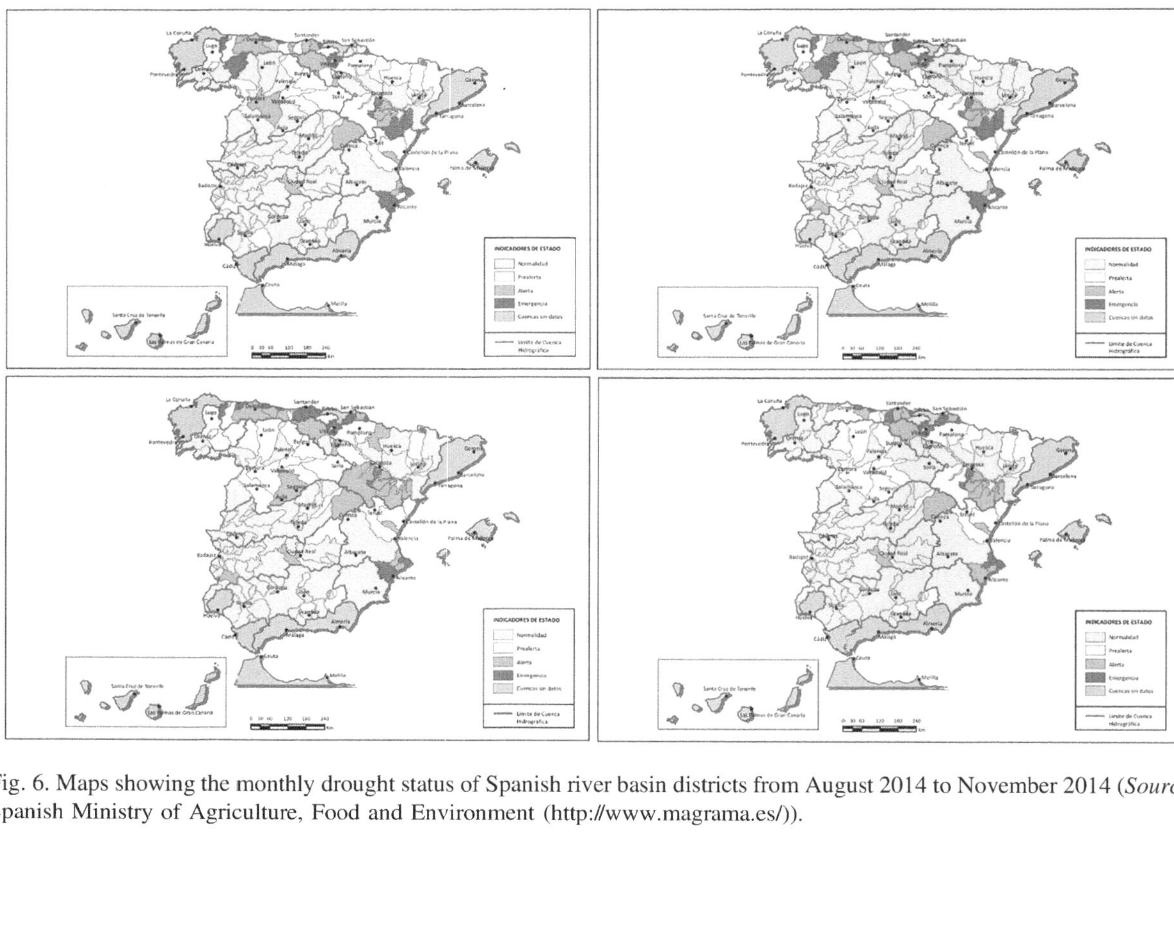

Fig. 6. Maps showing the monthly drought status of Spanish river basin districts from August 2014 to November 2014 (*Source:* Spanish Ministry of Agriculture, Food and Environment (http://www.magrama.es/)).

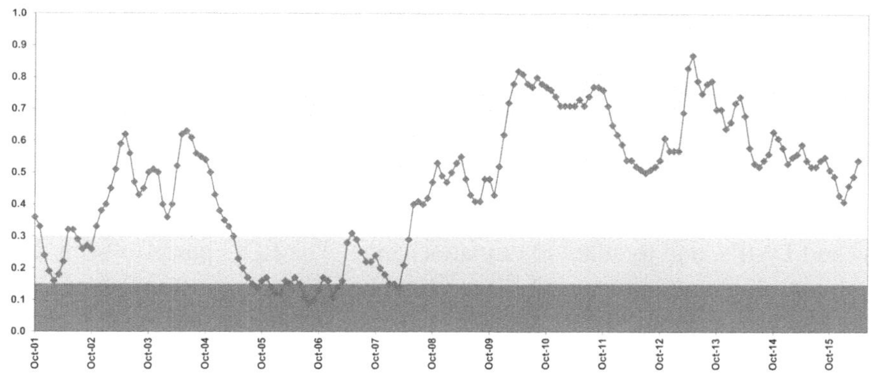

Fig. 7. Evolution of drought indicator in the Júcar river basin for the 2001–2016 period (elaborated with data taken from the Drought Indicator System of the Júcar River Basin Authority).

river basin management plans to be developed during normal phases, such as improving performances by increase of efficiency, incorporating new conventional and non-conventional resources, etc. Tactical measures are short-term actions activated at early warning and alert stages, such as small connections to activate alternative supply, changes in the management system, changes in demands, lower demand in compliance with environmental requirements, etc. Such measures are adopted when the 'pre-alert' situation of the system of indicators occurs. Emergency measures respond to droughts which are longer or more severe than expected, and when a reduction in the availability of water reaches an extreme situation. They must allow for the taking care of the urban supply and other priority water uses. Common measures of this phase are the construction of emergency wells, the establishment of restrictions on the water supply, prohibition of use, temporary modification of water rights, etc. Such measures are adopted when the system of indicators indicates the system is in emergency.

Figure 8 shows this type of measures included in the DMPs in the context of the values of the indicators for each phase or status of drought and of the general objectives corresponding to each of these phases.

During the follow-up to a drought event, information generated by the National Drought Indicator System is used by administrations and stakeholders, under the supervision of the governing and management bodies of the RBOs, to apply the various measures of the DMPs. When information about the assessment of a drought situation becomes publicly available, and objective and clear, citizens – and in particular water users – perceive the need to adopt measures with greater confidence. A proper diagnosis communicated in an appropriate and sufficient way, facilitates voluntary conservation measures (Arqued *et al.*, 2013).

TYPE OF MITIGATION MEASURES							
Indicator	1-0.5	0.5-0.4	0.4-0.3	0.3-0.2	0.2-0.15	0.15-0.1	0.1-0
Status	Normal	Pre-alert		Alert		Emergency	
Objective	Planning	Information-control		Conservation		Restriction	
Type of measure	Strategic			Tactics		Emergency	

Fig. 8. Typology of measures of drought mitigation and timing for application.

7. Conclusions

The EU has been historically affected by droughts. One of the most effective EU policy tools on droughts was the Communication issued in 2007 from the European Commission to the European Parliament and Council entitled 'Addressing the challenge of water scarcity and droughts in the European Union'. Since 2007, actions to integrate water scarcity and droughts into sector policies have been undertaken by EU Member States, in particular efforts to reduce water consumption, develop drought warning systems and DMPs, and to adapt to climate change. The EDO has also been developed providing inputs for the continuous monitoring of drought indicators across Europe since 2011.

Within the EU, Spain is a country characterised by a high variability and uneven water resources distribution, with the Mediterranean coast and southeast region being the areas most affected by water scarcity and droughts, with consequent socio-economic and environmental impacts. In the past, this situation has led to large investments in hydraulic works. Today, Integrated Water Resources Management (IWRM) policies are applied, using conventional (surface and groundwater) and non-conventional resources in a sustainable way, integrating hydraulic works developed during the second half of the 20th century into current IWRM policies.

As is well documented, Spain has periodically suffered prolonged droughts. Recently, a Spanish research centre, CEDEX, has elaborated a catalogue of historical droughts from three different types of historical information: reconstructed precipitation series, works and historical climatology, and dendrochronological analysis. This catalogue clearly shows the negative impacts of droughts in Spain throughout history: poor harvests, scarcity of food, hunger, plagues, emigration and, more recently, restrictions to public water supply and especially to economic water uses with great losses in agriculture and in hydropower production.

Emergency actions have been traditionally applied in Spain to manage drought situations, with actions leading to increasing water resources by developing hydraulic works, especially wells for groundwater abstractions, as well as reinforcing the use of non-conventional water resources. The socio-economic and environmental impacts produced by the drought which occurred in the period 1990–95 allowed appropriate experiences to be derived that have resulted in what became the DMPs. Spanish drought management policies, in accordance with European policies and legislation, have evolved in the last few years from emergency actions to face crisis situations to a planning approach. This was translated into the designing of a National Drought Indicator System to foresee these situations and elaborate DMPs for the Spanish river basin districts.

An indicator system that allows the foreseeing of extreme situations, establishing levels or thresholds depending upon the degree of the drought, and consequently developing actions aiming to delay or impede critical situations has been developed. DMPs approved in March 2007 contributed to alleviating the negative effects of the last drought which occurred during the years 2004–2008, resulting in water conservation, avoiding public water supply restrictions and improving aquatic ecosystem protection. Thus, the economic, social and environmental impacts of this last drought were quite attenuated compared with the drought of the previous decade, without causing public alarm. DMPs must be updated in 2017 and are now being revised taking into account the experience acquired since their approval.

The high impacts of droughts in Spain have raised awareness in administrations that drought is a normal and recurring fact of climate variability, shifting the drought management policies from emergency management actions to a planned approach in the last years with a high level of stakeholder participation.

Acknowledgements

The authors wish to thank the Ministry of Agriculture, Food and Environment of Spain and the Júcar River Basin Authority for providing the opportunity to take part in different national and international projects related to droughts.

References

Álvarez-Rodríguez, J., Potenciano, Á. & Villaverde, J. J. (2008). Evolución histórica y sequías en España. *Revista de Obras Públicas 3486*, 56–68.

Arqued, V., Martínez Cortina, L. & Ureta, J. (2013). La gestión del riesgo de sequía. Vallés, F. J., Andrés-Doménech, I., Escuder-Bueno, I., López Jiménez, P. A. & Marco, J. B. (eds), III Jornadas de Ingeniería del Agua. La protección contra los riesgos hídricos. Volumen 1, ix–xxviii. Ediciones Marcombo.

CEDEX (2012). Estudio de los impactos del cambio climático en los recursos hídricos y las masas de agua. Informe final. Clave CEDEX: 40-407-1-001. Madrid, España.

CEDEX (2013). Catálogo y Publicación sobre Sequías Históricas. Informe para Dirección General del Agua. Informe final. Clave CEDEX: 42-405-1-082. Madrid, España.

Custodio, E., Andreu-Rodes, J. M., Aragón, R., Estrela, T., Ferrer, J., García-Aróstegui, J. L., Manzano, M., Rodríguez-Hernández, L., Sahuquillo, A. & del Villar, A. (2016). Groundwater intensive use and mining in south-eastern peninsular Spain: hydrogeological, economic and social aspects. *Science of the Total Environment 559*, 302–316.

DGA-CEDEX (2016). Síntesis de los planes hidrológicos de cuenca del segundo ciclo (2015–2021). Draft report. Madrid.

Estrela, T. & Vargas, E. (2012). Drought management plans in the European Union. The case of Spain. *Water Resource Management 26*(6), 1537–1553.

Estrela, T., Fidalgo, A. & Pérez-Martin, M. A. (2006). Droughts and the European water framework directive: implications on Spanish river basin districts. In: *Water Resource Systems in Drought Management and Planning for Water Resources.* Rossi, G., Vagliasindi, F. & Vela, A. (eds). CRC Press. Taylor and Francis, Boca Raton, FL, pp. 169–192.

Estrela, T., Pérez-Martín, M. A. & Vargas, E. (2012). Impacts of climate change on water resources in Spain. *Hydrological Sciences Journal 57*(6), 1154–1167. IAHS Press.

European Commission (2007a). *Drought Management Plan Report.* Including Agricultural, Drought Indicators and Climate Change Aspects. Technical Report, 2008–023.

European Commission (2007b). Addressing the challenge of water scarcity and droughts in the European Union. Communication from the Commission to the European Parliament and the Council, COM (2007) 414 final.

European Commission (2008). First Follow up Report to the Communication on water scarcity and droughts in the European Union COM (2007) 414 final. Brussels, 19.12.2008 COM (2008) 875 final.

European Commission (2010). Second Follow up Report to the Communication on water scarcity and droughts in the European Union COM (2007) 414 final. Brussels, 18.5.2010 COM (2010) 228 final.

European Commission (2011). Third Follow up Report to the Communication on water scarcity and droughts in the European Union COM (2007) 414 final. Brussels, 21.3.2011 COM (2011) 133 final.

European Commission (2012). Communication to the European Parliament and the Council – Blueprint to Safeguard Europe's Water Resources, COM/2012/673 final.

European Commission (2013). Communication to the European Parliament and the Council – An EU Strategy on Adaptation to Climate Change. COM/2013/216 final.

European Environment Agency (2001). *Sustainable Water Use in Europe – Part 3: Extreme Hydrological Events: Floods and Droughts.* Environmental Assessment report No. 21. European Environment Agency, Copenhagen.

European Parliament and Council of the European Union (2000). Directive 2000/60/EEC of the European Union and of the council of 23 October 2000 establishing a framework for community action in the field of water policy. *Official Journal of the European Communities* 2000, L327/1.

Federman, D. K., Cortés, F. A. & Pérez, M. L. (2014). Constructing a framework for National Drought Policy: the way forward in Mexico. *Weather and Climate Extremes 3*, 90–94.

Garrote, L., Martín-Carrasco, F., Flores-Montoya, F. & Iglesias, A. (2007). Linking drought indicators to policy actions in the Tagus River drought management plan. *Water Resource Management 21*(5), 873–882.

Iglesias, A., Garrote, A., Cubillo, F. & Wilhite, D. A. (eds) (2009). *Coping with Drought Risk in Agriculture and Water Supply Systems. Drought Management and Policy Development in the Mediterranean.* Series: Advances in Natural and Technological Hazards Research, Springer, The Netherlands, Vol. 26. XVIII, p. 322.

Intergovernmental Panel on Climate Change (2007). Climate change 2007: synthesis report. In: *Contribution of Working Groups I, II and III to the Fourth Assessment Report of the Intergovernmental Panel on Climate Change.* Core Writing Team, Pachauri, R. K., Reisinger, A. (eds). IPCC, Geneva, Switzerland, pp. 104.

Intergovernmental Panel on Climate Change (2014). Climate change 2014: synthesis report. In: *Contribution of Working Groups I, II and III to the Fifth Assessment Report of the Intergovernmental Panel on Climate Change.* Core Writing Team, Pachauri, R.K., Meyer, L.A. (eds). IPCC, Geneva, Switzerland, pp. 151.

Kampragou, E., Apostolaki, S., Manoli, E., Froebrich, J. & Assimacopoulos, D. (2011). Towards the harmonization of water-related policies for managing drought risks across the EU. *Environmental Science and Policy 14*, 815–824.

Martínez Cachá, A. (2004). *Impacto económico de las sequías en el sureste agrario español.* Fundación Universitaria San Antonio, Murcia.

Ministerio de Medio Ambiente (2000). Libro Blanco del Agua en España, Madrid.

Ministerio de Medio Ambiente (2005). Guía para la redacción de los planes especiales de actuación en situación de alerta y eventual sequía. Madrid.

Ministerio de Medio Ambiente (2008). La gestión de la sequia de los años 2004 a 2007. Estrela T y Rodriguez Fontal A (coord.) Madrid.

Nelson, R., Howden, M. & Stafford Smith, M. (2008). Using adaptive governance to rethink the way science supports Australian drought policy. *Environmental Science & Policy 11*(7), 588–601.

Quevauviller, P. (2014). European water policy and research on water-related topics – an overview. *Journal of Hydrology 518*, 180–185.

Stein, U., Özerol, G., Tröltzsch, J., Landgrebe, R., Szendrenyi, A. & Vidaurre, R. (2016). Governance for Drought Resilience, Chapter 2 European Drought and Water Scarcity H. Bressers *et al.* (eds). Policies.

Stringer, L. C., Dyer, J. C., Reed, M. S., Dougill, A. J., Twyman, C. & Mkwambisi, D. (2009). Adaptations to climate change, drought and desertification: local insights to enhance policy in Southern Africa. *Environmental Science and Policy 12*, 748–765.

Urquijo, J., De Stefano, L. & La Calle, A. (2015). Drought and exceptional laws in Spain: the official water discourse. *International Environmental Agreements: Politics, Law and Economics. 15.* N° 2. May 2015.

Wilhite, D. A. & Buchanan-Smith, M. (2005) Drought as hazard: understanding the natural and social context. In: *Drought and Water Crisis. Science, Technology, and Management Issues.* Wilhite, D.A. (ed.). CRC Press Taylor and Francis, Boca Raton, FL, pp. 3–29.

Wilhite, D. A., Hayes, M. J., Knutson, C. & Smith, K. H. (2000). Planning for drought: moving from crisis to risk management. *Journal of the American Water Resources. Association. 36*, 697–710.

Wilhite, D. A., Sivakumar, M. V. K. & Pulwarty, R. (2014). Managing drought risk in a changing climate: the role of national drought policy. *Weather and Climate Extremes 3*, 4–13.

Drought planning and management: experience in the Seyhan River Basin, Turkey

Ismail Kaan Tuncok

Acin cd. 271. sk., No. 77/8 Cankaya, Ankara, Turkey. E-mail: ktuncok@solarisweb.com

Abstract

The focus of this study was to integrate drought planning and management into local and regional decision-making processes in the Seyhan River Basin, which is the second-largest basin after the Nile in the Eastern Mediterranean and agriculturally one of the most productive regions in Turkey and Europe. The methodological approach consisted of two steps: Step 1 – review and analyse historical data sets to evaluate and characterize water resources and drought-driven elements; Step 2 – evaluate drought indices to characterize drought conditions through use of the Standardized Precipitation Index and Standardized Precipitation Evapotranspiration Index. Historical and future expected drought periods were identified in the context of hydrologic, meteorologic and agricultural drought conditions.

Keywords: Agricultural drought; Climate change; Drought indices; Drought planning and management; Seyhan River Basin; Standardized Precipitation Index; Standardized Precipitation Evapotranspiration Index

1. Introduction

The Seyhan River Basin (Figure 1) has the second-largest river drainage basin area in the Eastern Mediterranean after the Nile, with an area of approximately 25,000 km^2. The basin is characterized midstream and upstream with rain-fed agriculture that spreads over the hilly regions, while the lower flat region of the basin is characterized by irrigated agriculture cultivating maize, wheat, fruits and other cash crops (Watanabe, 2007).

The Seyhan River is the longest river in Turkey flowing into the Mediterranean Sea. It is 560 km long and flows southwest from its headwaters in the Tahtali Mountains (in Sivas and Kayseri provinces) in the Anti-Taurus Mountains to the Mediterranean Sea via a broad delta. Its main tributaries are the Zamanti and Goksu, which unite in Aladag, Adana, to form the Seyhan River. The Zamanti River originates from the Uzun Plateau in Pinarbasi, Kayseri, and crosses the Tomarza, Develi and Yahyali districts in Kayseri. The main sectoral activities in the respective river systems include: agriculture

doi: 10.2166/wp.2016.019

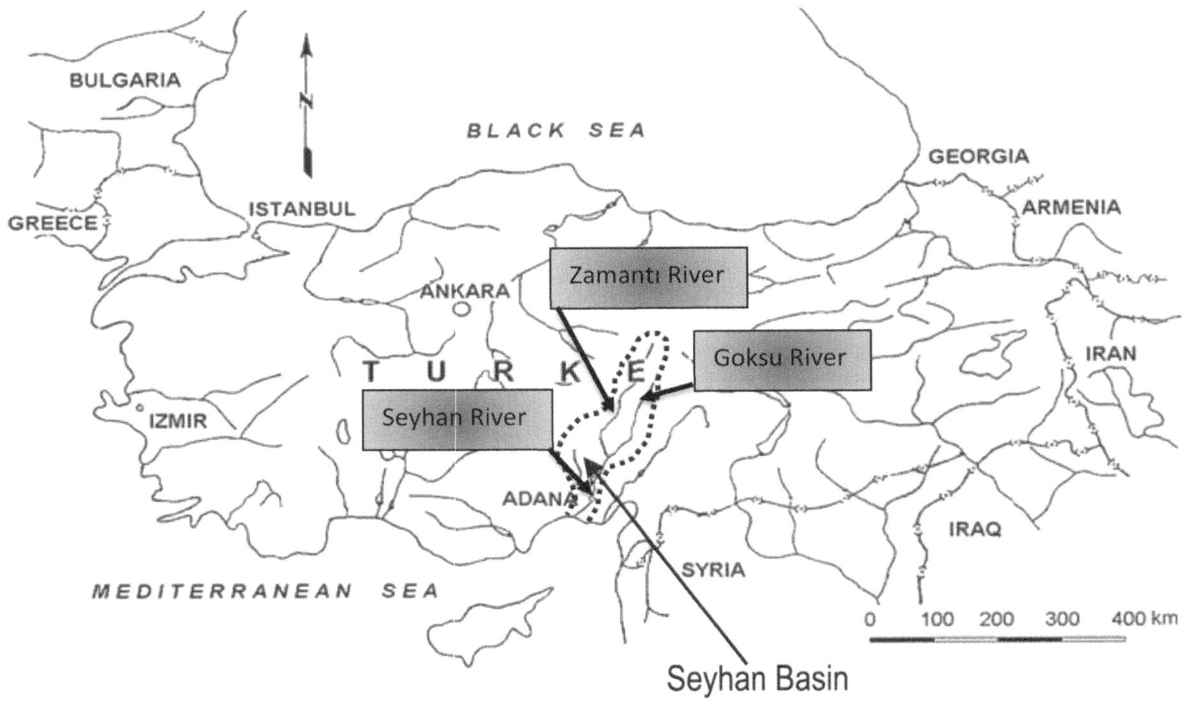

Fig. 1. The Seyhan River Basin.

on the Zamanti River; hydroelectric power generation on the Goksu River; and agriculture, hydroelectric power generation and an industrial sector on the Seyhan River.

1.1. Meteorologic and hydrologic characterization

Each of the three river systems (the Zamanti, Goksu and Seyhan rivers) has distinct meteorologic and hydrologic characteristics. In this context, it is important to evaluate dry and wet cycles within the respective river systems and to structure the basis of evaluating existing conditions for meteorologic and hydrologic events.

In order to set a representative timeframe for evaluation of existing conditions, various alternatives (i.e. 1960–2009, 1970–2000, etc.) were evaluated in the context of wet and dry periods and their associated impacts on meteorologic and hydrologic characteristics. The significance of dry and wet cycles within the respective river systems are depicted in Figure 2. The period 1970–2008 was selected as the reference, to allow the use of a conservative basis for characterizing meteorologic and hydrologic conditions. Associated rainfall and runoff characteristics are reflected through historical rainfall and naturalized runoff patterns within the three river systems, and are depicted in Figures 3 and 4, respectively.

The meteorologic and hydrologic (rainfall and runoff) characteristics in the river systems are summarized as follows (existing conditions runoff is represented by the 1970–2009 timeframe in each case).

1.1.1. Zamanti River.

- Average precipitation over the complete period of record: 628.5 mm.
- Average naturalized runoff during four distinct timeframes: 2,049 hm^3 (1960–2009), 2,150 hm^3 (1960–1990), 2,077 hm^3 (1970–2000), 1,984 hm^3 (1970–2008).

1.1.2. Goksu River.

- Average precipitation over the complete period of record: 494.2 mm.
- Average naturalized runoff during four distinct timeframes: 1,828 hm^3 (1960–2009), 1,948 hm^3 (1960–1990), 1,867 hm^3 (1970–2000), 1,776 hm^3 (1970–2008).

1.1.3. Seyhan River.

- Average precipitation over the complete period of record: 397.2 mm.
- Average naturalized runoff during four distinct timeframes (reflects contributions from Zamanti and Goksu river systems): 6,183 hm^3 (1960–2009), 6,589 hm^3 (1960–1990), 6,349 hm^3 (1970–2000), 5,994 hm^3 (1970–2008).

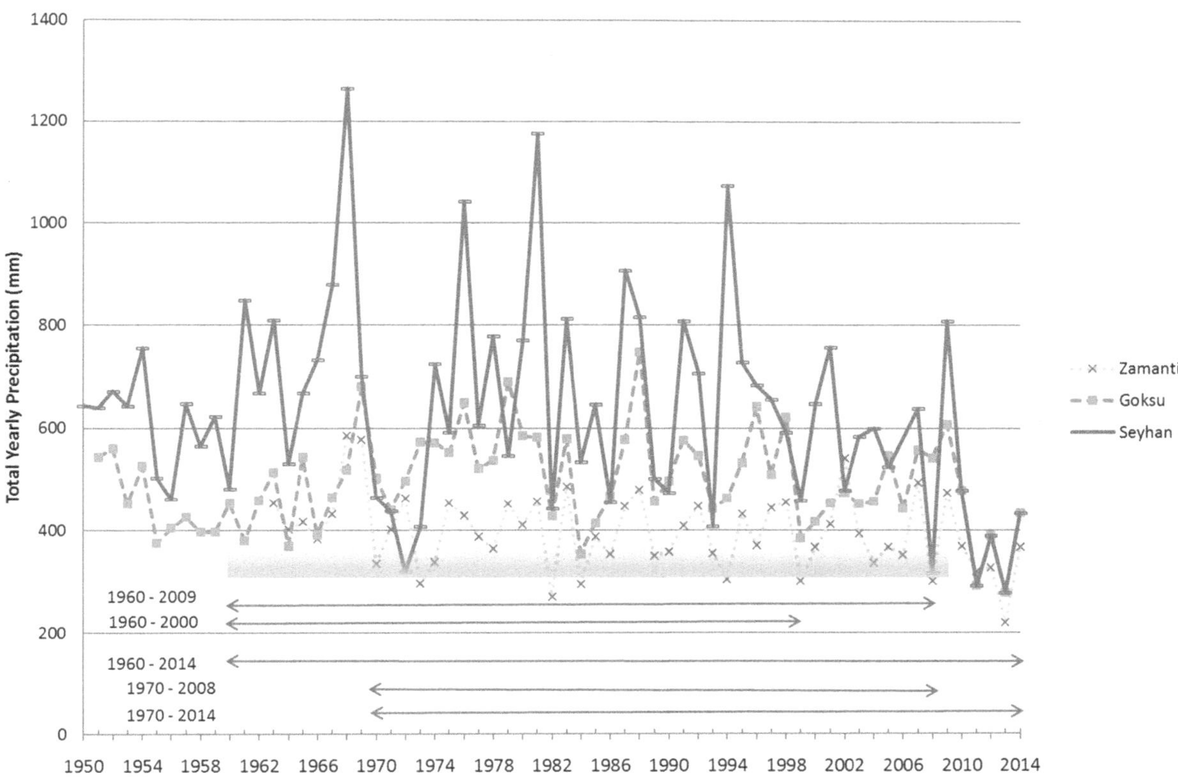

Fig. 2. Historical rainfall in the Seyhan River Basin.

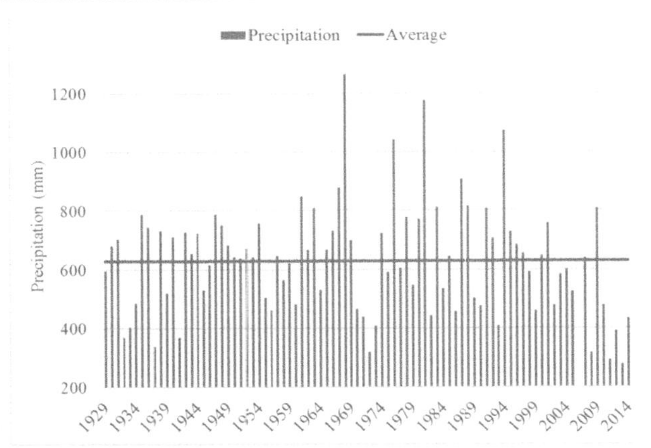

Zamantı River

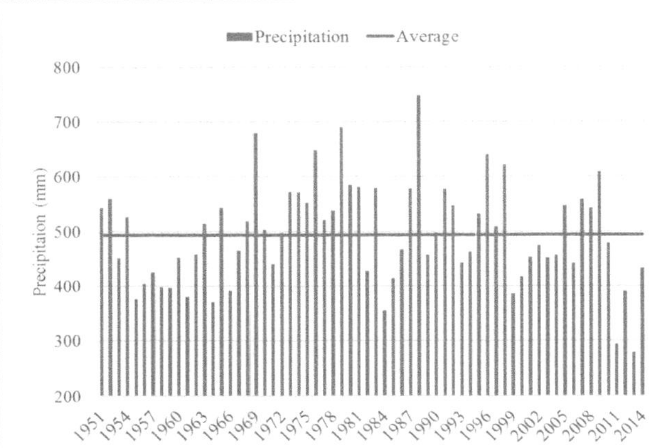

Goksu River

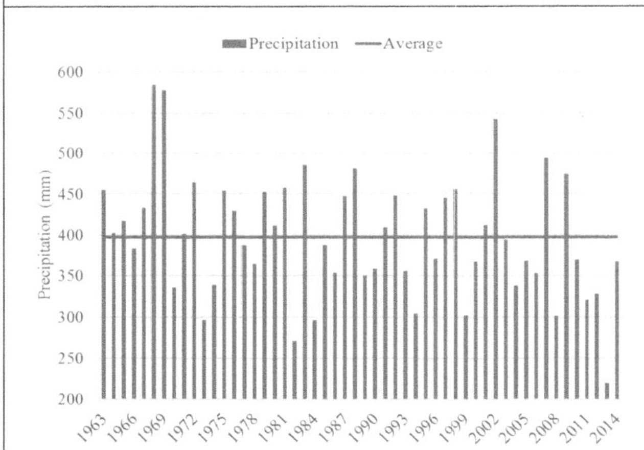

Seyhan River

Fig. 3. Historical rainfall in the Seyhan River and main tributaries.

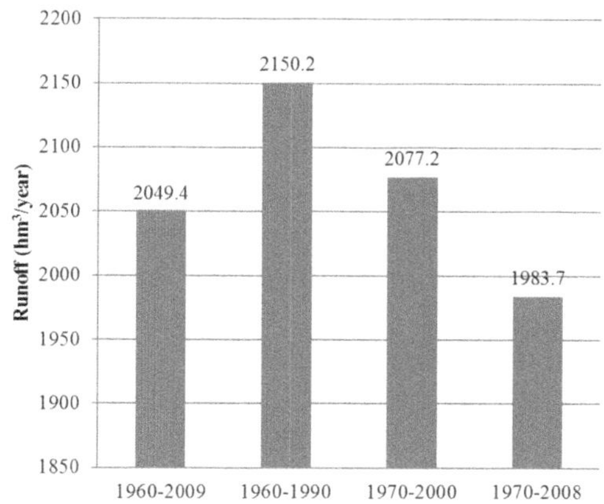

Zamantı River

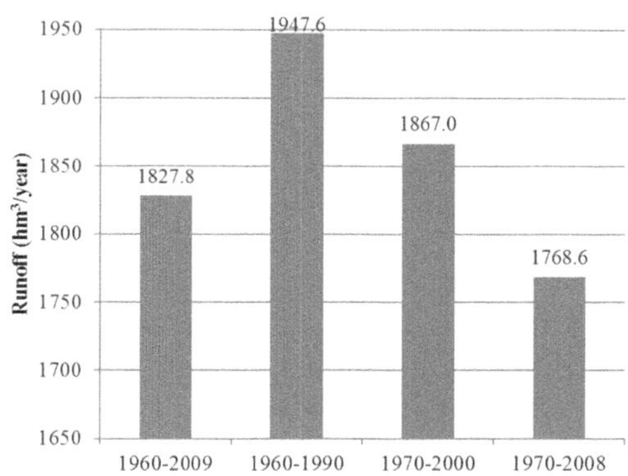

Goksu River

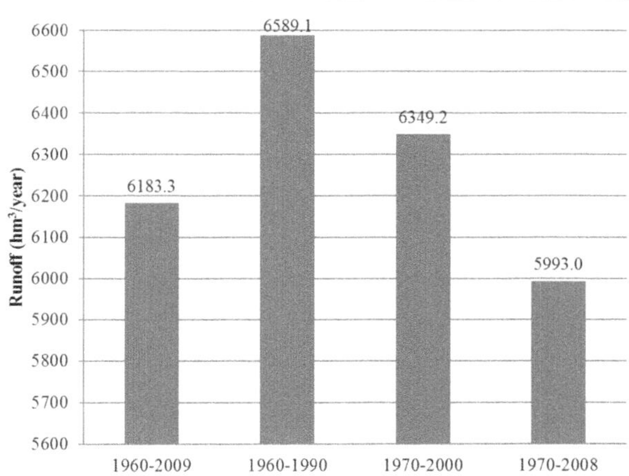

Seyhan River

Fig. 4. Historical runoff in the Seyhan River Basin.

1.2. Drought characterization

There is no single definition of drought (Wilhite & Glantz, 1985), and they can generally be grouped into meteorologic, agricultural and hydrologic droughts (Figure 5). Drought is a very complex hazard to define and detect. It spans multiple sectors and timescales. Just as there is no single definition of drought, there is no single drought index that meets the requirements of all applications.

All droughts begin with a deficiency of precipitation over some timeframe. These early stages of accumulating precipitation deficiencies are commonly referred to as 'meteorologic drought'. A continuation of these dry conditions over a longer period of time, sometimes in association with above-normal temperatures, high winds and low relative humidity, quickly results in impacts on agricultural and hydrologic sectors. Meteorologic droughts are driven by a change in the local meteorologic conditions. The geography and climatology of a region play an important role in what defines meteorologic drought,

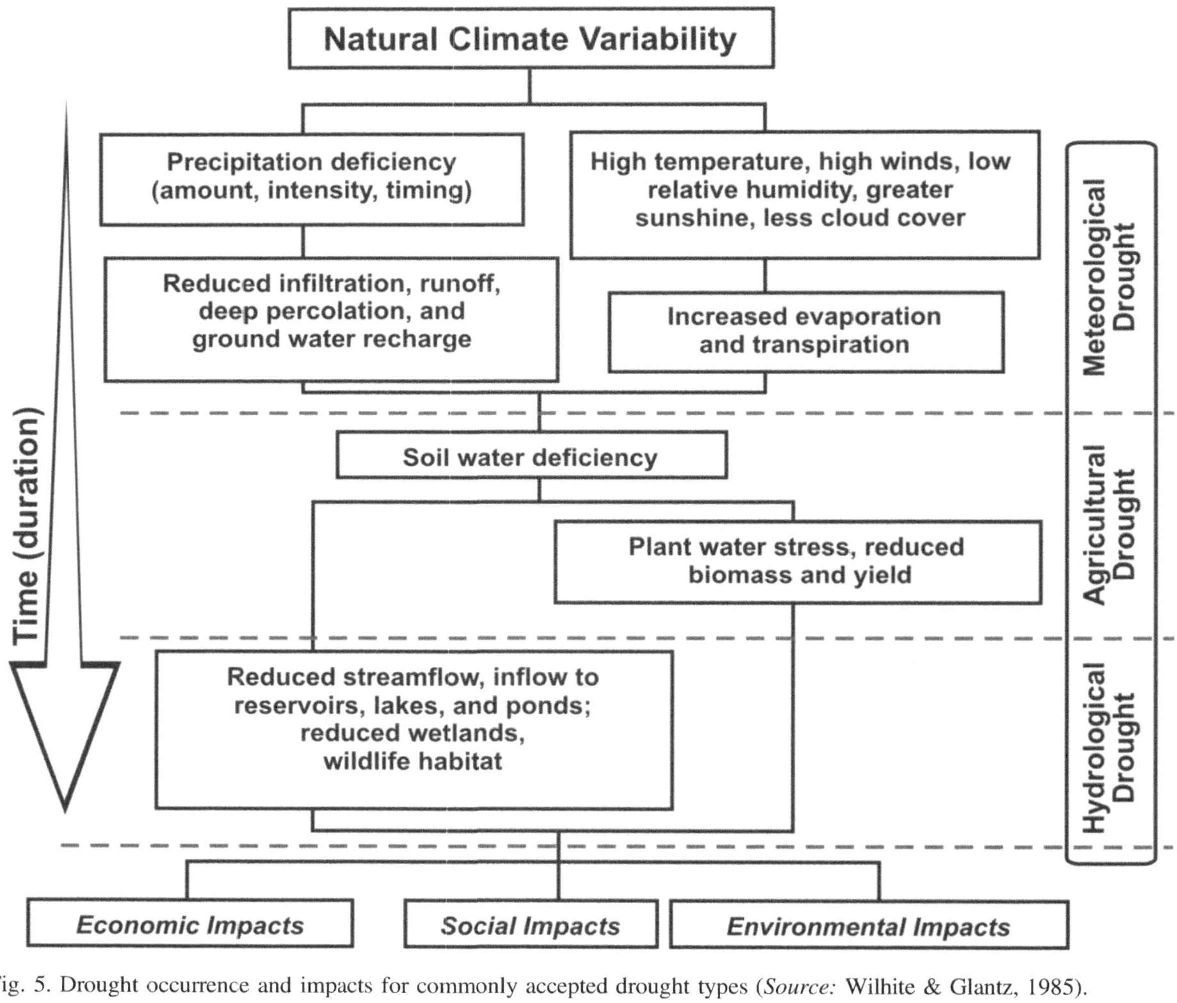

Fig. 5. Drought occurrence and impacts for commonly accepted drought types (*Source:* Wilhite & Glantz, 1985).

since regions have very different precipitation regimes. Meteorologic droughts can develop quickly, but they can also end just as quickly if the precipitation deficits are relatively small. However, these types of drought may also linger on into a multi-seasonal event and develop into one of the other types of drought (Wilhite & Glantz, 1987).

'Agricultural drought' can be characterized as an extended meteorologic drought in which there is a deficiency of precipitation during the growing season, such that crop growth and development are suppressed. Agricultural drought can also be associated with a dry period prior to the growing season in which soils were not fully recharged with moisture, which carries over to a growing season with inadequate precipitation, especially during critical crop growth stages. It may also include periods where available water for irrigation is lacking. Agricultural droughts are events that become the next phase of meteorologic drought. As the name implies, a drought of this nature will extend in duration to the point at which the agricultural concerns of a region are being impacted by dryness and lack of adequate soil moisture. Agricultural drought may extend into portions of more than one growing season, but the natural break between seasons is identified as a period where drought did not worsen or improve, because no agricultural production was taking place (such as in the winter season). Agricultural drought can also precede the actual start of the growing season in which conditions are not favourable for planting because of dryness and a lack of soil moisture, or not enough moisture available for pasture and rangelands to green up.

'Hydrologic drought', as the name implies, refers to drought events that curtail the amount of water available in rivers, streams, lakes, reservoirs, and groundwater. As meteorologic droughts are prolonged, the lack of runoff from precipitation events due to dry soils will begin to impact the hydrology of a region. There is typically a lag involved that will be unique to each region and to a particular time of year, as the impacts to hydrology will not be immediately after the start of the drought. With an extended dry period, soil moisture diminishes, surface runoff and subsurface recharge are reduced, and the amount of water in the hydrologic system of a region declines. During winter months, frozen precipitation is accumulated for future runoff, so a dry winter can induce hydrologic drought months later. Even with precipitation, dry soils can inhibit substantial runoff, as they will capture excess moisture before it is able to reach rivers, streams, and reservoirs. Dryness and heat (via evaporation) will work together to reduce the amount of water available in hydrologic systems. As some hydrologic systems are managed, water managers can choose to withhold water if hydrologic drought is of concern to try to lessen future impacts. Without proper recharge, a long-term drought will impact the hydrology of a region even after precipitation returns to more normal levels. It typically takes the longest time period for a hydrologic drought to develop and, in turn, the recovery time can also take months or even years (Wilhite, 1992).

In addition to these three components, it is important to evaluate socio-economic impacts and implications of socio-economic drought, which is a measurement of the impact of droughts, including supply and demand, usually expressed as an economic value (Wilhite & Glantz, 1985).

In a research report by the International Water Management Institute (Eriyagama *et al.*, 2009), a formulation for a Socio-economic Drought Vulnerability Index (SDVI) was presented to measure the vulnerability of individual countries to socio-economic drought. It was formulated on the consideration that higher GDP contributions from non-agricultural sectors, lower percentage employment in the agricultural sector and a higher crop diversity would collectively lower a country's chances of developing socio-economic drought when meteorologic drought occurs. Three sub-indices, namely the Income

Diversity Index (IDI), Employment Diversity Index (EDI) and the Crop Range Index (CDI), make up the composite SDVI.

According to Fleig *et al.* (2006), the development of droughts may initially start as a meteorologic drought, and thereafter become an agricultural drought and, if storage is not refilled, develop into a hydrologic drought. An analysis of meteorologic drought patterns may give information about initial stages of droughts.

In the context of this paper, drought characterization was performed using a Standardized Precipitation Index and Standardized Precipitation Evapotranspiration Index to evaluate meteorologic, agricultural and hydrologic drought conditions.

2. Methods

Three distinct processes were used to evaluate drought conditions in the Seyhan River Basin:

- an existing conditions drought assessment using drought indices and historical meteorologic parameters;
- a future conditions drought assessment, using drought indices and future conditions meteorologic parameters, calculated through regional climate models; and
- a comparison of the outcomes of the existing conditions drought assessment and historical streamflow gauge measurement.

2.1. Drought indices

2.1.1. Data sets. Meteorologic data sets, as recorded by the Meteorologic General Directorate (MGM) of Turkey, were used as the basis of drought calculations, including those for: precipitation, temperature, evaporation, relative humidity and solar radiation duration. The location and distribution of MGM stations used in the analyses are presented in Figure 6.

Each of the sub-basins are represented through use of the gauge network defined below:

- Sub-basin No. 1, represented by Pinarbasi and Tomarza MGM Gauges;
- Sub-basin No. 2, represented by Sariz MGM Gauge; and
- Sub-basin No. 3, represented by Adana Central and Karaisali MGM Gauges.

2.1.2. Processes. Two of the most commonly used drought indices, the Standardized Precipitation Index (SPI) and Standardized Precipitation Evapotranspiration Index (SPEI), were used to evaluate drought conditions in the Seyhan River Basin.

2.1.2.1. Standardized Precipitation Index. The SPI is a widely used index to characterize meteorologic drought on a range of timescales. Over short timescales, SPI is closely related to soil moisture, whilst over longer timescales, SPI can be related to groundwater and reservoir storage. SPI can be compared across regions with markedly different climates. It quantifies observed precipitation as a standardized

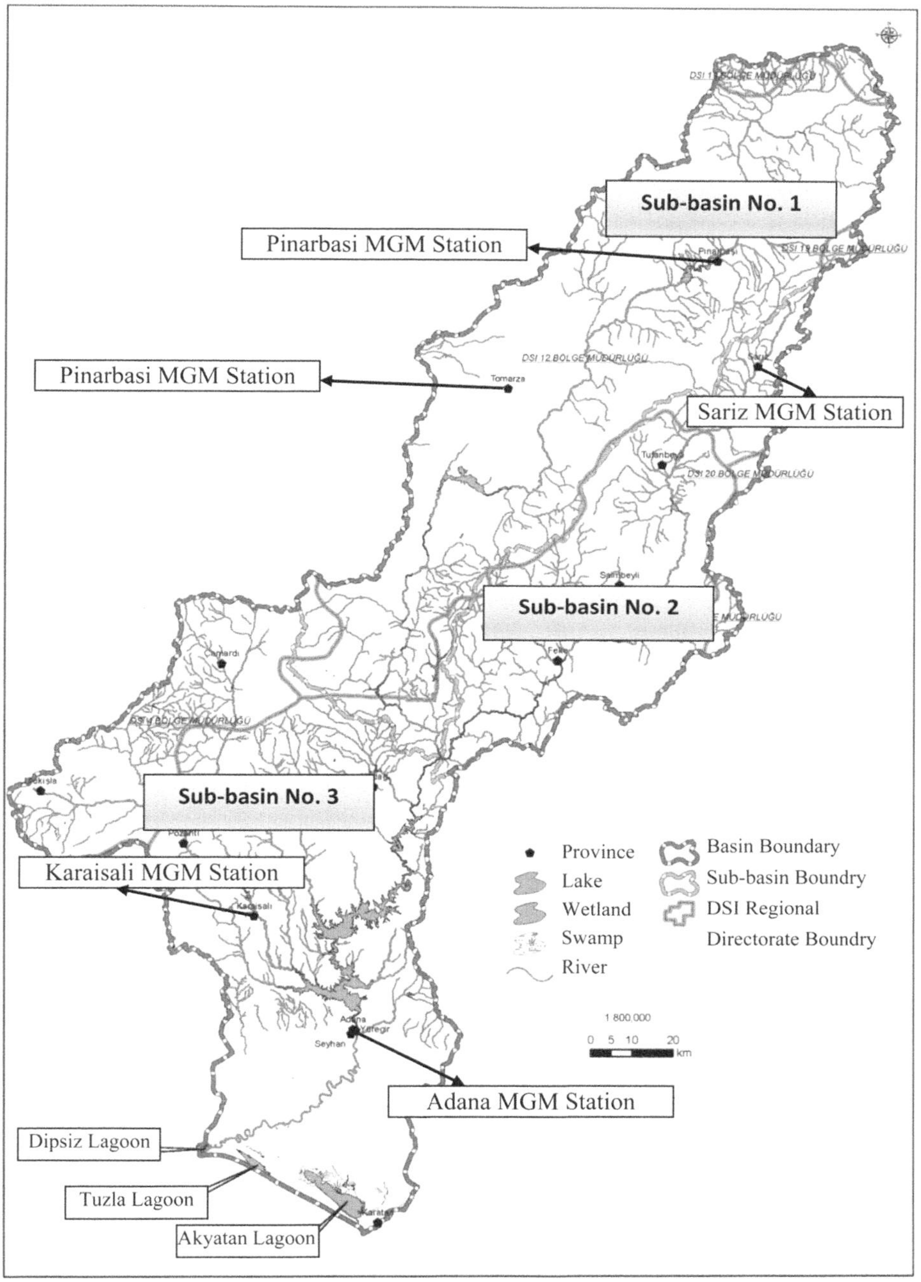

Fig. 6. Spatial distribution of MGM stations used in drought analysis (Source: DSI, 2013).

departure from a selected probability distribution function that models the raw precipitation data. The raw precipitation data are typically fitted to a gamma or Pearson Type III distribution, and then transformed to a normal distribution. The SPI values can be interpreted as the number of standard deviations by which the observed anomaly deviates from the long-term mean. The SPI can be created for differing periods of 1–36 months, using monthly input data (Keyantash & National Center for Atmospheric Research Staff, 2016).

SPI calculation for any location is based on the long-term precipitation record for a desired period at respective meteorologic gauge stations. This long-term record is fitted to a probability distribution, which is then transformed into a normal distribution so that the mean SPI for the location and desired period is zero (Edwards & McKee, 1997). Positive SPI values indicate greater than median precipitation, whilst negative values indicate less than median precipitation. Because the SPI is normalized, wetter and drier climates can be represented in the same way; thus, wet periods can also be monitored using the SPI. A classification system, shown in Table 1, was used to define drought classes resulting from the SPI.

2.1.2.2. Standardized Precipitation Evapotranspiration Index. The SPEI is an extension of SPI. SPEI is designed to take into account both precipitation and potential evapotranspiration (PET) in determining drought. Thus, unlike the SPI, the SPEI captures the main impact of increased temperatures on water demand. Like the SPI, the SPEI can be calculated on a range of timescales from 1 to 48 months (Vicente-Serrano & National Center for Atmospheric Research Staff, 2015).

SPI has no soil–water balance component, thus, no ratios of evapotranspiration/potential evapotranspiration (ET/PET) can be calculated. In this context, a variation of the index by Vicente-Serrano *et al.* (2010) attempted to address the PET issue by including a temperature component in calculation of their new index, called the Standardized Precipitation and Evapotranspiration Index.

Calculation of SPEI is based on the following steps: (i) calculation of potential evapotranspiration (ETo); (ii) accumulation of deficit or surplus using a climatic water balance at different timescales (PETo); and (iii) normalization of the water balance into a log-logistic probability distribution to obtain the SPEI index series.

In both SPI and SPEI, a variety of techniques can be used to convert point data to a continuous map of meteorologic drought. One such technique generates an interpolated surface of estimated values at locations between sites based on the mathematical relationships of the indicator or index between the original point data. Often this produces a map that appears 'natural', but is still based on the data from specific points and is only as accurate as the original data and the interpolation technique. No single interpolation method can be applied to all situations, and the most commonly used interpolation techniques include Kriging, Spline, and Inverse Distance Weighting.

Table 1. SPI classification.

Index range		Drought class	Colour code
2.0	Higher	Exceptionally moist	
	0.5	Normal–upper bound	
− 0.5		Normal–lower bound	
− 0.79	− 0.51	Abnormally dry	
− 1.29	− 0.8	Moderately dry	
− 1.59	− 1.3	Severely dry	
− 1.99	− 1.6	Extremely dry	
Lower	− 2.0	Exceptionally dry	

Soil moisture conditions respond to precipitation anomalies over a relatively short timescale. Groundwater, streamflow and reservoir storage reflect the longer-term precipitation anomalies. So, 1-month or 3-month SPI can be considered to evaluate meteorologic drought, 1–6-month SPI for agricultural drought, and 6–24-month SPI or more for hydrologic drought analyses and applications (WMO, 2012). In this context, the current study utilized a 3-month drought as a meteorologic drought, a 6-month drought as an agricultural drought, and a 12-month drought as a hydrologic drought.

2.2. Water resources potential

2.2.1. Data sources. Water resources potential for existing conditions were determined by evaluating the gauge stations depicted in Figure 7. The stations administered by MGM record meteorologic conditions. The stations administered by the General Directorate of State Hydraulics Works (DSI) of Turkey record hydrologic and hydraulic conditions. As shown in Figure 8, the distribution of the MGM and DSI networks enables accurate representation of both meteorologic conditions and rainfall–runoff processes.

2.2.2. Processes. Average surface and groundwater resources potential in the Seyhan River Basin within the reference timeframe of 1970–2008 are summarized below:

- Zamanti River: 1983.7 hm^3/year;
- Goksu River : 1768.6 hm^3/year;
- Seyhan River: 5993.0 hm^3/year.

The groundwater resources potential of the Seyhan Basin is 1259.6 hm^3/year.

In order to evaluate consistency between historical drought analysis and water resources potential, naturalized yearly streamflow conditions, downstream of the junction of the Zamanti River and the Goksu River, and in the vicinity of Adana Central, were used. Naturalized streamflow conditions, as documented in DSI (2013), are presented in Figure 8. In this context, the ability of SPI methodology to identify one of the worst historical dry periods (i.e. 1972–1974) was evaluated (as discussed in Section 3.2 below).

2.3. Regional climate models

Two recent regional climate model studies were undertaken in Seyhan Basin: Regional Climate No. 1 by MGM (2015), and Regional Climate No. 2 by SYGM (2015), as per guidelines provided in the IPCC 5th Evaluation Report (AR5).

2.3.1. Data sets. The model studies were performed using MGM meteorologic station data sets (precipitation and temperature) and European Centre for Medium-Range Weather Forecasts (ECMWF) analyses data sets (ERA Interim) (evaporation).

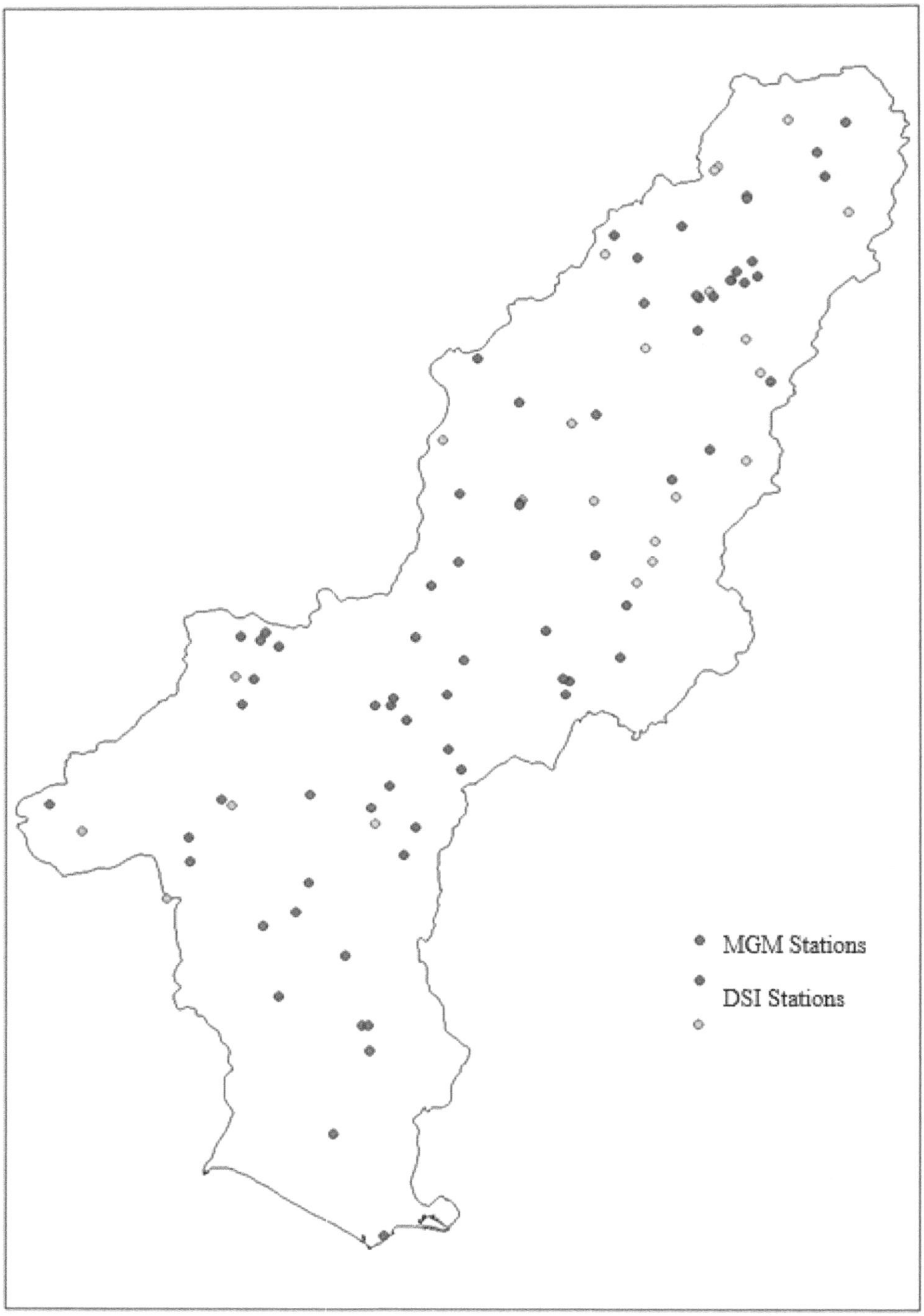

Fig. 7. Spatial distribution of MGM and DSI stations in the Seyhan Basin (*Source:* Selek & Tuncok, 2014).

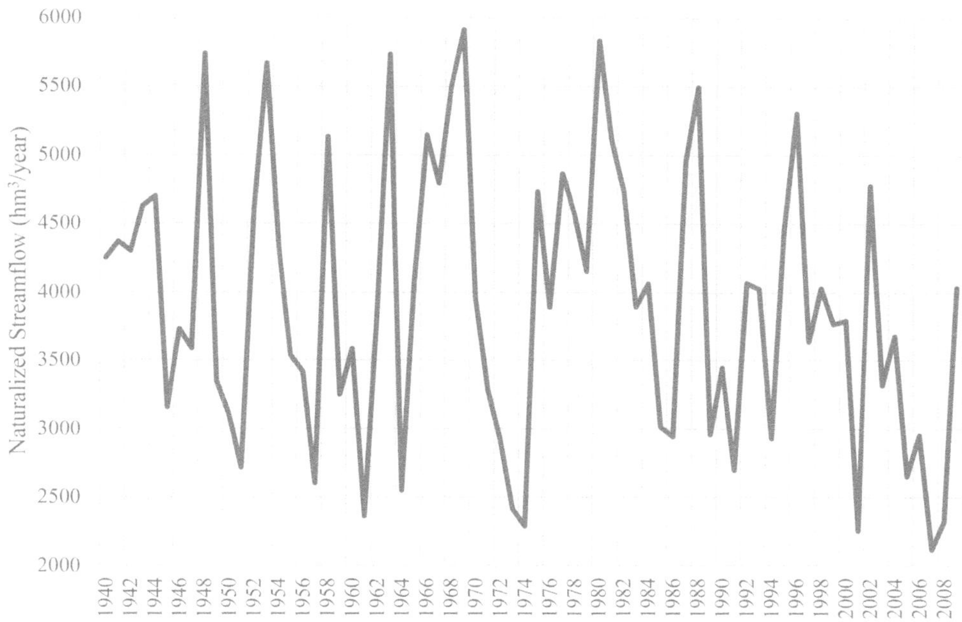

Fig. 8. Naturalized streamflow in the vicinity of Adana Central (*Source:* DSI, 2013).

2.3.2. Models. Three different general circulation models, representing climatic conditions in Turkey, were used:

- Met-Office Hadley Centre (England): HadGEM2-ES;
- NOAA Geophysical Fluid Dynamics Laboratory (USA): GFDL-ESM2M; and
- Max Planck Meteorology Institute (Germany): MPI-ESM-MR.

The outcomes from HadGEM2-ES, GFDL-ESM2M and MPI-ESM-MR were used as the boundary conditions to RegCM 4.3.5 (Regional Climate Model) at a grid size of 10 km in SYGM (2015) and at a grid size of 20 km in MGM (2015).

2.3.3. Scenarios. *A Representative Concentration Pathway (RCP) 4.5 scenario* represents balancing of the radiative forcing at a level of 4.5 W/m^2 by 2100.

An RCP 8.5 scenario represents an increase of radiative forcing to a level of 8.5 W/m^2 by the end of the century. It was determined that the HadGEM2-ES model was more in compliance with historical data sets, in comparison to the GFDL-ESM2M and MPI-ESM-MR models.

The outcomes of these models were used to define future conditions meteorologic parameters and associated drought indices.

3. Results and discussion

The impacts of drought both in existing conditions and future conditions were evaluated. The existing conditions were evaluated in the context of historical data sets and drought conditions, and the ability of SPI-based drought analysis to capture such conditions. The future conditions were evaluated.

3.1. Regional climate change models

The outcomes of two regional climate models reflect the following changes in precipitation and temperature.

3.1.1. Regional Climate Model 1 (MGM, 2015). The expected *decreases* in precipitation are summarized below.

As per RCP 4.5 scenario:

- 2016–2070 period: 5–10%; and
- 2071–2099 period: 10–15%.

As per RCP 8.5 scenario:

- 2016–2070 period: 5–10%; and
- 2071–2099 period: 15–20%.

The expected *increases* in temperature are summarized below.

As per RCP 4.5 scenario:

- 2016–2040 period: 0.5–1 °C; and
- 2041–2099 period: 1–1.5 °C.

As per RCP 8.5 scenario:

- 2016–2040 period: 1–1.5 °C;
- 2041–2070 period: 2–2.5 °C; and
- 2071–2099 period: 4–4.5 °C.

3.1.2. Regional Climate Model 2 (SYGM, 2015). The expected *decreases* in precipitation are summarized below.

As per RCP 4.5 scenario:

- 2015–2040 period: approximately 6%;
- 2041–2070 period: approximately 7%; and
- 2071–2100 period: approximately 10%.

As per RCP 8.5 scenario:

- 2015–2040 period: approximately 2%;
- 2041–2070 period: approximately 11%; and
- 2071–2100 period: approximately 13%.

The expected *increases* in temperature are summarized below.

As per RCP 4.5 scenario:

- 2015–2040 period approximately 1.7 °C;
- 2041–2070 period approximately 2.7 °C; and
- 2071–2100 period approximately 3.3 °C.

As per RCP 8.5 scenario:

- 2015–2040 period approximately 1.9 °C;
- 2041–2070 period approximately 3.4 °C; and
- 2071–2100 period approximately 5.4 °C.

The outcomes of Regional Climate Model No. 2 were used to evaluate future meteorologic conditions (mainly precipitation and temperature) to reflect conditions (in terms of decrease in precipitation and increase in temperature), which are worse than the ones resulting from Regional Climate Model No. 1.

3.2. Historical drought analysis

Drought analysis was performed in three sub-basins characterized by the Zamanti River, Goksu River and Seyhan River systems, respectively. In this context, 'drought periods' for historical events and future conditions were identified in respective sub-basins.

Sub-basin No. 1: Meteorologic data sets in Pinarbasi and Tomarza stations were used to conduct drought analysis using SPI and to identify dry periods.

The dry periods are summarized in Tables 2 and 3. Temporal variation of the SPI is depicted in Figures 9 and 10. It is important to note that 'dry periods' described in respective tables cover the entire duration for individual 3-, 6- and 12-month SPI index periods. As such, SPI index periods are observed within distinct durations during the dry periods.

Sub-basin No. 2: Meteorologic data sets in Sariz station were used to conduct drought analysis using SPI and to identify dry periods which are summarized in Table 4. Temporal variation of the SPI is depicted in Figure 11.

Table 2. Sub-basin No. 1: Pinarbasi station: historical dry periods.

SPI Index Periods	Dry Periods
3-months	1989 February–1989 July
6-months	1989 March–1989 October
12-months	1984 October–1986 May

Table 3. Sub-basin No. 1: Tomarza station: historical dry periods.

SPI Index Periods	Dry Periods
3-months	1994 April–1994 September
6-months	1993 November–1994 October
12-months	1994 April–1995 April

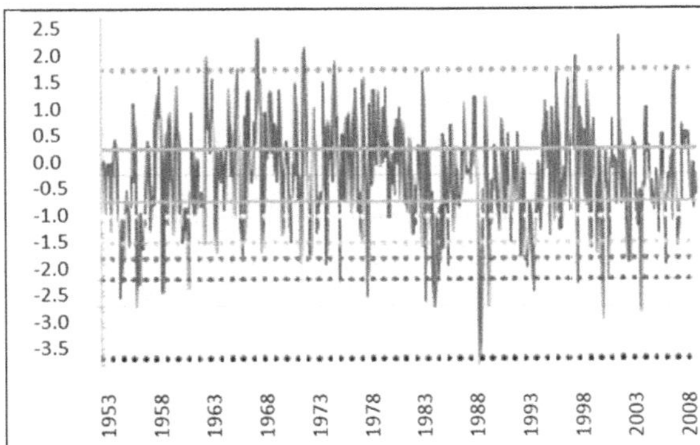

3-month:

Agricultural Drought

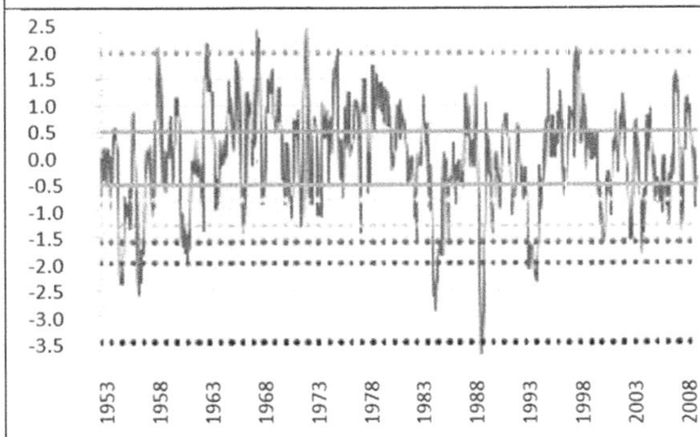

6-month:

Hydrological Drought

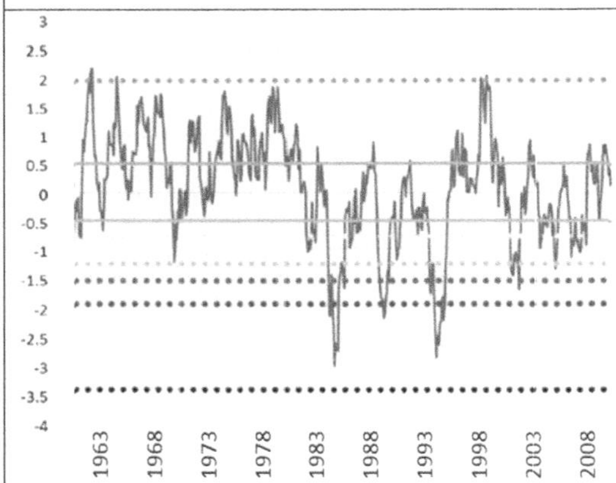

12-month:

Meteorological Drought

Fig. 9. Sub-basin No. 1: Pinarbasi: historical SPI drought index values (*Source:* MGM Drought Monitoring System 2.1, 2016).

Sub-basin No. 3: Meteorologic data sets in Adana Central and Karaisali stations were used to conduct drought analysis using SPI and to identify dry periods, which are summarized in Tables 5 and 6. Temporal variations of the SPI are depicted in Figures 12 and 13.

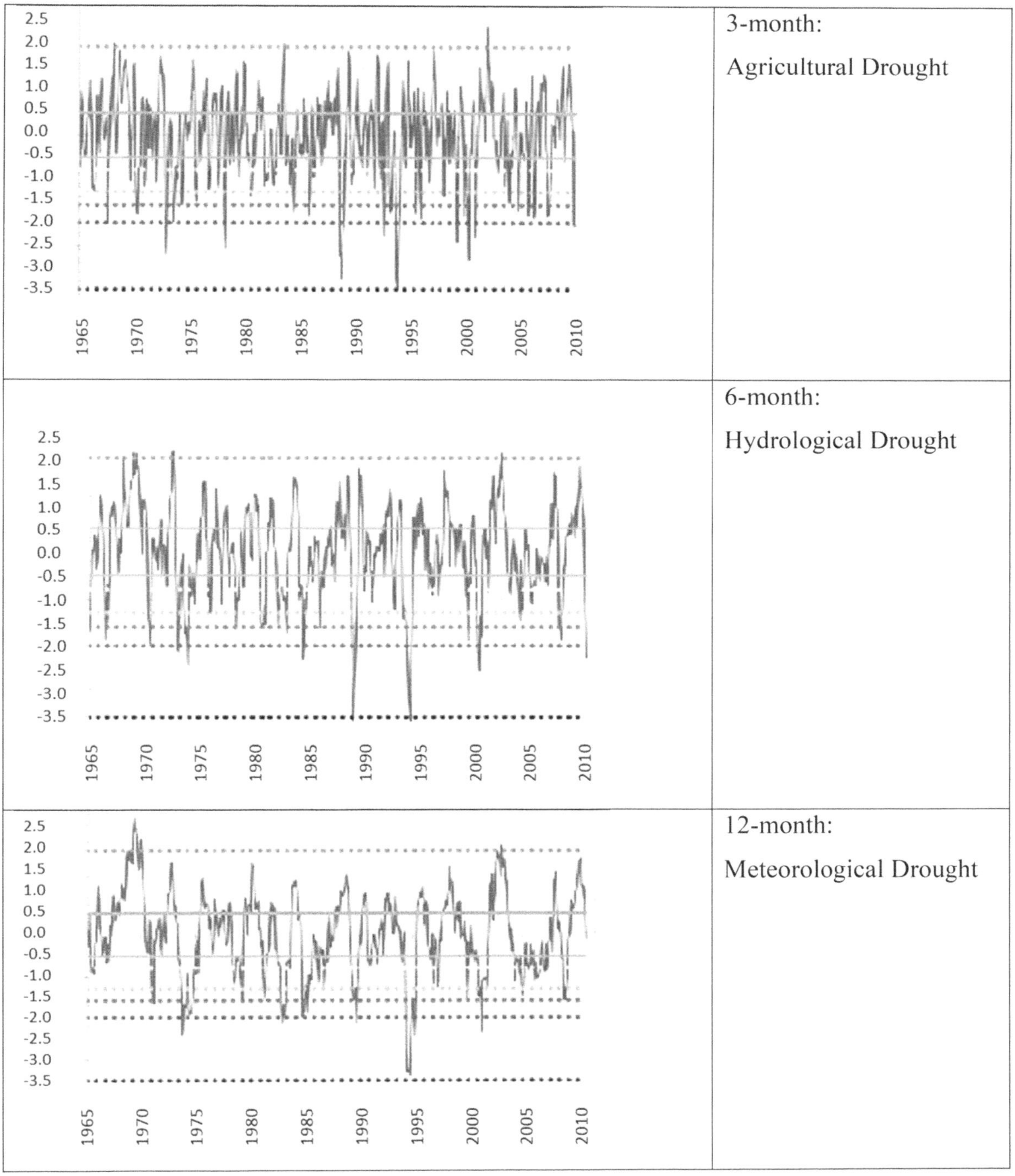

Fig. 10. Sub-basin No. 1: Tomarza: historical SPI drought index values (*Source:* MGM Drought Monitoring System 2.1, 2016).

Table 4. Sub-basin No. 2: Sariz station: historical dry periods.

SPI Index Periods	Dry Periods
3-months	1958 June–1958 December
6-months	1966 September–1967 August
12-months	2011 December–2013 December

3.3. Future conditions drought analysis

The future conditions analysis was undertaken using the outcomes of the climate change model studies presented in Section 3.1. The 12-month SPI drought index values, representing hydrologic drought conditions, were calculated by using meteorologic conditions driven by RCP 4.5 and RCP 8.5 scenarios. The associated dry periods are summarized in the following sections.

The distinguishing features of historical droughts and future expected droughts can be characterized as:

- historical droughts – characterized by less frequent but more severe droughts; and
- future droughts – characterized by more frequent but less severe droughts.

Sub-basin No. 1: Implications of meteorologic data sets in Pinarbasi and Tomarza stations were used to conduct drought analysis using SPI and to identify future dry periods. The dry periods identified using SPI 12-month for RCP 4.5 and RCP 8.5 conditions are documented in Tables 7 and 8.

Sub-basin No. 2: Implications of meteorologic data sets in Sariz station were used to conduct drought analysis using SPI and identify future dry periods. The dry periods identified using SPI 12-month for RCP 4.5 and RCP 8.5 conditions are documented in Table 9. The associated temporal variation for RCP 8.5 is depicted in Figure 14.

Sub-basin No. 3: Implications of meteorologic data sets in Adana Central and Karaisali stations were used to conduct drought analysis using SPI and identify future dry periods. The dry periods identified using SPI 12-month for RCP 4.5 and RCP 8.5 conditions are documented in Tables 10 and 11. The associated temporal variation for RCP 8.5 is depicted in Figures 15 and 16.

3.4. Standardized precipitation evapotranspiration index (SPEI)

Sub-basin No. 2 and the meteorologic station at Sariz were used to calculate dry periods using the SPEI approach. The outcomes are depicted in Figure 17. The drought periods with both the SPI and SPEI approaches identified the same dry periods of 1983–1986, 2000–2003 and 2011–2013. It was determined that SPEI analysis will not provide added value in terms of identifying drought conditions worse than the ones identified through the use of SPI.

4. Policy implications of the Seyhan drought

Turkey's Drought Management System has taken drought management to entail just response and improvement until very recent times. But in line with climatic changes and associated impacts on natural resources, government institutions have taken risk reduction, mitigation and preparedness into consideration just as importantly as response itself.

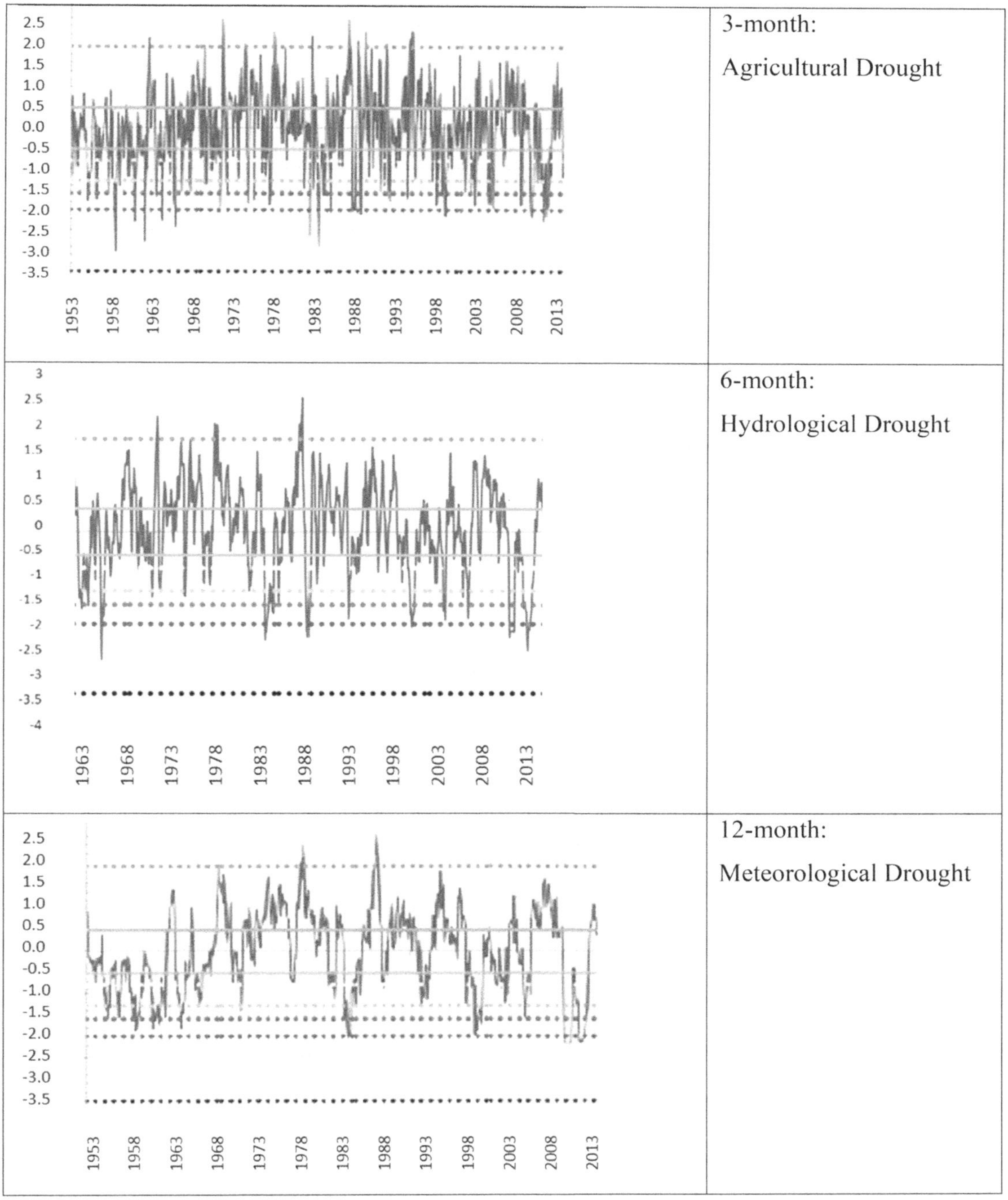

Fig. 11. Sub-basin No. 2: Sariz: historical SPI drought index values (*Source:* MGM Drought Monitoring System 2.1, 2016).

Table 5. Sub-basin No. 3: Adana central station: historical dry periods.

SPI Index Periods	Dry Periods
3-months	1972 November–1974 July
6-months	1972 August–1974 August
12-months	1972 April–1974 November

Table 6. Sub-basin No. 3: Karaisali station: historical dry periods.

SPI Index Periods	Dry Periods
3-months	1989 February–1989 June
6-months	1989 March–1989 September
12-months	1989 May–1990 February and 2012 November–2013 October

The Turkish Agricultural Drought Action Plan (TAKEP) aims at regulating all relations between government, non-government and civil organizations. But since the national regulations have not been totally reviewed and modified, there are some needs in coordinating the preparedness and response activities. There are several organizations studying drought management (i.e. meteorologic drought, hydrologic drought and agricultural drought) but coordination and cooperation are still the most important issues to be improved. Furthermore, a less structured level of authorization and information at the local level causes drought management initiatives to be less effective.

The national preparations developed as part of TAKEP indicate the following challenges:

- A need to strengthen legislative support for response at all levels.
- A need for hazard and risk assessments at national and local levels, considering types of droughts (i.e. hydrologic, meteorologic and agricultural).
- A need for vulnerability and capacity assessments at national and local level.
- A need for strategic plans, master plans and land use plans for risk reduction and mitigation, including measures according to types of droughts.
- A need for an efficient coordination mechanism between the institutions involved at national and local levels.

In order to address these challenges, strategic integration of the following units (see Figure 18), formed as part of TAKEP, can serve as the basis of strategic actions. The mandates and responsibilities of three of these strategic units are summarized below.

4.1. Agricultural drought management coordination council

The agricultural drought management coordination council has two purposes. First, it supervises and coordinates development of the Drought Management Plan. Second, after the plan is developed and during times of drought when the plan is activated, a Task Force coordinates actions, and implements mitigation and response programmes.

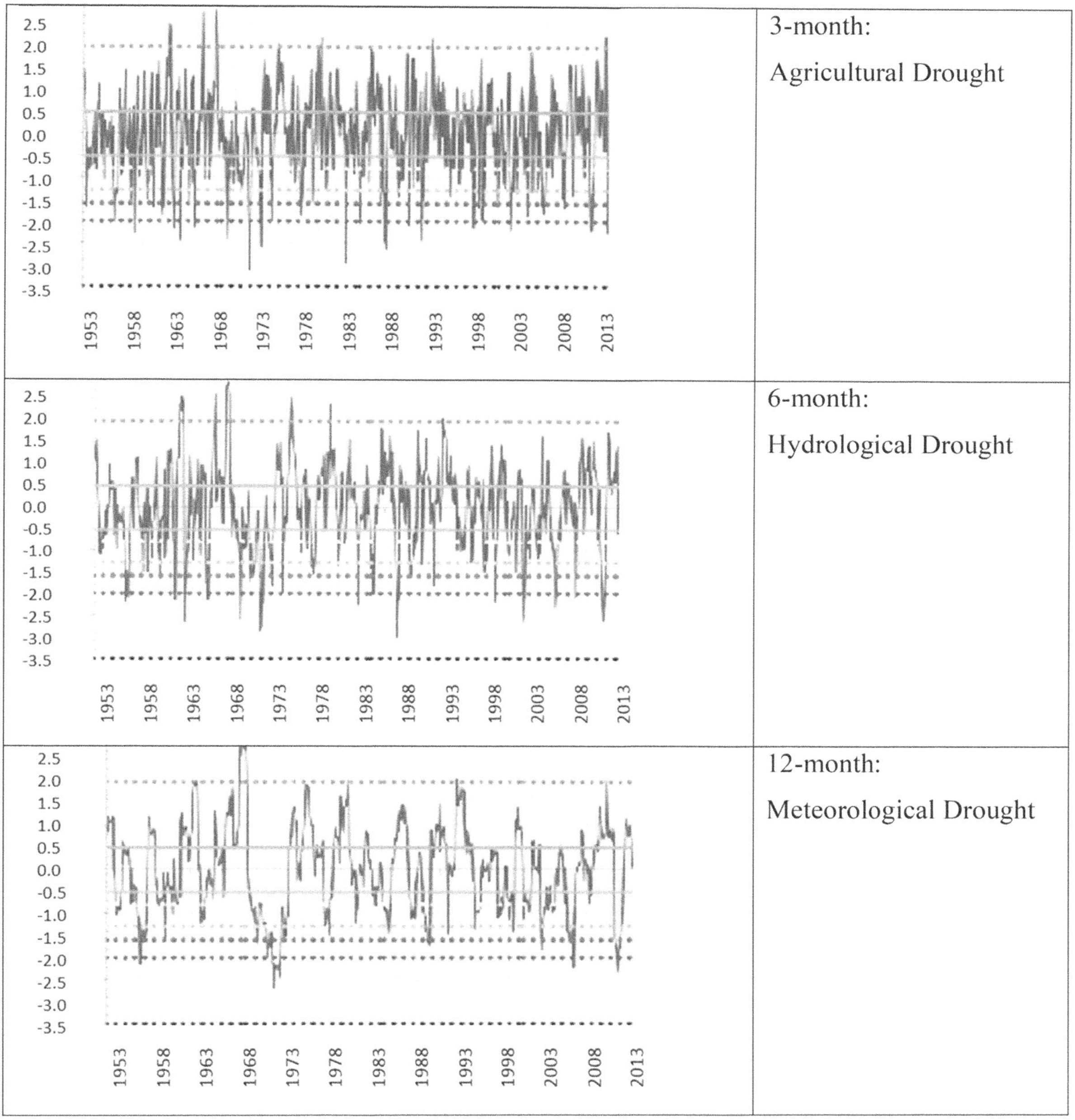

Fig. 12. Sub-basin No. 3: Adana Central: historical SPI drought index values (*Source:* MGM Drought Monitoring System 2.1, 2016).

4.2. Monitoring, early warning and forecasting committee

A reliable assessment of water availability and its outlook for the near and long term is valuable information in both dry and wet periods. During drought, the value of this information increases markedly.

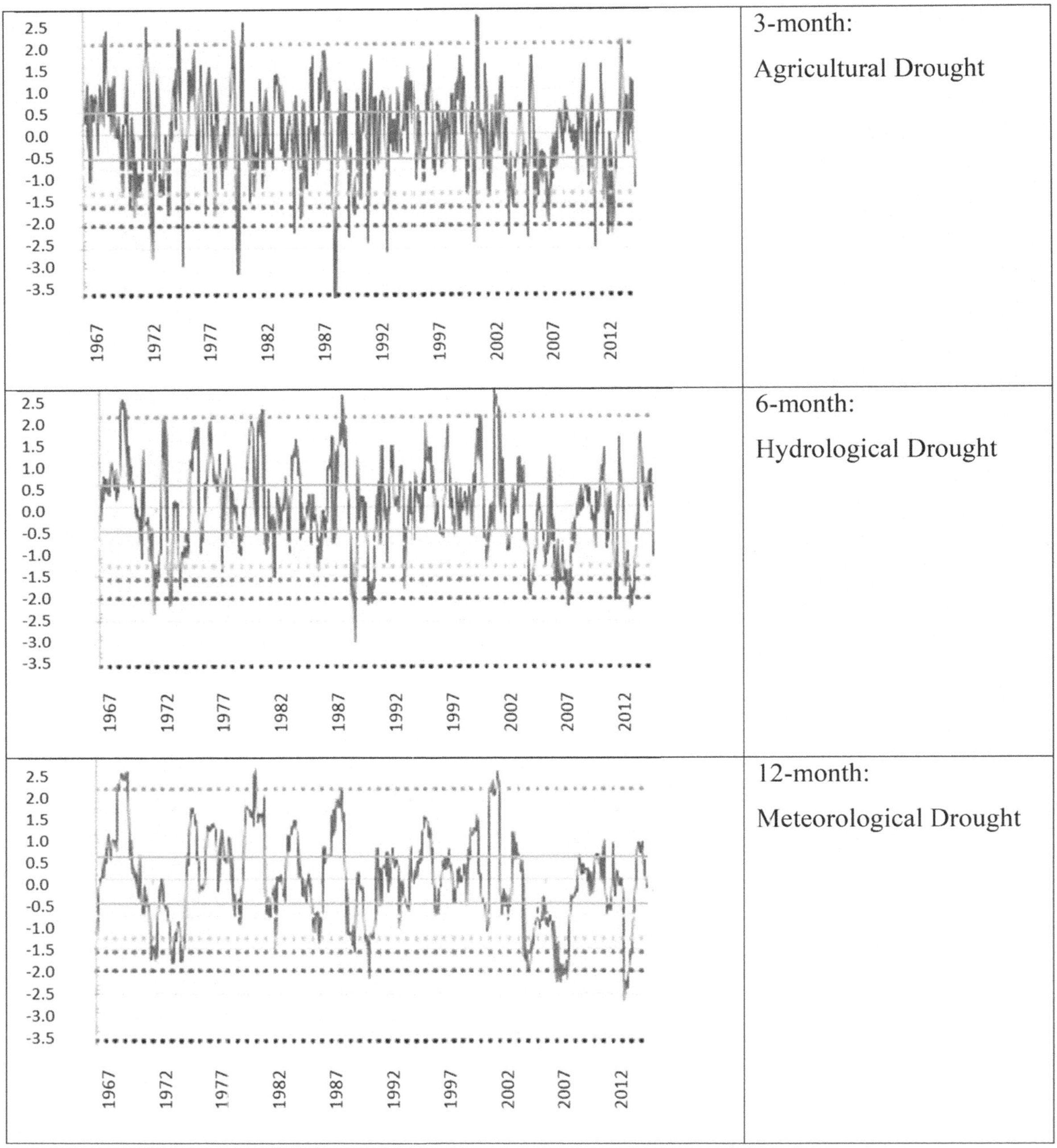

Fig. 13. Sub-basin No. 3: Karaisali: historical SPI drought index values (*Source:* MGM Drought Monitoring System 2.1, 2016).

The monitoring committee includes representatives from agencies with responsibilities for monitoring climate and water supply. Data and information on each of the applicable indicators (e.g. precipitation, temperature, evapotranspiration, seasonal climate forecasts, soil moisture, streamflow, groundwater

Table 7. Sub-basin No. 1: Pinarbasi station: future dry periods.

SPI Index Periods	Dry Periods	Climate Scenario
12-months	2037 March–2037 December	RCP 4.5
12-months	2094 May–2096 December	RCP 8.5

Table 8. Sub-basin No. 1: Tomarza station: future dry periods.

SPI Index Periods	Dry Periods	Climate Scenario
12-months	2066 April–2066 December	RCP 4.5
12-months	2095 March–2095 December	RCP 8.5

Table 9. Sub-basin No. 2: Sariz station: future dry periods.

SPI Index Periods	Dry Periods	Climate Scenario
12-months	2066 April–2066 December	RCP 4.5
12-months	2095 March–2095 December	RCP 8.5

levels, reservoir and lake levels, and snowpack) ought to be considered in the committee's evaluation of the water situation and outlook.

4.3. Risk assessment committee

Risk is the product of exposure to the drought hazard (i.e. probability of occurrence) and societal vulnerability, represented by a combination of economic, environmental and social factors. Therefore, to reduce vulnerability to drought, the most significant impacts must be identified and their underlying causes assessed. Drought impacts cut across many sectors and across normal divisions of government authority. Experience has demonstrated that the most effective approach to follow in determining vulnerability to and impacts of drought is to create a series of working groups under the aegis of the risk assessment committee. The responsibility of the committee and working groups is to assess sectors, population groups, communities and ecosystems most at risk, and identify appropriate and reasonable mitigation measures to address these risks.

Vulnerability assessment provides a framework for identifying the social, economic and environmental causes of drought impacts. It bridges the gap between impact assessment and policy formulation by directing policy attention to underlying causes of vulnerability rather than the result – the negative impacts – which follow triggering events, such as drought. For example, the direct impact of precipitation deficiencies may be a reduction of crop yields. The underlying cause of this vulnerability, however, may be that the farmers did not use drought-resistant seeds, because they did not believe in their usefulness, the costs were too high, or there was some commitment to cultural beliefs. Another example could be farm foreclosure. The underlying causes of this vulnerability might include

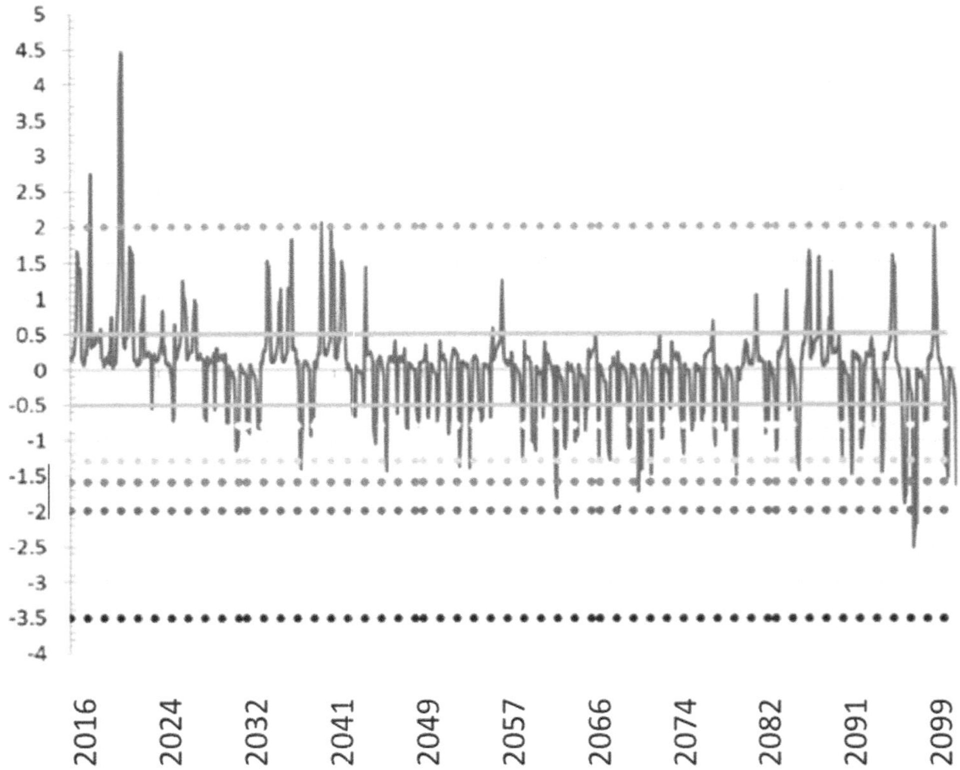

Fig. 14. Sub-basin No. 2: Sariz: SPI 12-month future drought index values – RCP 8.5.

Table 10. Sub-basin No. 3: Adana central station: future dry periods.

SPI Index Periods	Dry Periods	Climate Scenario
12-months	2032 October–2032 December	RCP 4.5
12-months	2049 September–2049 December	RCP 8.5

Table 11. Sub-basin No. 3: Karaisali station: future dry periods.

SPI Index Periods	Dry Periods	Climate Scenario
12-months	2042 June–2042 December	RCP 4.5
12-months	2066 April–2066 December	RCP 8.5

small farm size because of historical land appropriation policies, lack of credit for diversification options, farming on marginal lands, limited knowledge of possible farming options, a lack of local industry for off-farm supplemental income, or government policies.

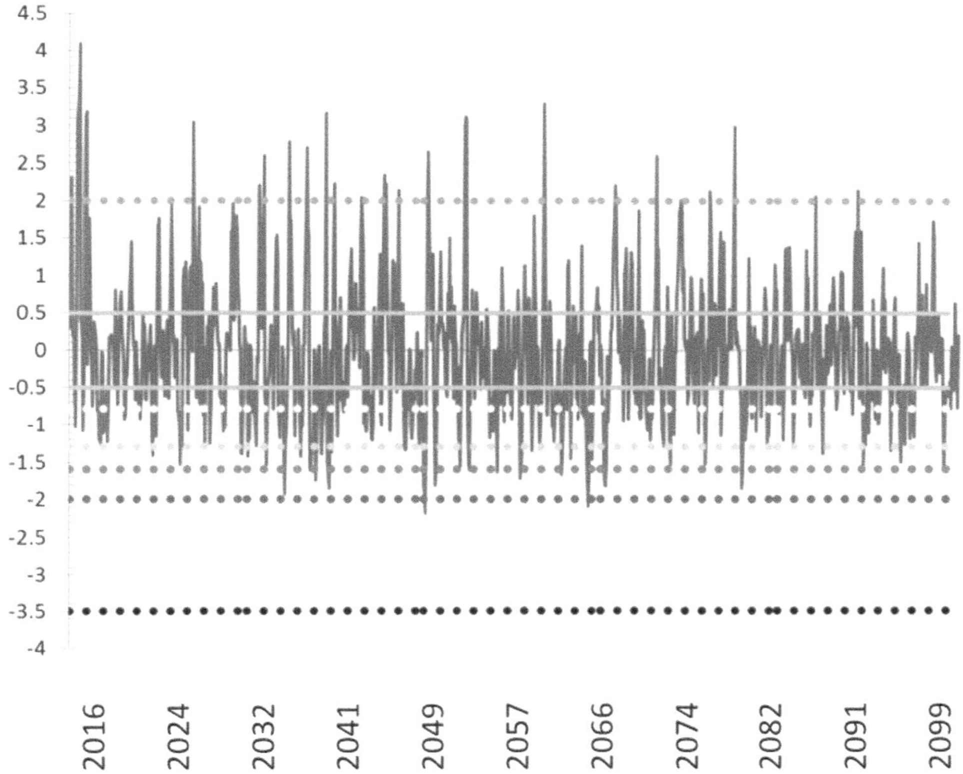

Fig. 15. Sub-basin No. 3: Adana Central: SPI 12-month future drought index values – RCP 8.5.

4.3.1. Integration of science and policy. An essential aspect of the proposed policy initiatives to adapt to drought dynamics in the Seyhan River Basin will be the ability to integrate science and policy aspects of drought management. The policy maker's understanding of the scientific issues and technical constraints involved in addressing problems associated with drought is often limited. Likewise, scientists generally have less interest in and knowledge of existing policy constraints for responding to the impacts of drought. In many cases, communication and understanding between the science and policy communities must be enhanced if the planning process is to be successful.

Good communication is required between the two groups in order to distinguish what is feasible from what is not achievable for a broad range of science and policy issues. Integration of science and policy during the planning process will also be useful in setting research priorities and synthesizing current understanding. As such, non-governmental organizations (NGOs) can play a vital role in the development of a communication and coordination platform to develop efficient and sustainable solutions by respecting both political and scientific requirements. Therefore, NGOs should consider various alternatives to bring these groups together and maintain a strong working relationship.

This platform should also be used to develop a broad-based education programme to raise awareness of short- and long-term water supply issues, which will help ensure that people know how to respond to

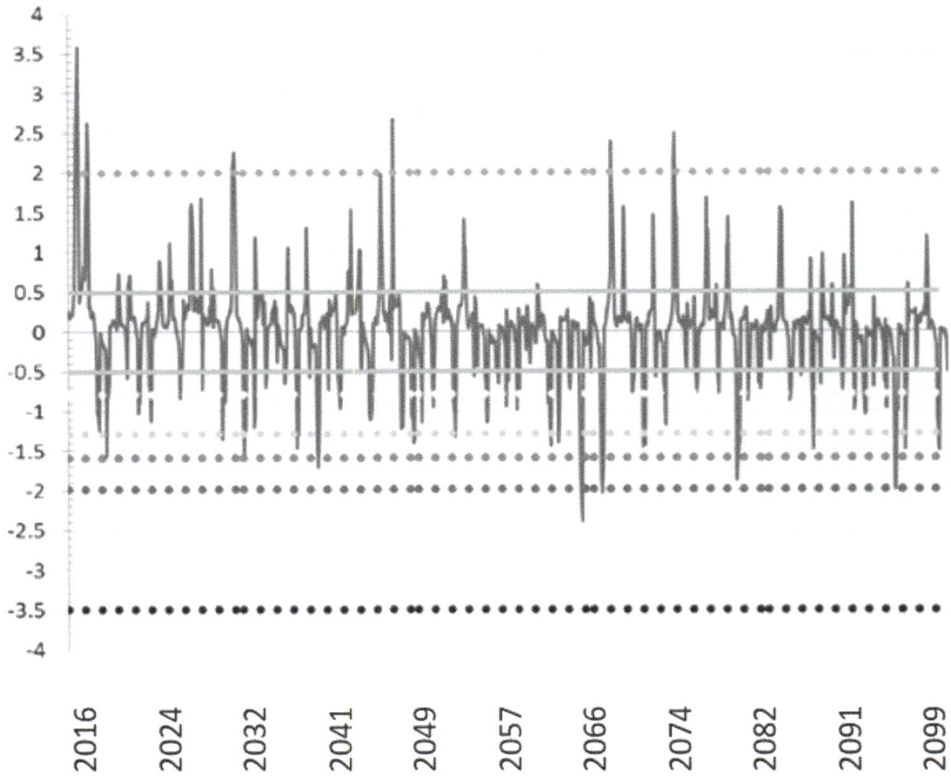

Fig. 16. Sub-basin No. 3: Karaisali: SPI 12-month future drought index values – RCP 8.5.

drought when it occurs and that drought planning does not lose ground during non-drought years. It is also important to tailor information to the needs of specific groups (e.g. elementary and secondary education, small business, industry, homeowners, utilities).

During drought, work groups identified by policy makers, universities and NGOs should work with public information professionals to keep the public well informed of the status of water supplies, whether conditions are approaching 'trigger points' that will lead to requests for voluntary or mandatory use restrictions, and how drought can limit assistance and access to vulnerable communities.

4.3.2. The political economy of drought. A crucial step for success lies in a decentralization policy, whereby decision-making on development issues is divested to local levels – which means drought management at the local and provincial level. However, gaps still exist in political commitment to address issues of drought management institutions, governance, risk and vulnerability identification, and local stakeholders' capacity (i.e. knowledge and technical skills). While there have been clear perceptions regarding the urgent need for drought relief and rehabilitation of vulnerable communities, there has been a notable lack of longer-term development plans established in the period between droughts to alleviate the vulnerability and apparent dependency syndrome among the target groups.

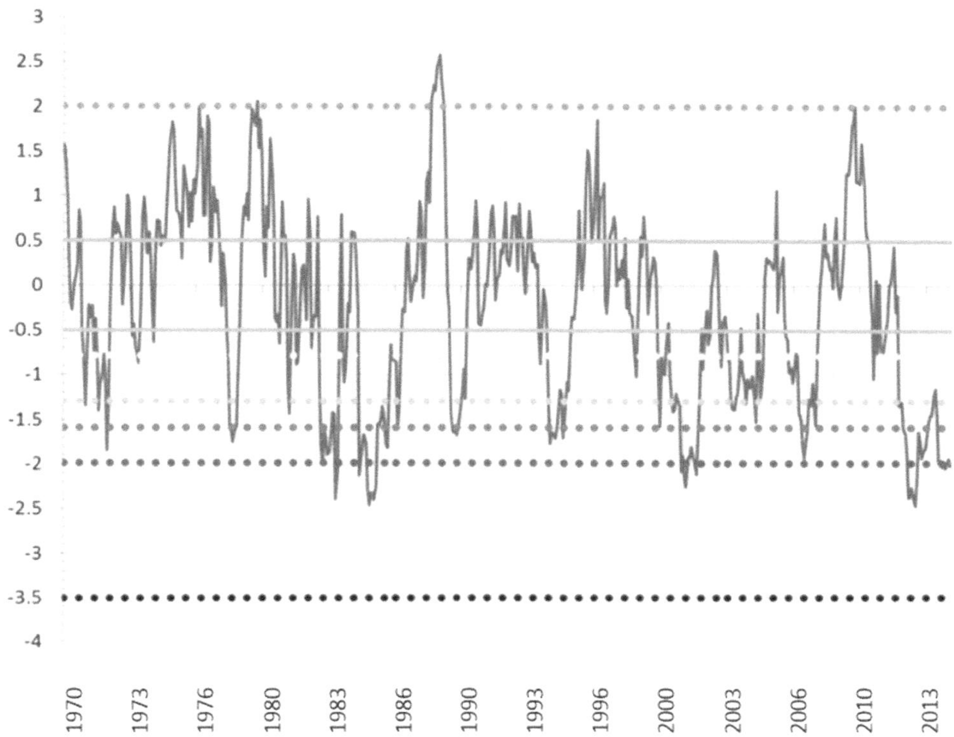

Fig. 17. Sub-basin No. 2: Sariz: Historical SPEI drought index values.

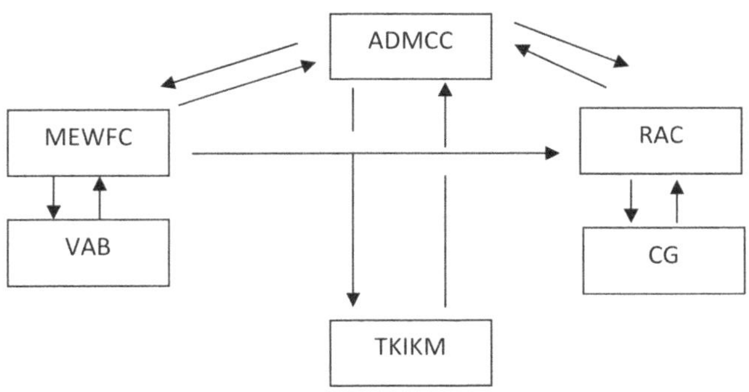

Fig. 18. Agricultural drought management in Turkey.

Political initiatives have usually advocated for emergency drought relief and rehabilitation measures, with less commitment to development aimed at reducing inherent risks and building vulnerable communities' resilience to drought. In this context, a lack of a budgetary framework and disbursement of funds for long-term drought preparedness and development programmes in the country's approach to drought risk reduction can put national development gains at risk.

5. Institutional structure

Drought management operates in a three-dimensional framework, usually constructed at the national scale, which includes the enabling environment, institutions and management. In this context, the institutional framework may be broadly defined with reference to three functions:

- constitutional function, to establish laws and framing policies;
- organizational function, to structure management at river basin scale; and
- operational function, to set up the functional relationships between stakeholders.

In order to integrate these functions within a 'drought management cycle' framework, it is important to implement a structured process of drought planning and management (Figure 19). This means that how a drought is managed can adapt to changing circumstances, e.g. to political changes, natural catastrophes and changes in demography.

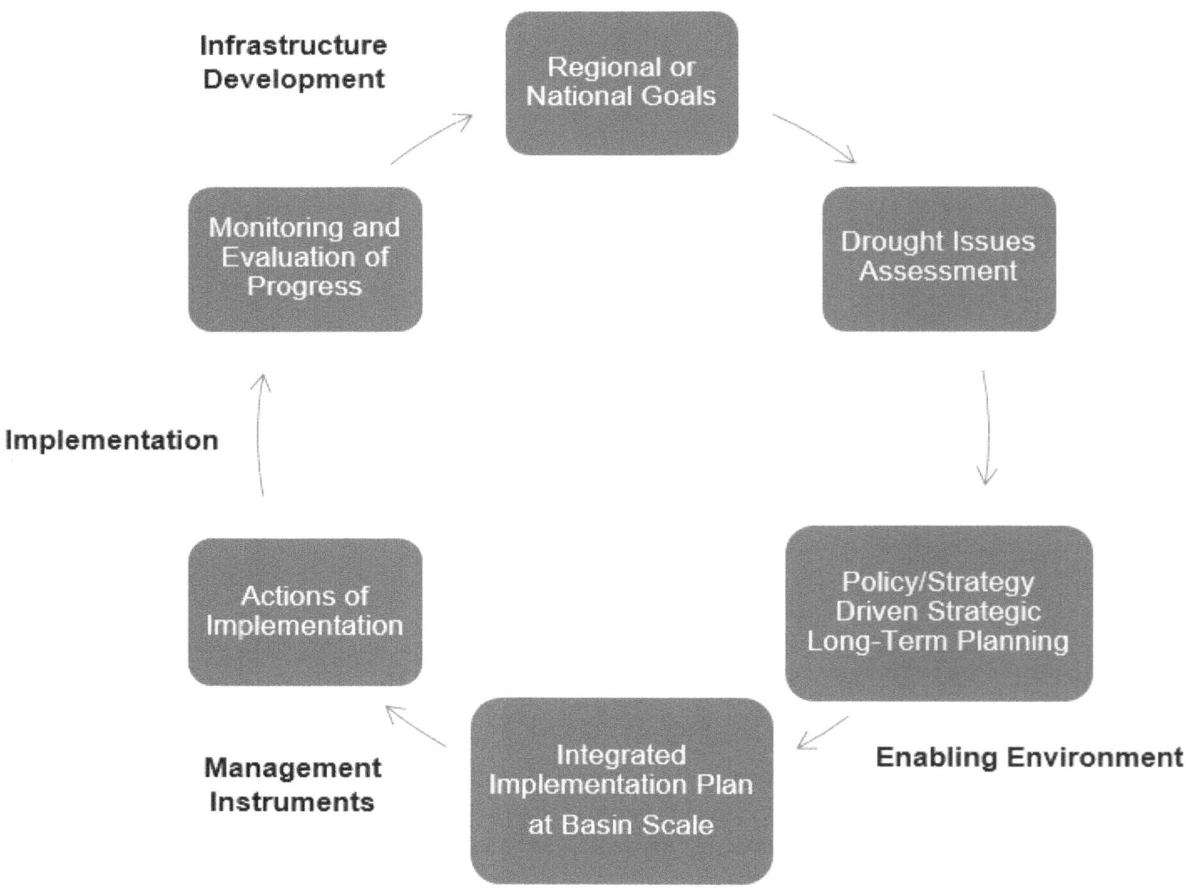

Fig. 19. Drought management cycle for planning and implementation.

In order to implement the respective components of the drought management cycle effectively, institutions that are in charge of river basin management activities can select the level of their actions in the following order:

- Local level (sub-basin plan, local government plan).
- Implementation level (basin- or provincial-scale management plan).
- Policy level (national and international processes for developing drought policies, treaties, and laws).

A key issue is how the basin administration fits with and relates to other administrative levels – national, provincial, district, community. This needs to be resolved in order to avoid duplication and confusion of responsibilities with other administrative bodies. What is needed is a clear legal framework that specifies the roles and responsibilities, rights and obligations of stakeholders, the levels of decentralization, and the processes and means for good water governance.

As per recent developments in Turkish legislation, water management was established at the basin scale by including various strategic committees (Figure 20). The basis for the establishment of the main institutional body, the Water Management Coordination Council (WMCC), provides the basis for sustainable monitoring and evaluation of site-specific measures, which includes drought-related

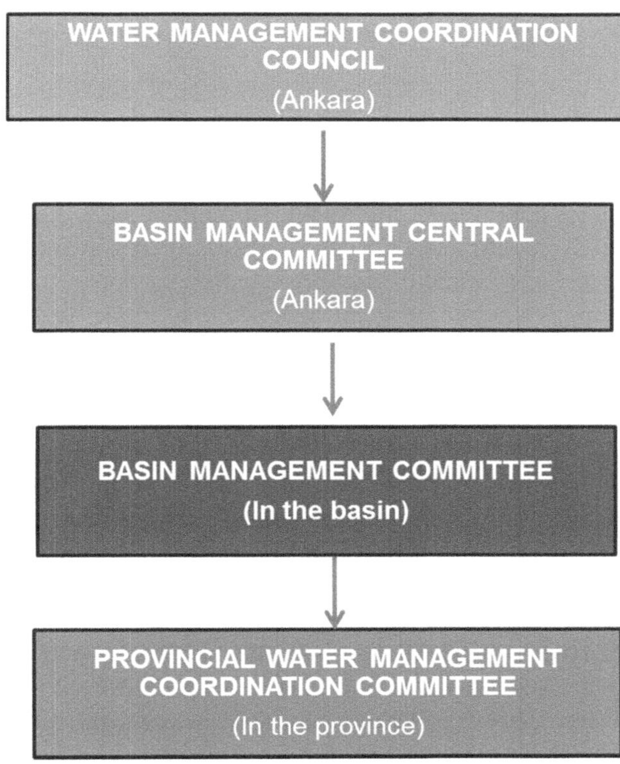

Fig. 20. Institutional structure of Basin Management Committee.

tasks and activities. One of the key bodies in this structure is the Basin Management Committee, which has the following roles and responsibilities:

- provide inter-sector coordination, collaboration and accelerated investment in water provision for effective water management;
- ensure a high level of coordination and cooperation between all stakeholders;
- determine necessary measures to conserve water resources within the framework of integrated river basin management; and
- develop strategy, policy and plan for achieving objectives set out in national and international documents.

One of the key committees, River Basin Management Committee, is in charge of developing basin management plans, and monitoring/evaluation of basin-scale initiatives and practices. Coordinator Governors coordinate the Basin Management Committees.

Establishing linkages between respective agencies to avoid any overlaps and/or redundancies is a challenge in this institutional context. As such, all governmental authorities, institutes, various lower-level public authorities, and laboratories have responsibilities in the system, and are stakeholders. Civil society stakeholders should also be encouraged to take active part in the implementation of plans and programmes at local level concerning water resources and drought management. In particular, cooperatives and unions, organized in the areas of agriculture and forestry, should participate actively in the implementation and monitoring of policy initiatives set by government institutions.

6. Conclusions

The main focus of this study has been to evaluate drought conditions in the Seyhan River Basin, in the context of historical conditions and future expected dry periods. The analyses were performed by using meteorologic gauge stations within the basin and using the most commonly used indices, SPI and SPEI. Meteorologic, agricultural and hydrologic drought conditions were then identified by using 3-month, 6-month and 12-month index values, respectively. It was determined that SPEI analysis would not bring additional value to identify drought conditions worse than those periods identified through use of SPI. A future conditions drought analysis was implemented using the outcomes of recent regional climate models to estimate impacts of climate change on the main meteorologic parameters (precipitation and temperature). The distinguishing features of historical droughts and future expected droughts are summarized as:

- historical droughts are characterized by less frequent but more severe droughts; and
- future droughts are characterized by more frequent but less severe droughts.

The differences in these characteristics will have an impact on respective sectoral activities along the river systems and associated adaptation activities at two main scales: the river system and the entire Seyhan basin. In the Goksu river system, for example, hydroelectric power generation will have to be aligned for more frequent reductions in average flow volume, and hence facilities without storage volumes will face more severe reductions in power generated. On the other hand, agricultural activities – the leading sectoral activity along the Seyhan River system – will have to adapt to the changes in flow volume through a combination of tools including optimized crop cycles/patterns and technology-driven solutions (drip and sprinkler irrigation schemes, remote sensing and soil moisture, etc.). As such, a structured drought management plan needs to be developed around the core requirements of a drought cycle including pre-drought, drought and post-drought oriented activities. Some of the strategic activities in the drought cycle are depicted in Figure 21, whilst the core activities within this drought cycle are summarized below.

Pre-drought activities:

(1) 'Loss reduction' should consider risk-based evaluation of restructuring activities.
(2) 'Preparation' activities prior to the impact are implemented in line with drought planning scenarios.

Drought activities (lessons learned provide added value for future activities):

(3) Lessons learned from emergency conditions contribute to effective development of communication strategies and planning of activities prior to the impact.
(4) Prioritization of emergency action management activities.
(5) 'Improvement' activities in relation to restructuring, development of an effective and sustainable plan and review of all functional services.

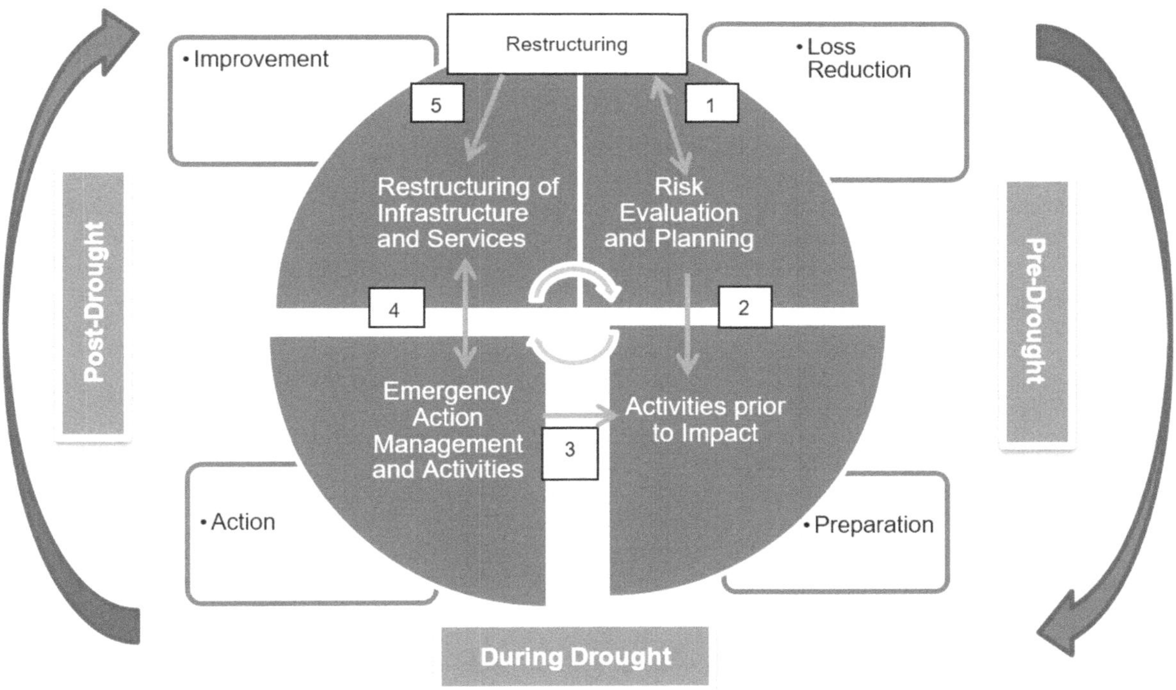

Fig. 21. Drought cycle management.

In addition to the elements documented as part of the drought cycle, some of the strategic components of a Drought Management Plan should include the following elements:

- An emergency water allocation strategy needs to be adopted for implementation during severe drought conditions.
- Recommendations need to support water suppliers to develop drought plans.
- Worst-case drought scenarios need to be evaluated for possible further actions.
- Interaction between public and policy actions need to be elevated through strategic and public/government bodies, such as natural hazard mitigation council and public advisory committee.
- Strategic consideration of pricing and market trading is essential by considering the fact that water is largely a public good, and water rights rest with a governing body, while individuals essentially have 'use' rights. In addition, water is not a standard commodity, rather water supply is stochastic and flows through complex natural and man-made systems. Transaction costs for water trades can be high because of the need to physically transport water and the required administrative approvals, which may not be granted because of externalities to third parties. Additionally, institutional features affect transaction costs, such as the structure of the water district, the water rationing mechanism, and other rules such as environmental return-flow requirements.

References

Edwards, D. C. & McKee, T. B. (1997). *Characteristics of 20th Century Drought in the United States at Multiple Time Scales*. Climatology Report 97-2, Department of Atmospheric Science, Colorado State University, Fort Collins, CO.

Eriyagama, N., Smakhtin, V. & Gamage, N. (2009). *Mapping Drought Patterns and Impacts: A Global Perspective, International Water Management Institute (IWMI) Research Report No. 133*, pp. 36. IWMI, Colombo, Sri Lanka.

Fleig, A. K., Tallaksen, L. M., Hisdal, H. & Demuth, S. (2006). A global evaluation of streamflow drought characteristics. *Hydrological Earth System Sciences Discussions 10*, 535–552.

General Directorate of State Hydraulics Works (DSI) (2013). *Seyhan River Basin Master Plan*. DSI, Ankara.

Keyantash, J. & National Center for Atmospheric Research Staff (eds) (2016). *The Climate Data Guide: Standardized Precipitation Index (SPI)*. Available at: https://climatedataguide.ucar.edu/climate-data/standardized-precipitation-index-spi (accessed 11 November 2016).

MGM (General Directorate of Meteorology) (2015). *Climate projections for Turkey using new scenarios*. MGM, Ankara, pp. 68.

MGM (General Directorate of Meteorology) (2016). Drought Monitoring System 2.1. Available at: http://212.175.180.197/bitkidon/kuraklikizleme.aspx (accessed 14 March 2016).

Selek, B. & Tuncok, I. K. (2014). Effects of climate change on surface water management of Seyhan Basin, Turkey. *Environmental and Ecological Statistics 21*(3), 391–409.

SYGM (General Directorate of Water Management) (2015). *Impacts of Climate Change on Water Resources*. Interim Report on Climate Change Model in Seyhan River Basin. SYGM, Ankara, pp. 256.

Vicente-Serrano, S. M. & National Center for Atmospheric Research Staff (eds) (2015). *The Climate Data Guide: Standardized Precipitation Evapotranspiration Index (SPEI)*. Available at: https://climatedataguide.ucar.edu/climate-data/standardized-precipitation-evapotranspiration-index-spei (accessed 11 November 2016).

Vicente-Serrano, S. M., Beguería, S. & López-Moreno, J. I. (2010). A multi-scalar drought index sensitive to global warming: The Standardized Precipitation Evapotranspiration Index – SPEI. *Journal of Climate 23*(7), 1696–1718. doi:10.1175/2009JCLI2909.1.

Watanabe, T. (2007). *Summary of ICCAP: Framework, Outcomes and Implication of the Project: ICCAP Project Report*. Available at: http://www.chikyu.ac.jp/iccap/finalreport.htm (accessed 11 November 2016).

Wilhite, D. A. (1992). Drought. In: *Encyclopedia of Earth System Science,* Vol. 2. Nierenberg, W. A. (ed.). Academic Press, San Diego, CA, pp. 81–92.

Wilhite, D. A. & Glantz, M. H. (1985). Understanding the drought phenomenon: the role of definitions. *Water International 10* (3), 111–120.

Wilhite, D. A. & Glantz, M. H. (1987). Understanding the drought phenomenon: the role of definitions. In: *Planning for Drought.* Wilhite, D. A., Easterling, W. E. & Wood, D. A. (eds). Vestview Press, Boulder, CO, pp. 11–27.

World Meteorological Organization (WMO) (2012). *Standardized Precipitation Index User Guide*. WMO, Geneva, Switzerland, pp. 16.

South Africa and the drought that exposed a young democracy

Anthony Turton

Centre for Environmental Management, University of Free State, Bloemfontein, Republic of South Africa
E-mail: tony@anthonyturton.com

Abstract

South Africa is a young democracy currently going through a crisis of leadership. The worst drought in recorded history has played out at regional level but against the backdrop of complex political dynamics. The government has lost significant capacity at the technical level, largely the result of political priorities driven by the need to decolonise society and the institutions of higher learning. This has manifest in the water sector as systemic failures of key instrumentation systems, rendering the El Niño event invisible until it hit. This case study of the El Niño event shows that drought management is embedded within a broader political process and is not simply a technical management issue. The Vaal River system sustains 60% of the national economy and 45% of the total population of the country, but water security in this system has been placed at risk because of political dynamics.

Keywords: DDT; Decolonization of science; El Niño; Indian Ocean Dipole; Inter-Tropical Convergence Zone; Instrumentation; Patronage; South Africa; Urogenital birth defects

Introduction

'*Southern Africa's worst drought in decades has left 40 million people, 14 per cent of the region's total population, food insecure this year. The World Food Programme is significantly scaling up operations to reach over 13 million by January 2017, but funding shortfalls remain acute: the Southern African Development Community estimates a gap of $2.5 billion*' (CSIS, 2016).

South Africa, as but one country in southern Africa, is an anomalous place: economically diversified beyond its natural water endowment and with the largest city in the world not on a river, lake or

doi: 10.2166/wp.2016.020

waterfront – Johannesburg – straddling a continental watershed divide (Turton *et al.*, 2006). This means that economic development is clustered around one major river basin (the Orange/Vaal) and the well-being of an entire nation rests in the hands of a highly competent technocratic elite. Water security at both the national and regional level is a core strategic issue (Oosthuizen *et al.*, 2015). Drought is a natural occurrence with the essence of strategic water resource planning being the mitigation of the impact of the next event (Middelton & Bailey, 2008). Hydrologically the country is squeezed between four major planetary systems – the El Niño Southern Oscillation (ENSO), the frontal systems of the Southern Ocean, the Inter-Tropical Convergence Zone and the Indian Ocean Dipole. This paper delves into the politics and decision-making processes that have been exposed by the 2015/16 El Niño event, which is the worst in recorded history. Analytically, three independent drivers will be assessed in order to gain a deeper understanding about drought management in a young democracy that is currently in crisis (Feinstein, 2007; Johnson, 2015). The policy-related linkages between hydrological reality and the legitimate aspirations of a nation with a long history of oppression and abuse by political elites will be examined.

Drivers of the 2015/16 El Niño water resources drought disaster

It is logical to assume that policy experiments take place in a young democracy and in this regard South Africa is no exception. After all, a liberation movement that fought for almost a century faces a different set of challenges when it becomes the party of government. In order to understand how the broad policy positions have played out since the advent of democracy in 1994, we can see three distinct phases, each different from the others in a fundamental way, but also linked directly to the executive leadership at a given moment in time. By understanding these policy dynamics, we can assess the degree to which the impact of the current drought is a human-induced disaster by virtue of the politicization of water resource management decisions.

Driver 1: overarching political dynamics

The history of apartheid is well documented so little needs to be said here apart from the fact that the desire for liberation from oppression is universal and deeply rooted in South Africa (Johns, 1989; Karis & Carter, (1972); Liebenberg, 1994; Meredith, 2007; Simpson, 2016). The transition to democracy took place in 1994 when a negotiated peace emerged from the inability of either side to gain a military victory (Gottschalk, 1994; SASS, 1996; Spitz & Chaskalson, 2000). For the purposes of this analysis, three phases are considered since the democratic transition in 1994, each driven by a different executive president.

The first phase of democracy is directly associated with Mandela (1994). Highly acclaimed for his leadership and wisdom, his term in office is associated with the redrafting of all legislation in order to align the new democratic order with the Constitution (1996). This was followed shortly afterwards by the National Water Act (1998), which separated water rights from land rights as a key element of land redistribution (Movik, 2008). This is important, because the African National Congress (ANC) came into existence as a liberation movement in response to the 1913 Land Act that gave the majority of land to persons of European origin, while reserving small tracts of land held under tribal authority in what became known as Bantustans (Giliomee, 1981; Liebenberg, 1994). This was the genesis of

Apartheid, so the linkage between water and land was fundamental to this policy reform (Reed & de Wit, 2003; Funke *et al.*, 2007). The National Waste Management Act (1998) and the National Waste Management: Air Quality Act (1998), were both radical departures from previous policy that had evolved to protect strategic industries in the face of comprehensive economic sanctions (Findlater *et al.*, 2007). These reforms in waste management reversed the externalization of liabilities that had been used to sustain the Apartheid state (Gutteridge, 1984; Adler *et al.*, 2007).

This era of legislative and policy reform was driven by charismatic leaders of the armed struggle against Apartheid, many of whom were internationally recognized and respected, including Nelson Mandela, winner of the Nobel Peace Prize, and Professor Kader Asmal, an international lawyer, Stockholm Water Prize laureate and winner of the South African Water and Energy Forum prize for outstanding leadership. This era can be summarized by Mandela's emphasis on non-racial nation-building and the deepening of democracy under the popular banner of the Rainbow Nation (a phrase coined by another Nobel laureate, Archbishop Desmond Tutu). In summary, this period was driven by leaders of enormous integrity with a vision of reconciliation for a country that had emerged from a 350-year history of oppression, violence and bigotry (Meredith, 2007).

The second phase of democracy is directly associated with Thabo Mbeki (Gevisser, 2007), heir apparent to Nelson Mandela and a second generation freedom fighter (Mbeki, 1984). Thabo Mbeki was educated in exile with a post-graduate degree in development studies from the University of Sussex. A deeply intellectual man, his presidency was associated with two major events. The first was the emergence of the concept of the African Renaissance, which effectively ended all public reference to nation building and the so-called Rainbow Nation. In effect this ended the period of non-racialism noted above. The second was the emergence of a dark period in which western science was shunned as the HIV-AIDS pandemic took hold of the country. Dubbed AIDS Denialism, this period unlocked internal tensions in the ruling ANC party, resulting ultimately in the removal of Mbeki from office during an event known as the Polokwane Conference. Science, perceived to be a western concoction imposed on Africa by colonial powers, became the subject of suspicion, so the ban on the use of DDT for malaria eradication was set aside (Cone, 2009; see also: http://www.southafrica.info/about/health/malaria-190906.htm#.V9P5_cEaJqM) and indigenous treatment for AIDS became government policy (Visser *et al.*, 2004; see also: http://www.health24.com/Diet-and-nutrition/Nutrition-basics/Beetroot-garlic-onions-and-Aids-20120721).

The third phase of democracy is directly associated with the consolidation of personal power by Jacob Zuma. Much has been written about the Polokwane Conference, which was effectively a bloodless coup, meted out swiftly with the alleged covert support of the National Intelligence Agency (NIA) (Diakanyo, 2009). The remilitarization of politics has become increasingly apparent by the growing presence of camouflage uniformed members of the ANC militia at political events, sometimes carrying weapons without any attempt by the police to intervene (Hosken, 2016). Zuma is unlike any of his predecessors. He has only limited formal education, rising to prominence as head of intelligence in the armed wing of the ANC (Mkonto we Sizwe – Spear of the People – abbreviated as MK) (Simpson, 2016) with a grassroots political intuition that is legendary. Under his presidency, still in existence at the time of writing, the South African democratic experiment has been put to a series of extreme tests. For analytical purposes, the Zuma phase is characterized by seven major trends or events.

1. The first is the consolidation of political power, which many have labelled a criminalization of the state (Feinstein, 2007; Johnson, 2015; Blair, 2016; CCT, 2016). The liberation movement that

brought democracy to South Africa has effectively been captured by an elite (Munusamy, 2016). This so-called 'state capture' is a central element of many daily newscasts and media outputs. At its core, it is about control over power to entrench the politics of patronage and tribalism, while protecting key collaborators. (February, 2016).

2. The South African Police Service was remilitarized (Dintwe, 2013). The national security paradigm was also changed to one of state security, in which the increasingly embattled political elite need protection from its own citizens (Turton, 2010), bringing the State Security Agency (SSA) in as an instrument of power by replacing the more benign NIA.
3. Zuma's presidency's consolidation of power has engendered racial intolerance and genocidal rhetoric (see http://mg.co.za/article/2010-11-04-motlanthe-malemas-cockroach-comment-bad-manners), both of which are linked to the high levels of xenophobic violence that has marred South Africa during this phase of democracy (Johnston & Wolmarans, 2008; Turton, 2009; Grootes 2011).
4. The AIDS issue has taken an about face, with the denialism of the Mbeki-era being actively swept away, at the same time that President Zuma became embroiled in a rape case that went to court, but drove the alleged victim into exile (Times Live, 2014). The decolonization of science is now playing out as a nation wide student protest in demand for free and decolonised tertiary education that has placed the 2016 academic year in jeopardy.
5. The once proud veterans of the armed struggle are now squabbling. A recent example of this was a court battle against the Chairman of the MK Veterans Association and Deputy Minister of Defence (Kebby Maphatsoe) by the former Chief of Staff of MK, Minister of Intelligence and subsequent Minister of Water Affairs (Ronnie Kasrils) (Mabuza, 2016). This centred on the victim of the Zuma rape trial (noted above), with Kasrils pledging the money awarded by the court to a trust set up to protect the exiled victim, who has subsequently died.
6. The Constitutional Court has affirmed the primacy of the constitution, ruling that Zuma has failed to uphold his constitutional obligations (CCT, 2016; see also: http://cdn.24.co.za/files/Cms/General/d/3834/24efe59744c642a1a02360235f4d026b.pdf). The Constitutional Court has also ruled on the legitimacy of the Public Protector's office, openly flaunted by Jacob Zuma (see https://www.enca.com/south-africa/full-judgment-concourt-ruling-nkandla-matter#). This indicates that the judiciary is still functional and effective despite the many attempts to manipulate it. However, the recently appointed new Public Protector has links to the SSA (Joubert, 2016).
7. Finally, the most significant trend has been the unintended consequence of Zuma's leadership, manifesting as a unification of the plethora of opposition parties, to such an extent that during the 2016 municipal elections, the ruling ANC lost its overwhelming national majority with opposition coalitions taking over all major metropolitan councils.

So how is Driver 1, overarching political dynamics, relevant to water resource management in general, and drought management in particular? To answer this question, empirical data underpinning the technical capacity of the South African state to manage water resources have been examined. While state capture has been taking place, a number of important technical issues have remained invisible from media scrutiny. The first part of this narrative is driven by the emerging suspicion of science as a western machination, most aptly manifested by Mbeki's AIDS denialism (Natrass, 2010), and the dogged determination by the ANC to ignore global evidence regarding human health risks to DDT exposure (Bornman *et al.*, 2005, 2008, 2009; Aneck-Hahn *et al.*, 2007; Cone, 2009). The general disrespect for science is starkly illustrated in the water sector by the number of gauging stations that have been

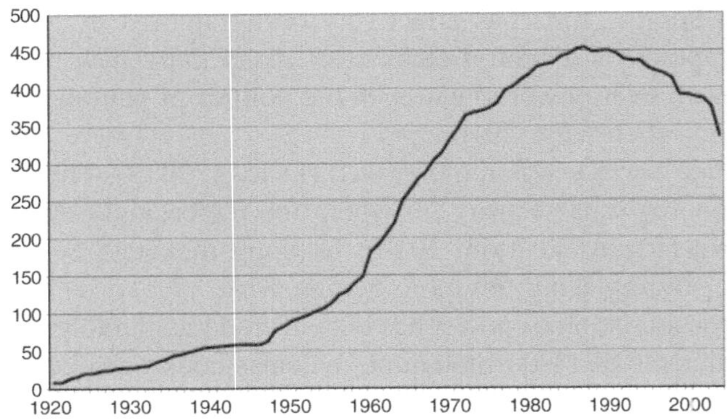

Fig. 1. Number of useful streamflow gauges open each year in South Africa (Pitman, 2011).

functional over time. Figure 1 shows the number of useful stream flow gauging stations while Figure 2 shows the number of rainfall gauging stations that have been functional over time (Pitman, 2011).

Commenting on the data shown in Figure 1, Pitman (2011) notes: 'This figure shows a rapid growth after a relatively slow start before 1960, and then a flattening off to a peak of around 450 in the late 1980s. Since 1990 there has been a steady decline such that the number open in 2004 has dropped to about 350. If this trend continues into the future it will be a serious cause for concern.' With regard to Figure 2, Pitman (2011) said: 'When the use of time series of flow was first adopted in the 1981 survey, the period prior to 1920 was rejected owing to the relative scarcity of stations open before that date. However, we are now in a similar position with the decline in numbers from a high around 1970 to only about half of that in 2004, which is roughly the same number of stations as in 1920. This is a major problem as rainfall is the primary input to WRSM2000 (or any other rainfall runoff model) – not only for simulating runoff but also for calculating irrigation demands and losses from reservoirs and wetlands. Some water management areas (WMAs) are worse off; for example, WMAs No. 4 (Olifants) and No. 13 (Upper Orange) now have less than half the number of stations open than was the case in 1920.'

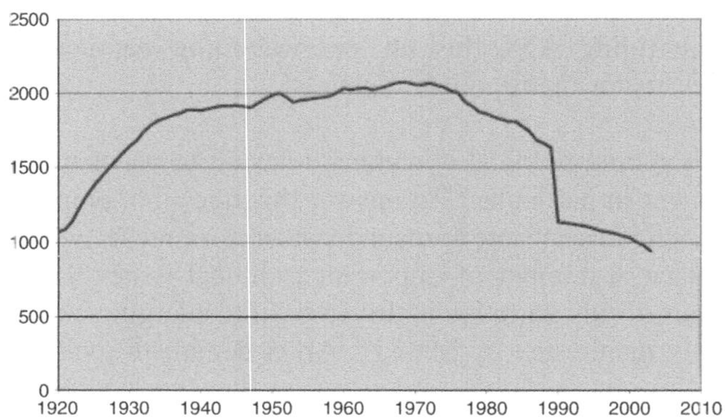

Fig. 2. Number of rainfall gauging stations open each year in South Africa (Pitman, 2011).

In Figure 2, our attention is drawn to the vertical drop in gauging capacity in 1994, when approximately 500 gauges went off line simultaneously. This happened at the same time that a decision was made to disband the Computing Centre for Water Research (CCWR), previously hosted at the University of Natal, and a significant technical resource underpinning a wide range of water resource decision-making platforms. The CCWR used to capture primary data, verify and collate it for onward supply to a range of data users in research, management and regulatory structures. In this one decision, a major element of the technical capacity of the state to manage water, most notably in the capture, storage and processing of metadata, was lost forever. This single event has led to the capture of state data by private consultants, a process exacerbated by the outmigration of technically skilled specialists as the ideological purges began to take hold under Mbeki and accelerated under Zuma.

If we look wider, a similar trend is noted in the capacity of the Council for Scientific and Industrial Research (CSIR) (Walwyn & Scholes, 2006). Figure 3 was part of a review conducted by two senior fellows within the CSIR (Turton, 2008a). The vertical lines show revenues over time, with a peak around 1989. The employee peak of around 5,000 happened in 1988. Therefore, technical capacity was already in decline during the decade before the formal transition to democracy in 1994. This coincides with the external military campaigns, most notably the culmination of Cold War hostilities in Angola that occurred in 1987 with the Battle of the Lomba River (sometimes called the Battle of Cuito Cuanavale) (Steenkamp & Heitman, 2016). It must be noted that the transition to democracy merely culminated in 1994, but actually began in 1987 after the Battle of the Lomba (Turton, 2010). In effect then, the era of science denialism triggered by Thabo Mbeki, merely exacerbated a process already underway and is still playing out as the call by students to decolonize university education.

The same trend is evidenced by the loss of engineering skills. This has become so critical that the South African Institute of Civil Engineers (SAICE) did a presentation to Parliament in 2008, during the transition from Mbeki to Zuma, in which they alerted national policy-makers to the implications of this trend (SAICE, 2008). SAICE's data showed that many of the rural municipalities no longer had a single qualified engineer on their staff, with some of the larger cities also showing a deficit. Furthermore, the outmigration of skilled engineers after 1994, in response to changing employment equity policies, has resulted in the loss of a mentoring cohort capable of guiding new engineers. Similar briefings were given by the CSIR during that period

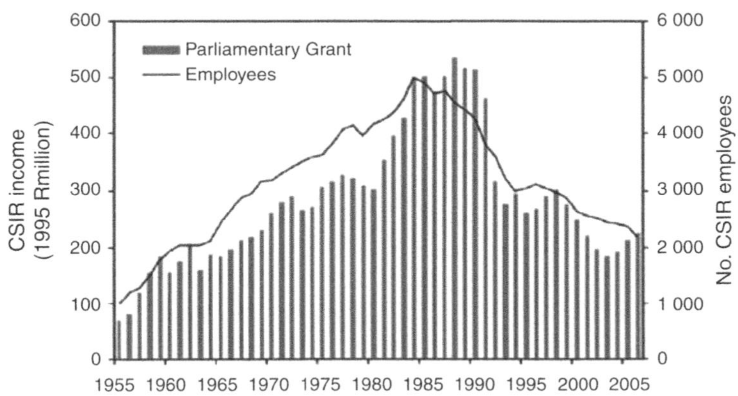

Fig. 3. Number of employees and revenues for the South African CSIR over time (Walwyn & Scholes, 2006).

of great uncertainty (Genthe & Steyn, 2007; Oberholster, 2008; Turton, 2008b) at considerable personal risk to senior scientists making a stand (Nordling, 2008; Power, 2008).

The conclusion regarding Driver 1 (overarching political dynamics) is that there has been a general loss of technical capacity, commencing in 1987 but accelerating under the Mbeki regime's suspicion of science in general, acutely manifest by the reversal on the global position regarding the use of DDT for malaria control. This loss of technical capacity has its origins in the general political environment which has framed the policy debates and driven a range of decisions that have collectively resulted in a loss of resilience in the face of a major drought. While this is understandable, if we accept that the transition from being a liberation movement to becoming a coherent government is challenging, the unfortunate outcome is a human-induced disaster that could have otherwise been avoided, or at least minimized.

Driver 2: climatic reality

Politics is one thing but climate is another indeed. As noted earlier, South African hydrology is driven by major global systems (Schulze, 1990) of which the ENSO is but one (Scholes *et al.*, 2015). Figure 4 shows ENSO data as a deviation from the mean over time (Meissner, 2010).

This dataset shows a number of important trends. First, there is a general cyclical oscillation between positive and negative deviations from the norm over time. Second, a typical trend is an oscillation between a positive and a negative deviation, but this is not always the case when one deviation in a given direction can be followed by another in the same direction. This is unusual however, but not impossible. Third, and most importantly, the peak values on both sides are on a clear upward trend. This is evidence of global warming but also evidence that suggests future events will be more extreme than in the past. These data were used by the author in 2012 (for commercial clients) to predict a major

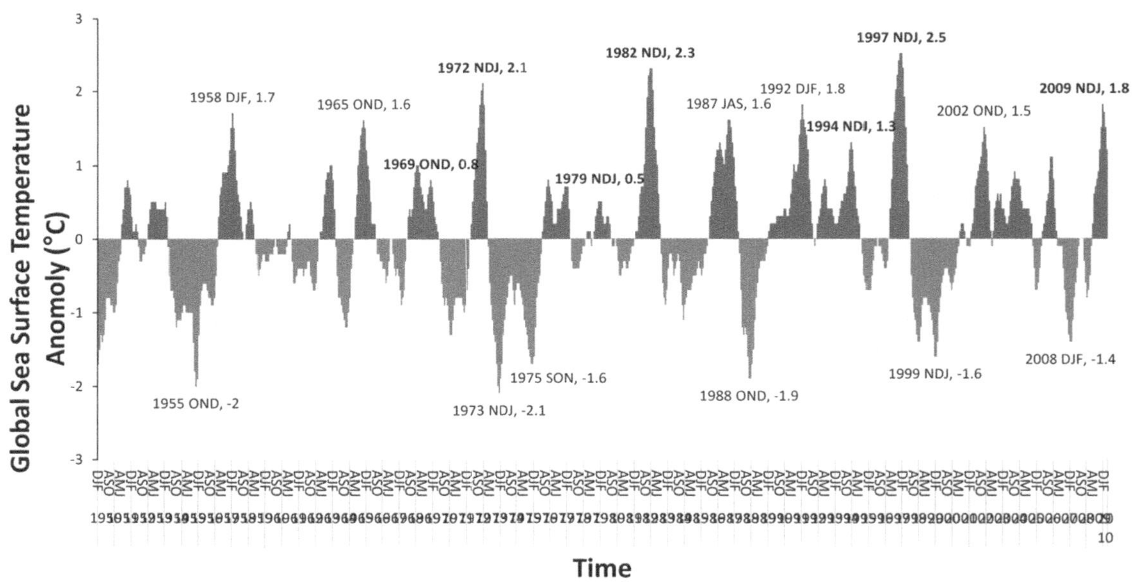

Fig. 4. Global sea surface anomaly for the Pacific Ocean showing El Niño/La Niña events over time (Meissner, 2010).

regional El Niño event 'in the foreseeable future'. This event took place in 2015–16 and was the worst in recorded memory (Crilly, 2016), triggering famine and requiring emergency response at regional level (see http://www.fews.net/southern-africa).

The significance of this El Niño drought event, in the context of government response, is that it remained invisible to government (even though it was known to the CSIR) until it was too late to prepare for it. The rainfall and streamflow gauging stations (shown in Figures 1 and 2) were in such a state of disrepair and the data management systems that they fed into had become so dysfunctional that the El Niño of 2015–16 approached unnoticed. This was best illustrated in the Mvoti River of the Kwazulu Natal (KZN) province.

The Mvoti River supplies water to a large rural community in the iLembe District. Potable water is supplied from a treatment plant that abstracts and treats about 17 Ml/D directly from the Mvoti River. There is no dam for storage. The river carries a heavy silt load and is used by sand miners to harvest aggregate for the construction industry. While some of these sand miners are illegal, not all fall into this category.

In 2014 an employee of the potable water plant run by Umgeni Water sent a report to head office that the water supply had disappeared and that sand miners had diverted the flow of the river. This triggered an aggressive response from the state, with a raid under armed protection, of all legal mining operations. One sand mining operation with lawful mineral rights was served with a directive and told to cease mining immediately.

The mining rights owner, as a legal operation, approached the courts for judgement in their specific case, citing the rapid escalation from *status quo* to enforcement without first going through the customary engagement to determine the level of compliance. The Water Tribunal was dysfunctional and had been abandoned, so the mining rights holder approached the Durban High Court for an interdict (Case number 14236/14 & 14621/14). The High Court ruled that the only legitimate structure capable of hearing such a matter was the Water Tribunal, so they passed an order of court mandating the Minister of Water and Sanitation. This was done in a Cabinet meeting on 15 March 2015 with letters of appointment being signed on 16 June 2015.

In the interim the mining rights holder was being financially prejudiced, because they had been forced to shut down their operation without the chance to engage with the authorities to determine the compliance status (and reasonable actions that might have been needed to ensure full compliance). Their case was based on the logical step of compliance monitoring before the more drastic step of enforcement by shutting down the operation. The matter was heard by the newly reconstituted Water Tribunal, but it was clear that the Chairman was unfamiliar with water resource management principles, and she therefore made specific rulings that effectively reinforced the unilateral decision to shut down the mine.

This meant that the tribunal was forbidden to hear evidence of the lack of cause and effect, because the central allegation was that mining had diverted the flow of the river, which was not supported by any hard evidence. This meant that the logic or legality of compliance monitoring before enforcement through shut-down was not heard. The location and status of gaging stations is shown in Figure 5. There is incomplete information on which management decisions can be made as a direct result of this. The Umvoti River has not had a Comprehensive Reserve Determination Study called for by the National Water Act, so there is no robust scientific data on which to base any management decisions. There is a major El Niño event unfolding, with the catchment of the Hazelmere Dam – the largest water storage infrastructure for the entire KZN North Coast – in the worst drought in 33 years. The dam had insufficient water in it to supply the KZN North Coast until the next rainy season. There was an almost

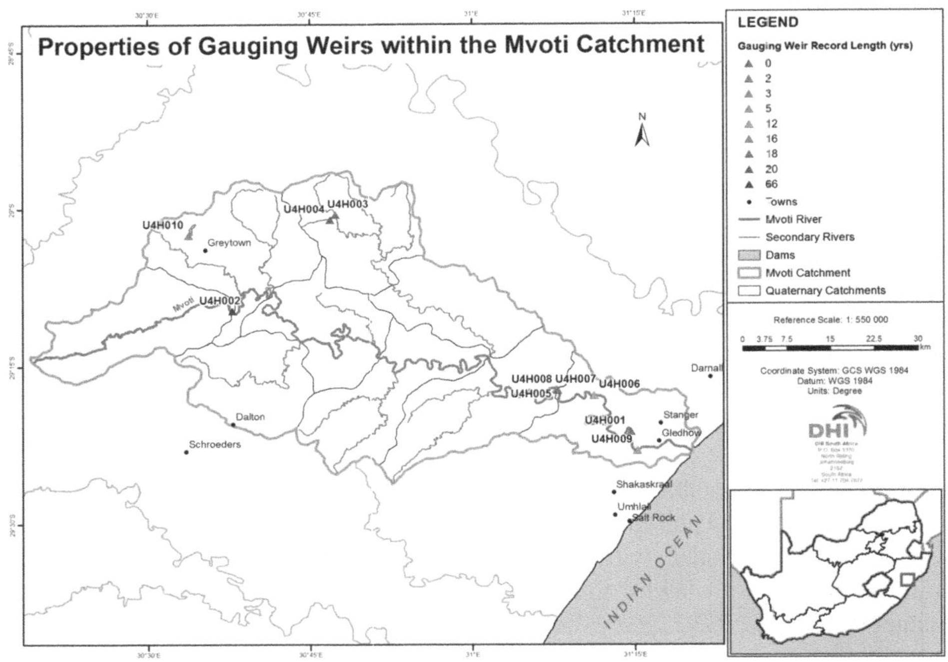

Fig. 5. The Mvoti River Basin north of Durban showing the location and status of gauging weirs (courtesy of the Danish Hydrological Institute).

100% certainty that the water supply infrastructure would fail in the next few months and there was an engineering race underway to build a pipeline from the adjacent oThongathi River as an additional source of supply into Hazelmere Dam (see Figure 6). The oThongathi River was also dry so this pipeline was tenuous at best.

None of this factual evidence was heard by the Water Tribunal, simply because the Chairman ruled such evidence as being inadmissible.

The conclusion regarding Driver 2 (climatic reality) is that the El Niño drought remained invisible to government until it hit. This was felt in the Mvoti River basin when the Umgeni Water potable treatment plant supplying the iLembe District literally ran dry. In the absence of scientifically verifiable data, sand mining operations were wrongly blamed for being the primary diverter of water. The Water Tribunal was dysfunctional, so application was made to the High Court that resulted in an order of court to the Minister of Water to reconstitute that body. This was done, but the Chairman ruled that no evidence could be led on the state of systemic failure of the rainfall and streamflow gauging stations.

Driver 3: transition from an extractive economy

What happened in the epicentre of the El Niño drought event as experienced by the loss of supply for the iLembe District Municipality was not an isolated incident. Its impacts can also be seen in the Vaal River which sustains 60% of the national economy and 45% of the total population of South Africa.

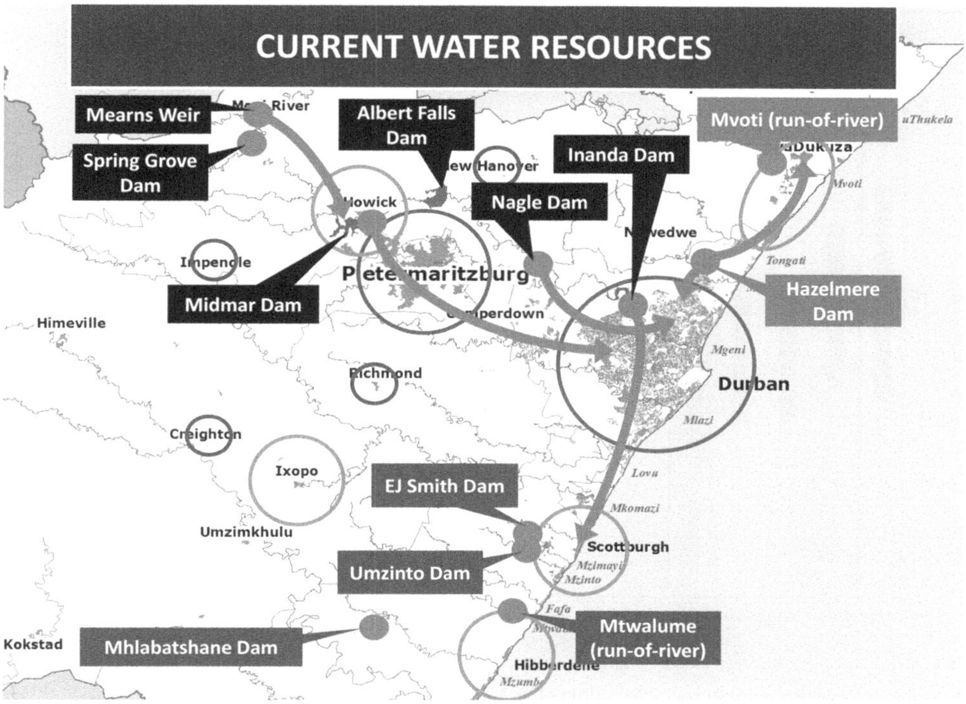

Fig. 6. Water resources for the greater Durban area showing the outlying run-of-river schemes including the Mvoti River site, all of which were distressed (Umgeni Water).

The facts in this matter are grounded on the growing levels of salinity in the overall system as shown in Figure 7. These data represent the situation as it was in 2006, at the end of the Mbeki regime and thus prior to the aggressive phase of patronage characteristic of the Zuma regime. Salinity is generally low along the upper reaches of the Vaal River (gauging stations VS1–7). There is a small spike in salinity at VS6 and a large spike again at VS8, coinciding with the point where surface water return flows come from the Witwatersrand Goldfields (DWA, 2012). From VS8 onwards, salinity remains high until the Vaal enters the Orange River. It must be noted that this dataset was generated before the 2015–16 El Niño event, so total dissolved solids (TDS) levels were lower by virtue of the dilution capacity still in existence at that time.

In response to the increasing levels of salinity in the Vaal, the Inter-Ministerial Task Team proposed the strategic water management scenario shown in Figure 8. System yield is represented on the vertical axis, with time on the horizontal axis. System yield drops dramatically after 2010, resulting in a growing but substantial deficit over time. The demand curve grows in a linear manner, even in response to an aggressive water conservation and demand management regime. The reason for the loss of yield is the need to release fresh water from the Lesotho Highlands Water Project (LHWP) to achieve the resource water quality objectives mandated by law. In effect, economic development is placed at risk because of this dramatic reduction in yield. A white arrow on the horizontal axis coinciding with 2020 represents the date by which Phase 2 of the LHWP needs to be online. It must be noted that even if this target date is met, there will still be a deficit with obvious implications for water security needed to sustain economic development for 60% of the national economy.

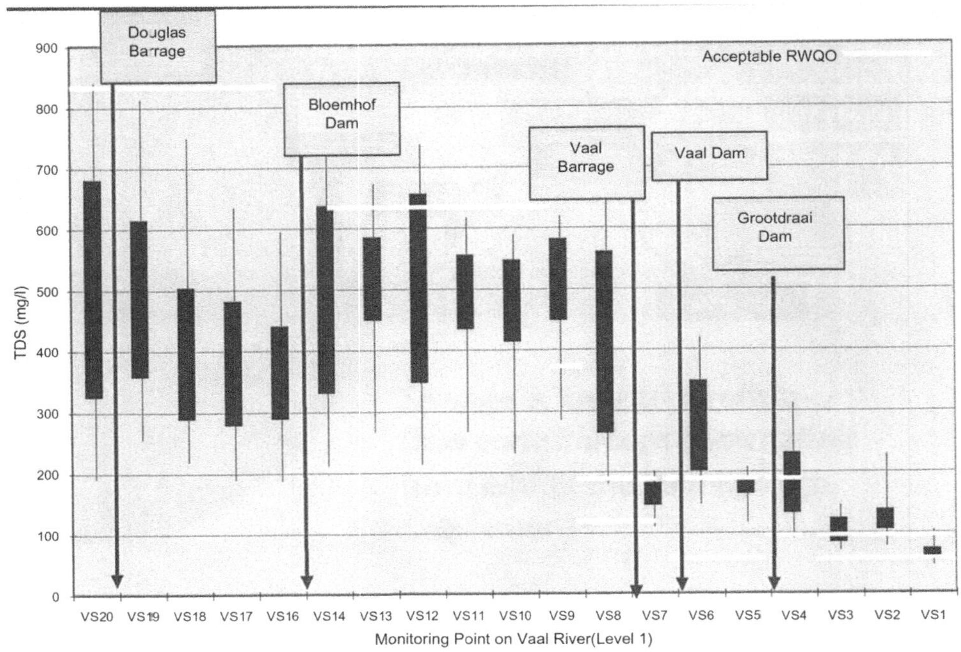

Fig. 7. Salinity values at various gauging stations along the Vaal River expressed as TDS in 2006 (DWA, 2012).

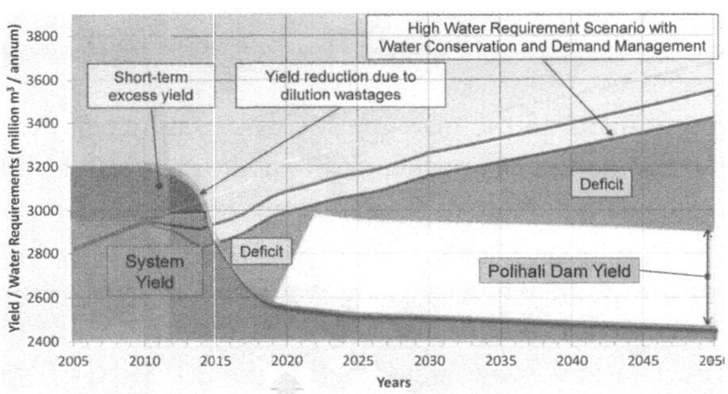

Fig. 8. Vaal River system short-term water balance showing the role played by Phase 2 of the LHWP (Polihali Dam).

Figure 9 shows the long-term water balance of the Vaal River system, with the desalination of acidic mine water having a dramatic effect on the yield. In effect then, the technical members of the Inter-Ministerial Task Team have developed a solution to a pressing problem of national strategic importance. But how important is the mining issue in the context of the national drought?

The importance derives from the fact that the Witwatersrand Goldfields (Davenport, 2013), once the largest contributor of gold to the world (Turton *et al.*, 2006; Hart, 2013), is now in a state of rapid decline (Hartnady, 2009; Turton, 2015), as shown in Figure 10. With a production peak in 1970, there are three distinct sub-cycles evident. The first peak in the 1930s coincided with relatively shallow mining to a depth of 1,000 metres below surface. The second peaked in 1970 and was driven by

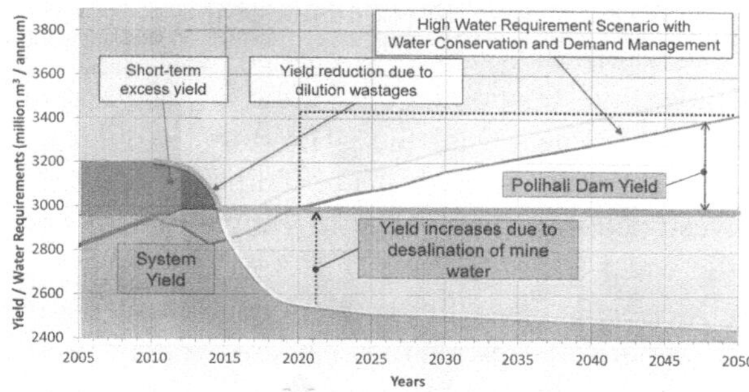

Fig. 9. Vaal River system long-term water balance showing the impact of desalinating mine water (DWA, 2012).

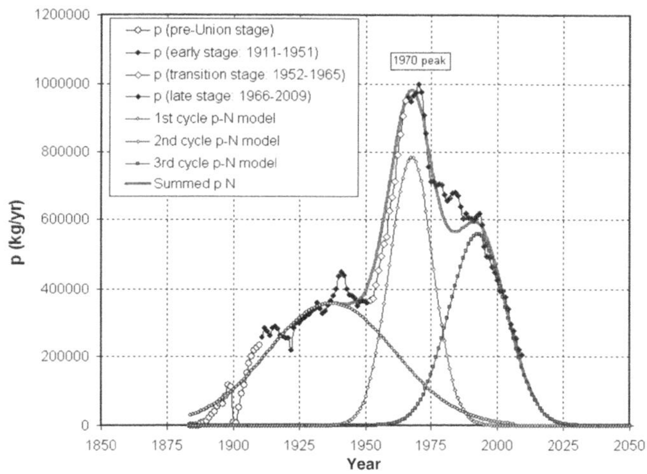

Fig. 10. Production from the Witwatersrand Goldfields showing three distinct sub-cycles with a near linear decline by 2020 (Hartnady, 2009; Turton, 2015).

advances in engineering needed for deep level mining (>4,000 metres below surface). The third peak coincided with the transition to democracy in 1994 and is driven by improvements in metallurgical engineering that enables residual gold found in old tailings dams to be recovered.

While gold mining is in a known state of decline, and the impact of flows out of abandoned workings has been assessed by the Inter-Ministerial Task Team on Acid Mine Drainage to be of major concern, the Ministry of Water did not seem to prioritize this matter.

The current Minister of Water was previously Premier of Gauteng Province. The Gauteng Provincial Legislature commissioned a report on mining residues to assess the risk (Hartnady *et al.,* 2011). This report clearly spelt out the multiple risks arising from the demise of the gold sector, all of which are exacerbated by the absence of a coherent strategy and policy. It could reasonably be expected that the Minister knew about the contents, given the strategic importance in both roles as Premier of Gauteng and subsequently as Minister of Water.

Rand Water Board commissioned a technical study, centred on a high confidence mathematical model, designed specifically to assess risk arising from the flow of acidic mine water (Hartnady *et al.,* 2012). The Minister would be aware of the contents of this report by virtue of the fact that Rand Water Board reports directly to the Ministry.

The Minister of Water is under investigation by the Public Protector regarding an attempted intervention in the procurement process for Phase 2 of the LHWP that would award a lucrative contact to a company owned by a financial backer of President Jacob Zuma (Timse & Ntaote, 2016). This particular company is already under investigation by the Special Investigating Unit (dealing with organized crime) for alleged tender fraud in the water sector.

At the same time, the Minister is also defying the National Treasury regarding the appointment of a controversial Chairman of an about-to-be mega Water board, arising from the merger of Umgeni and Mhlathuze water boards in KwaZulu-Natal (Masondo, 2016). This is significant because it is here that the El Niño drought is at its severest, but this merger has nothing to do with the drought. The Chairman designate is also Chairman of the Zuma Foundation and is also the embattled Chairman of the bankrupt South African Airways (see http://mg.co.za/article/2015-11-27-00-how-myeni-broke-saa). So the water board itself appears publicly to have more to do with the patronage politics which many see so deeply entrenched under the Zuma regime. It also demonstrates a deterioration in the dialogue needed between technical and political in meeting drought uncertainties so necessary to assuring water security.

Implications for economic development and job creation

It is reasonable to conclude from the analysis presented above that the ruling party in South Africa, as it is currently constituted, is less interested in matters such as drought management than in the politics of patronage. The general suspicion of western-based science, ushered in by the Mbeki regime and manifesting as AIDS-denialism and the politicization of DDT for the control of malaria, has resulted in a slow but steady decline in data capture, processing and management systems needed for water resource management. The outcome of the Mbeki-era's short-sighted policy approach driven by science scepticism is now manifesting as a significant loss of human lives, as people living with AIDS were unnecessarily denied access to anti-retroviral medication. Sadly, it is also manifesting as a cohort of children born with urogenital defects (Bornman *et al.,* 2005, 2008, 2009) and men with impaired semen quality (Aneck-Hahn *et al.,* 2007) due to the reintroduction of DDT (Cone, 2009). This science scepticism has merely been accelerated by the Zuma regime.

When the priority is self-enrichment, with collaboration by a select few, then it is easy to understand how water boards and other parastatals become significant vehicles for the dispensing of patronage. In the Vaal River system, a deliberate delay to the implementation of Phase 2 of the LHWP has exposed 60% of the national economy and 45% of the total population to the risk of acute water scarcity exacerbated by the El Niño event.

Indeed, the restructuring of two water boards in the epicentre of the worst drought on record, without any benefit to those affected by the drought, deserves closer scrutiny (Timse & Ntaote, 2016). In strictly institutional terms, the Minister of Water has no reason to defy the National Treasury. However, this action is consistent with the battle currently waging between the President and the Minister of Finance (Whittles & van Wyk, 2016). Unfortunately, this places the Minister of Water primarily supporting the President and his patronage base. However, most see such action as inconsistent with the legal mandate of the Minister of Water as defined in the National Constitution and National Water Act.

The implications are stark. The South African economy has been stagnant for a decade, with high levels of unemployment and the ever present risk of a ratings agency downgrade to junk status. The El Niño drought event has merely exacerbated what is already a national disaster. The actions by the state, as evidenced in the sand mining case along the Mvoti River, show that systemic dysfunction is deeply entrenched, with the Water Tribunal itself having been resuscitated only because of an order of the High Court.

The delay of Phase 2 of the LHWP, in spite of an extremely strong case developed by the competent technocrats who have survived various purges, suggests that water resource management is playing a secondary role to tribal and patronage politics. Yet in spite of this, the drought persists because it takes no sides in any power struggle between political elites.

The only conclusion that can be drawn with any confidence is that South Africa now faces the gravest drought in recorded history. The impact of that drought on the economy, livelihoods and human suffering cannot be assessed at this time. The part played by political decay has yet to be evaluated by scholars.

Policy implications

The most significant policy implication is that measurement systems for all parameters related to water resource management need to be managed and adequately resourced if the impact of drought is to be mitigated. The impact of climate change, most notably manifest as the rising peaks in Figure 4, have strategic implications. We can reasonably anticipate a deep and protracted drought, possibly of decadal duration, across the entire Southern African area. Work done by South African members of the IPCC indicate that while a global average increase in temperature of 2 °C is survivable, the actual case in South Africa is likely to be closer to 5 °C in certain parts of the country (Scholes *et al.*, 2015). The impact of this startling fact on food security needs to be carefully considered (Reeve & Entholzner, 2016).

Conclusion

The El Niño drought of 2015–16 has provided an excellent opportunity to analyse the response by the South African government. A growing body of evidence suggests that the state response has been inadequate. This has been driven by the broader political milieu in which water resources drought policy and decision making is embedded. Science has become politicized, with an unfortunate linkage having been made to colonial rule. This has undermined the scientific community and has meant that evidence-based policy reform has failed. Instrumentation systems have been neglected and allowed to fall into a state of dysfunction, which means that the early onset of the El Niño event in 2014 was not noticed until the people of iLembe suddenly ran out of water, threatening social stability. Nothing that the government can do now will have major impact on capacities to adapt to the drought impacts. Indeed, the main adaptation policies have become battening down the hatches and surviving as best one can.

The big lesson to be learned is that the El Niño severity is likely to increase, so the next event can be better planned for. This will only be possible once the institutions of state are reformed. The prognosis for this is good, because the Constitutional Court and Chapter 9 Institutions like the Public Protector have been tested and found to be robust in spite of the many attempts to circumvent them.

References

Adler, R. A., Claassen, M., Godfrey, L. & Turton, A. R. (2007). Water, mining and waste: an historical and economic perspective on conflict management in South Africa. In: *The Economics of Peace and Security Journal* 2(2), 32–41.

Aneck-Hahn, N. H., Schulenburg, G. W., Bornman, M. S., Farias, P. & de Jager, C. (2007). Impaired semen quality associated with environmental DDT exposure in young Men living in a Malaria Area in the Limpopo Province, South Africa. *Journal of Andrology 28*(3), 423–434.

Blair, D. (2016). Jacob Zuma 'should' Face 783 Criminal Charges, Declares South African Court. In: The Telegraph, 29 April, 2016. Available online at http://www.telegraph.co.uk/news/2016/04/29/south-african-court-clears-way-for-jacob-zuma-to-face-783-crimin/.

Bornman, M. S., Delport, R., Becker, P., Risenga, S. S. & de Jager, C. P. (2005). Urogenital birth defects in neonates from a high-risk malaria area in Limpopo Province, South Africa. *Epidemiology 5*(16), S126–S127.

Bornman, R., de Jager, T., Worku, Z. & Farias, P. (2008). Indoor residual spraying (IRS) with DDT is associated with urogenital birth defects in Newborn Venda boys from South Africa. In: *Epidemiology 19*(6), 167–168. Available online at http://journals.lww.com/epidem/fulltext/2008/11001/Indoor_Residual_Spraying__IRS__with_DDT_is.479.aspx.

Bornman, M. S., de Jager, C. P., Worku, Z., Farias, P. & Reif, S. (2009). DDT and Urogenital Malformations in Newborn Boys in a Malaria Area, in British Journal of Urology (BJU) International.

CCT (2016). Economic Freedom Fighters v Speaker of the National Assembly and Others; Democratic Alliance v Speaker of the National Assembly and Others [2016] ZACC 11. South Africa: Constitutional Court. Available online at http://cdn.24.co.za/files/Cms/General/d/3834/24efe59744c642a1a02360235f4d026b.pdf.

Cone, M. (2009). Should DDT Be Used to Combat Malaria? In: Environmental Health News, 4 May 2009. Available online at http://www.scientificamerican.com/article/ddt-use-to-combat-malaria/.

Constitution (1996). *The Constitution of the Republic of South Africa. Act 108 of 1996*. Government Printer, Pretoria.

Crilly, R. (2016). El Niño sends records tumbling – with the worst weather yet to come. In: The Telegraph. 10 September 2016. Available online at http://www.telegraph.co.uk/news/worldnews/northamerica/usa/12083969/El-Nino-sends-records-tumbling-with-the-worst-weather-yet-to-come.html.

CSIS (2016). *The CSIS Global Food Security Project and the CSIS Africa Program: Sounding the Alarm on Southern Africa's Food Crisis*. Center for Strategic and International Studies, Washington, DC.

Davenport, J. (2013). *Digging Deep: A History of Mining in South Africa (1852–2002)*. Jonathan Ball, Johannesburg.

Diakanyo, S. (2009). The Spy Who Caused all Trouble. In: Mail & Guardian, Thought Leader. Available online at http://thoughtleader.co.za/sentletsediakanyo/2009/08/31/the-spy-who-caused-all-trouble/.

Dintwe, S. (2013). The survival of community policing in a remilitarized police approach: a paradoxical case of South Africa. Chapter 12. In: *Global Community Plocing: Problems and Challenges*. Verma, A., Das. D. K. & Abraham, M. (eds). CRC Press (Taylor & Francis), London.

DWA (2012). *Feasibility Study for a Long-Term Solution to Address the Acid Mine Drainage Associated with the East, Central and West Rand Underground Mining Basins*. Inter-Ministerial Task Team of Acid Mine Drainage, Department of Water Affairs, Pretoria.

February, J. (2016). Why Gordhan? Because he is a bulwark between SA and state capture. And they need him out of the way. In: Daily Maverick, 24 August 2016. Available online at http://www.dailymaverick.co.za/opinionista/2016-08-24-why-gordhan-because-he-is-a-bulwark-between-sa-and-state-capture.-and-they-need-him-out-of-the-way./#.V-ekLMEaJqM.

Feinstein, A. (2007). *After the Party: A Personal and Political Journey Inside the ANC*. Jonathan Ball, Johannesburg.

Findlater, K., Funke, N., Adler, R. & Turton, A. R. (2007). *South Africa's Hydropolitical History: Actors, Actions, Roles and Responsibilities*. Report No. CSIR/NRE/IR/WR/2007/0064/A. Council for Scientific and Industrial Research, Pretoria.

Funke, N., Nortje, K., Findlater, K., Burns, M., Turton, A. R., Weaver, A. & Hattingh, H. (2007). Redressing Inequality: South Africa's New Water Policy. In: *Environment: Science and Policy for Sustainable Development*, April, 2007, pp. 10–23.

Genthe, B. & Steyn, M. (2007). *Parliamentary Briefing Paper on Endocrine Disrupting Chemicals*. CSIR Report No. CSIR/NRE/WR/IR/2008/0079/C. Council for Scientific and Industrial Research (CSIR), Pretoria.

Gevisser, M. (2007). *Thabo Mbeki: The Dream Deferred.* Jonathan Ball, Johannesburg.

Giliomee, H. (1981). Processes in development of the Southern African Frontier. In: *The Frontier in History: North America and Southern Africa Compared.* Lamar, H. & Thompson, L. (eds). Yale University Press, New Haven & London.

Gottschalk, K. (1994). United Democratic Front, 1983–1991: Rise, impact and consequences. In: Liebenberg, I., Lortan, F., Nel, B. & van der Westhuizen, G. (eds). *The Long March: The Story of the Struggle for Liberation in South Africa.* Pretoria: Haum.

Grootes, S. (2011). 'Kill the Boer' Ruling: Malema's Loss is also SA Freedom's Loss. In: Daily Maverick. 13 September 2011. Available online at http://www.dailymaverick.co.za/article/2011-09-13-kill-the-boer-judgement-malemas-loss-is-also-sa-freedoms-loss/#.V9Ufx8EaJqM.

Gutteridge, W. (1984). Mineral Resources and National Security, in Conflict Studies, No. 162; 1-25. Reprinted in Gutteridge, W. (ed.) (1995). South Africa: From Apartheid to National Unity, 1981–1994. Page 59–83. Dartmouth Publishing, Aldershot, Hants & Brookfield, VT.

Hart, M. H. (2013). *Gold: The Race for the World's Most Seductive Metal.* Simon & Schuster, London.

Hartnady, C. J. H. (2009). South Africa's gold production and reserves. In: *South African Journal of Science 105*, 328–329. September/October 2009.

Hartnady, C. J. H., Mlisa, A. & Turton, A. R. (2011). *Report on Mine Residue Areas Reclamation Strategy.* Gauteng Department of Agriculture and Rural Development (GDARD), Johannesburg.

Hartnady, C., Mlisa, A., Turton, A. R., Blake, D., Goyns, A., Simpson, G., von Scherenberg, L., Khudzai, A., Burgher, K. & Seyler, H. (2012). *Research Project to Investigate Acid Water Plumes, Decants and Intersects with Rand Water's Potable Water Pipelines: Phase 1.* Report No. 810-04/01/2012. Rand Water, Johannesburg.

Hosken, G. (2016). MK Vets Force Journalists Back from Luthuli House. In: Times Live. 5 September 2016. Available online at http://www.timeslive.co.za/politics/2016/09/05/MK-vets-force-journalists-back-from-Luthuli-House?platform=hootsuite.

Johns, S. (1989). Volume 1: protest and hope 1882–1934. In: *From Protest to Challenge: A Documentary History of African Politics in South Africa 1882-1964.* Karis, T. & Carter, G. (eds). Hoover Institution Press, Stanford.

Johnson, R. W. (2015). *How Long will South Africa Survive? The Looming Crisis.* Jonathan Ball, Jeppestown.

Johnston, N. & Wolmarans, R. (2008). Xenophobic Violence Grips Johannesburg. Mail & Guardian, 23 May 2008. http://www.mg.co.za/articlePage.aspx?articleid=339509&area=/breaking_news/breaking_news__national/.

Joubert, J. J. (2016). Busisiwe Mkhwebane nominated as Public Protector by Parliament. In: Time Live. 7 September, 2016. Available online at http://www.timeslive.co.za/politics/2016/09/07/Busisiwe-Mkhwebane-nominated-as-Public-Protector-by-Parliament.

Karis, T. & Carter, G. M. (eds) (1972). *From Protest to Challenge: A Documentary History of African Politics in South Africa 1882–1964.* Hoover Institution Press, Stanford.

Liebenberg, I. (1994). Resistance by the SANNC and the ANC, 1912–1960. In: *The Long March: The Story of the Struggle for Liberation in South Africa.* Liebenberg, I., Lortan, F., Nel, B. & van der Westhuizen, G. (eds). HAUM, Pretoria, pp. 8–21.

Mabuza, E. (2016). Maphatsoe Apologizes to Kasrils, Agrees to Pay R500,000 – Money to Go to Fund for Kwezi. In: Times Live, 23 August, 2016. Available online at http://www.timeslive.co.za/politics/2016/08/23/Maphatsoe-apologises-to-Kasrils%E2%80%9A-agrees-to-pay-R500%E2%80%9A000---money-to-go-to-fund-for-Khwezi.

Mandela, N. R. (1994). *The Long Walk to Freedom: The Autobiography of Nelson Mandela.* Little, Brown & Co., London.

Masondo, S. (2016). Nomvula Defies Treasury. In: News 24. 9 September 2016. Available online at http://www.news24.com/SouthAfrica/News/nomvula-defies-treasury-20160910-2.

Mbeki, G. (1984). *South Africa: The Peasants' Revolt.* IDAF, London.

Meissner, R. (2010). *El Niño Southern Oscillation Data Trends over Time.* Council for Scientific and Industrial Research (CSIR), Pretoria.

Meredith, M. (2007). *Diamonds, Gold and War: The Making of South Africa.* Jonathan Ball, Johannesburg.

Middleton, B. J. & Bailey, A. K. (2008). *Water Resources of South Africa, 2005 Study (WR 2005). Water Research Commission Report No. TT 381/08.* Water Research Commission, Pretoria.

Movik, S. (2008). Fluid Rights: South Africa's Water Allocation Reform. Unpublished D.Phil. Thesis. University of Sussex.

Munusamy, R. (2016). Original Sin: The arms deal and South Africa's sullied political story. In: Daily Maverick, 22 April 2016. Available online at http://www.dailymaverick.co.za/article/2016-04-22-original-sin-the-arms-deal-and-south-africas-sullied-political-story/#.V-fZc8EaJqM.

National Environmental Management Act (1998). *The National Environmental Management Act. Act 107 of 1998.* Government Gazette, Pretoria.

National Environmental Management: Air Quality Act (1998). *The National Environmental Management: Air Quality Act.* Act 39 of 2004. Government Gazette, Pretoria.

National Water Act (1998). *The National Water Act.* Act 36 of 1998. Government Gazette, Pretoria.

Natrass, N. (2010). Still crazy after all these years: the challenge of AIDS denialism for science. In: *AIDS and Behaviour 14*(2), 248–251.

Nordling, L. (2008). South Africa Suspends Water Scientist, in Nature News, available online at http://www.nature.com/news/2008/081125/full/news.2008.1255.html.

Oberholster, P. (2008). *Parliamentary Briefing Paper on Cyanobacteria in Water Resources of South Africa. Annexure 'A' of CSIR Report No. CSIR/NRE/WR/IR/2008/0079/C.* Council for Scientific and Industrial Research (CSIR), Pretoria.

Oosthuizen, M., Visser, W., Pooe, K. & Mathu, K. (2015). *The Future of Water Security in SA and SADC.* Gordon Institute of Business Science (University of Pretoria), Johannesburg.

Pitman, W. V. (2011). Overview of water resource assessment in South Africa: current state and future challenges. In: *Water SA 37*(5), Pages 659–664. Water Research Commission 40 Year Celebration Special Edition.

Power, M. (2008). CSIR Suspends Top Water Scientist. The Times. 23 November 2008. Available online at http://www.the-times.co.za/PrintEdition/Insight/Article.aspx?id=890387.

Reed, D. & De Witt, M. (eds) (2003). *Towards A Just South Africa: The Political Economy of Natural Resource Wealth.* World Wildlife Fund Macroeconomics Program Office, Washington, DC.

Reeve, C. & Entholzner, A. (eds) (2016). *Building Climate Resilience through Virtual Water and Nexus Thinking in the Southern African Development Community.* Springer Verlag, Berlin.

SAICE (2008). *Presentation to Parliamentary Portfolio Committee on Water Affairs and Forestry.* Cape Town. 4 June 2008. South African Institute of Civil Engineers (SAICE), Cape Town.

SASS (1996). SASS Newsletter: Special Edition. (Farewell Messages for M. J. M. Louw). South African Secret Service, Pretoria.

Scholes, R. J., Scholes, M. & Lucas, M. (2015). *Climate Change: Briefings from Southern Africa.* Wits University Press, Johannesburg.

Schulze, R. E. (1990). Climate change and hydrological response in Southern Africa: heading towards the future. *South African Journal of Science 86*(7), 373–381.

Simpson, T. (2016). *Umkhonto we Sizwe: The ANC's Armed Struggle.* Penguin Books, Cape Town.

Spitz, R. & Chaskalson, M. (2000). *The Politics of Transition: A Hidden History of South Africa's Negotiated Settlement.* Witwatersrand University Press, Johannesburg.

Steenkamp, W. & Heitman, H. R. (2016). *Mobility Conquers: The Story of 61 Mechanised Battalion Group 1978–2005.* Helion & Co, Solihull.

Times Live (2014). http://www.timeslive.co.za/local/2014/06/08/zuma-s-rape-accuser-s-mother-opens-up-about-fleeing-sa-after-case.

Timse, T. & Ntaote, B. (2016). Nomvula Mokonyane's alleged interference in Lesotho water project cited as causing delays. In: Daily Maverick. 5 August, 2016. Available online at http://www.dailymaverick.co.za/article/2016-08-05-amabhungane-nomvula-mokonyanes-alleged-interference-in-lesotho-water-project-cited-as-causing-delays/#.V9Vpt8EaJqM.

Turton, A. R. (2008a). *Three Strategic Water Quality Challenges that Decision-Makers Need to Know About and How the CSIR Should Respond. CSIR Report No. CSIR/NRE/WR/EXP/2008/0160/A. Keynote address, A Clean South Africa, presented at the CSIR Conference 'Science Real and Relevant', 18 November 2008.* Council for Scientific and Industrial Research (CSIR), Pretoria. Available online at http://www.bosparra.com/wp-content/uploads/2008/11/banned_csir2008.pdf and http://playpen.meraka.csir.co.za/~acdc/education/CSIR%20conference%202008/Proceedings/CPA-0001.pdf.

Turton, A. R. (2008b). *Three Strategic Water Quality Challenges that Parliamentarians Need to Know About. CSIR Report No. CSIR/NRE/WR/IR/2008/0079/C prepared for the October Parliamentary Briefing.* Council for Scientific and Industrial Research (CSIR), Pretoria.

Turton, A. R. (2009). Resource allocation and xenophobic violence. In: *African Environmental and Human Security and AFRICOM in the 21st Century.* Purkitt, H. (ed.). Cambria Press, New York, pp. 111–123.

Turton, A. R. (2010). *Shaking Hands with Billy: The Private Memoirs of Anthony Richard Turton.* Just Done Publications, Durban. http://www.shakinghandswithbilly.com.

Turton, A. R. (2015). Untying the Gordian knot: unintended consequences of environmental and water policy for the gold mining industry in South Africa. In: *Water International 41*(3), 330–350 (Special issue on mining).

Turton, A. R., Schultz, C., Buckle, H., Kgomongoe, M., Malungani, T. & Drackner, M. (2006). Gold, scorched earth and water: the hydropolitics of Johannesburg. In: *Water Resources Development 22*(2), 313–335.

Visser, J., Labadarios, D. & Labuschagne, I. (2004). Alternative nutritional remedies for people living with HIV/AIDS. Food Review, July 2004, pp. 10–12.

Walwyn, D. & Scholes, R. J. (2006). The impact of a mixed income model on the South African CSIR: a recipe for success or disaster? In: *South African Journal of Science 102*, 239–243.

Whittles, G. & van Wyk, P. (2016). The Zuma vs Gordhan showdown: What are the key battles? In: Mail & Guardian, 25 August 2016. Available online at http://mg.co.za/article/2016-08-25-the-presidents-daring-power-play-here-comes-the-zuma-vs-gordhan-showdown.

Drought management policies – from global collaboration to national action

Frederik Pischke[a] and Robert Stefanski[b]

[a]*Corresponding author. Global Water Partnership (GWP), Linnégatan 87D, Stockholm, PO Box 24177, Stockholm 104 51, Sweden. E-mail: frederik.pischke@gwp.org*

[b]*World Meteorological Organization (WMO), 7 bis, avenue de la Paix, Case postale 2300, Genève 2 CH-1211, Switzerland*

Abstract

The article provides an overview of the development of national drought management policies (NDMP). It explores collaborative efforts that were started at the High-level Meeting on National Drought Policy – whose declaration provides the backdrop to this article – and are implemented through the World Meteorological Organization (WMO)/Global Water Partnership (GWP) Integrated Drought Management Programme (IDMP) and related initiatives. Early outputs, e.g., the *National Drought Management Policy Guidelines – A template for action* and the *Handbook on Drought Indices and Indicators*, provide an indication of how expert-reviewed guidance is brought together and applied by drought practitioners for developing NDMP and applying drought indices/indicators. Regional examples from Central and Eastern Europe and the Horn of Africa, including a brief country example from Kenya, highlight how these guidelines and collaborative efforts and outputs are applied. The role of IDMP is to provide a framework and commensurate technical support to countries but the actual development and implementation of national drought plans and policies still needs to be done by governmental ministries and national stakeholders. The article emphasizes how information from different sources is used to support countries to shift from only reacting to droughts when they occur to adopt proactive national drought policies that focus on improved collaboration and the mitigation of drought impacts.

Keywords: Drought; Drought policies; Integrated drought management

Introduction

Droughts have generally been addressed in a reactive manner, only responding after drought impacts have occurred. This reactive or crisis management approach is untimely, poorly coordinated, disintegrated and provides negative incentives for adapting to a changing climate.

Despite recognition to move away from crisis management to risk management, no concerted efforts have been made in the past to initiate a dialogue on the formulation and adoption of national drought

doi: 10.2166/wp.2016.022

policies. In addition, since there are different impacts of drought across economic sectors and society as well as different providers of information and solutions, there must a collaborative effort to effectively manage drought. For example, at country level, the National Meteorological and Hydrological Services (NMHSs) provide drought monitoring and early warning of the weather and climate variables, the water resource management agency provides information on reservoir levels, the Ministry of Agriculture provides information on crop yields and estimates production, the Ministry of the Environment provides data on environmental flows and the Ministries of Planning and Finance are often key actors in the overall development planning. All of these institutions must collaborate to develop a coherent drought management policy for the country.

To address these issues, the High-level Meeting on National Drought Policy (HMNDP) was organized by the World Meteorological Organization (WMO), the Secretariat of the United Nations Convention to Combat Desertification (UNCCD) and the Food and Agriculture Organization (FAO) of the United Nations, in collaboration with a number of UN agencies, international and regional organizations, in Geneva from 11 to 15 March 2013. Four hundred and fourteen participants from 87 countries as well as representatives of international organizations, regional organizations and UN agencies participated in the HMNDP.

The HMNDP encouraged in its Final Declaration (WMO, 2013) all Governments to develop and implement national drought management policies (NDMP) guided by the following principles:

- Develop proactive drought impact mitigation, preventive and planning measures, risk management, fostering of science, appropriate technology and innovation, public outreach and resource management as key elements of effective national drought policy.
- Promote greater collaboration to enhance the quality of local/national/regional/global observation networks and delivery systems.
- Improve public awareness of drought risk and preparedness for drought.
- Consider, where possible within the legal framework of each country, economic instruments and financial strategies, including risk reduction, risk sharing and risk transfer tools in drought management plans.
- Establish emergency relief plans based on sound management of natural resources and self-help at appropriate governance levels.
- Link drought management plans to local/national development policies.

In addition, the Policy Document of the HMNDP (UNCDD, FAO; WMO, 2013) stated the essential elements of a national drought policy, namely:

- Promoting standard approaches to vulnerability and impact assessment
- Implementing effective drought monitoring, early warning and information systems
- Enhancing preparedness and mitigation actions
- Implementing emergency response and relief measures that reinforce national drought management policy goals.

These elements are considered to be the key pillars of national drought management policy. These pillars have been used in many different initiatives including the Integrated Drought Management Programme (IDMP), the UN-Water Initiative on Capacity Development to Support NDMP and the

Windhoek Declaration of the African Drought Conference (UNCCD, 2016). One of the successes of HMNDP is that it has drawn the attention of the international organizations and national governments to focus on proactive policies.

According to UN/ISDR (2007), the main elements for a drought risk reduction framework should comprise: (1) risk identification and early warning, which include drought hazard and vulnerability analysis, monitoring, impact assessment and communication; (2) mitigation and preparedness, referring to the application of effective and affordable practices; (3) political commitment and responsibilities; and (4) awareness and education for a well-informed public and participatory process.

The strong call for a framework in the form of a policy that combines different approaches that have been considered key in moving from a crisis management approach to a risk management approach has led to the launch of the IDMP by WMO and GWP at the HMNDP in March 2013. With the objective to support stakeholders at all levels by providing policy and management guidance and by sharing scientific information, knowledge and best practices for an integrated approach to drought management the IDMP aims:

- To shift the focus from reactive (crisis management) to proactive measures through drought mitigation, vulnerability reduction and preparedness.
- To integrate the vertical planning and decision-making processes at regional, national and community levels into a multi-stakeholder approach including key sectors, especially agriculture and energy.
- To promote the evolution of the drought knowledge base and to establish a mechanism for sharing knowledge and providing services to stakeholders across sectors at all levels.
- To build capacity of various stakeholders at different levels.

This article will give an overview of the guidance for NDMP and approaches that have taken place in Central and Eastern Europe (CEE) and the Horn of Africa (HOA) where the IDMP, its partners and other initiatives have been active.

Early outputs of global collaboration

The strength of the IDMP is to leverage activities of its various partners to determine the status and needs of countries and move forward collectively to support addressing these needs. To date, over 30 organizations have agreed to support and provide input to the goals of the IDMP. The IDMP also uses the network of NMHS and related institutions affiliated with WMO, the United Nations specialized agency for weather, climate and water, and the Regional and Country Water Partnerships of the GWP as the multi-stakeholder platform to bring together actors from government, civil society, the private sector and academia working on water resources management, agriculture and energy. In addition, the IDMP liaises with related initiatives which are not formally part of IDMP but which are contributing to WMO and GWP.

Against the background of the High-Level Meeting on National Drought Policies the IDMP developed guidance for national drought policy development and implementation. Based on one of the tools that has been instrumental for the development of drought preparedness plans in the United States, which comprises a 10-step planning process by Wilhite (1991) and subsequently modified to emphasize drought mitigation in the planning process (Wilhite *et al.*, 2000, 2005), Wilhite developed within the framework of the IDMP an adapted version. These guidelines, the *National Drought Management Policy Guidelines – A template for action* (WMO & GWP, 2014) focus on a national policy context and draw on experiences from different countries. The purpose of these guidelines is to provide countries with a template that they can use and modify for their own purposes. Countries should not blindly use the 10-step process. They should be modified by local experiences. For example, in CEE, they have adapted the guidelines to the context of the EU Water Framework Directive (see section later in this paper) and Mexico has also modified the 10-step process to its national context (WMO & GWP, 2014).

The 10 steps that are meant to provide a framework for the development of a national drought management policy are detailed as follows (WMO & GWP, 2014):

Step 1: Appoint a national drought management policy commission.
Step 2: State or define the goals and objectives of a risk-based national drought management policy.
Step 3: Seek stakeholder participation; define and resolve conflicts between key water use sectors, considering also transboundary implications.
Step 4: Inventory data and financial resources available and identify groups at risk.
Step 5: Prepare/write the key tenets of the national drought management policy and preparedness plans, which would include the following elements: monitoring; early warning and prediction; risk and impact assessment; and mitigation and response.
Step 6: Identify research needs and fill institutional gaps.
Step 7: Integrate science and policy aspects of drought management.
Step 8: Publicize the national drought management policy and preparedness plans and build public awareness and consensus.
Step 9: Develop educational programmes for all age and stakeholder groups.
Step 10: Evaluate and revise national drought management policy and supporting preparedness plans.

Another early accomplishment of the IDMP is the publication of the *Handbook of Drought Indicators and Indices* which will provide options for identifying the severity, location, duration onset and cessation of such conditions (WMO & GWP, 2016). The purpose of this handbook is to provide to drought practitioners some of the most commonly used drought indicators/indices that are being used across drought-prone regions, with the goal of further advancing monitoring, early warning and information delivery systems in support of risk-based drought management policies and preparedness plans. The indicators and indices are categorized by their ease of use and grouped into the following classifications: (a) meteorology, (b) soil moisture, (c) hydrology, (d) remote sensing and (e) composite or modelled. Table 1 provides selected examples of the meteorology drought indicators and indices. Currently, only physical indicators and indices have been reviewed and listed.

Indicators are variables or parameters used to describe drought conditions. Examples include precipitation, temperature, streamflow, soil moisture, groundwater and reservoir levels, and snowpack. Indices are typically computed numerical representations of drought severity, assessed using climate or hydrological inputs including the indicators listed above. They aim to measure the qualitative state of droughts on the landscape for a given time period. Indices can simplify complex relationships and provide useful communication tools for many diverse audiences and users, including the public. Indices are used to provide quantitative assessment of the severity, location, timing and duration of drought events.

Table 1. Selected examples of meteorology drought indicators and indices (WMO & GWP, 2016).

Drought indicator	Ease of use*	Input parameters	Additional information
Deciles	Green	Precipitation	Easy to calculate; examples from Australia are useful
SPI	Green	Precipitation	Highlighted by the WMO as a starting point for meteorological drought monitoring
Palmer Drought Severity Index	Yellow	Precipitation, temperature, available soil water content	Not green due to complexity of calculations and the need for serially complete data
Crop-specific Drought Index	Red	Precipitation, temperature, dewpoint temperature, wind data, solar radiation, crop data, modelled parameters, available soil water content	Quality data of many variables needed, making it challenging to use

*For more details on the 'ease of use' classification and the full breadth of indicators please consult WMO & GWP (2016) directly.

Therefore, drought indicators and indices, in combination with additional information on exposure information and vulnerability assessments, are essential for tracking and anticipating drought-related impacts and outcomes. Indices may also play another critical role, depending on the index, in that they can provide a historical reference for planners or decision-makers. This provides users with a probability of occurrence, or recurrence, of droughts of varying severities.

The handbook is a reference book with details of over 50 drought indices and indicators that includes information on their ease of use, origins, characteristics, input parameters, applications, strengths, weaknesses, resources including access to software code, and references. Selected examples of meteorology drought indicators and indices are given in Table 1.

Information derived from indicators and indices is useful in planning and designing applications (such as risk assessment, drought early warning systems and decision-support tools for managing risks), provided that the climate regime and drought climatology is known for the location.

The handbook is aimed at those persons who want to generate indicators and indices by themselves and for those who simply want to obtain and use products that are generated by other organizations. It is intended for use by general drought practitioners (for example, meteorological/hydrological services and ministries, resource managers and other decision-makers at various levels) and aims to serve as a starting point to these practitioners, by highlighting which indicators/indices are available and being put into practice around the world. Most importantly, the handbook has been designed with the drought risk management processes in mind and thus represents a readily available resource when a drought management policy is developed and implemented. It does not aim to recommend the best indicators or indices since the choice of indicators/indices is based on the specific characteristics of droughts closely associated with the impacts of concern to the stakeholders.

The handbook points out that the impacts of droughts can be as varied as the causes of droughts. Droughts can adversely affect agriculture and food security, hydropower generation, industry, human and animal health, livelihoods, ecology of natural and human-made environments, security and access to education. Such impacts are largely dependent on the socioeconomic contexts in which droughts occur, in terms of who or what are exposed to the droughts and the specific vulnerabilities of the exposed activities or persons. Therefore, the type of impacts that are relevant in a particular drought monitoring and early warning context is a crucial consideration in determining the selection of drought indicators.

One example of using an output from a partner organization is the use of a WMO survey to assess the status of the agricultural meteorological products and services provided by the NMHSs of the world. In the most recent survey (2010–2014), the NMHSs were requested to list their current drought indices in use in their country and whether the country has a national drought policy or plan. While this is not an all-inclusive list, out of the 52 countries that responded, 17 indicated that they have some sort of national drought policy or plan. The results of this survey are only a starting point. The IDMP uses various sources of information such as this survey to provide a status of national drought policies around the world. The work of the IDMP and its partners is then to liaise with these countries to see if these policies or plans are actually implemented. The IDMP then collaborates with its partners to fulfil its objectives.

In order to support countries in developing and implementing drought management actions an Integrated Drought Management Help Desk is in the process of being established. The Help Desk approach draws on the sister programme of the IDMP, the WMO/GWP Associated Programme for Flood Management and the concept of its Help Desk active since 2009 providing support in the implementation of the principles of integrated flood management. The Help Desk currently developed for the IDMP will consist of a 'Find' section – providing existing knowledge resources, an 'Ask' section – offering a point of contact to expertise, and a 'Connect' facility – providing an overview and a connection to ongoing initiatives. The resources that have been developed and the regional initiatives of the IDMP will populate the Help Desk. It should be stressed that these Help Desks are available to any government agency, national institutions or individuals. Requests for assistance can be done via email, correspondence or phone call. Depending on the type of request, the kind of expected assistance would include pointing out relevant resource material from the website, providing detailed information on a procedure and, in some cases, the development of training or a country visit.

Based on the discussion of the IDMP partners at the annual IDMP Advisory and Management Committees meetings, several publications and projects have been approved. Besides the ones noted below, the IDMP partners are working on a framework document on integrated drought management, which will serve as the intellectual framework of the concept of integrated drought management. It will be a synthesis of existing publications around the topic to succinctly define the concept and approaches to integrated drought management with a focus on water management.

Related initiatives

The IDMP liaises with other similar initiatives to assist its work and mandate. A related activity that also originated from the HMNDP was the UN-Water Initiative on Capacity Development to Support NDMP. This was a collaborative initiative of several UN-Water organizations: the WMO, the UNCCD, the FAO of the United Nations, the Convention on Biological Diversity and the UN-Water Decade Programme on Capacity Development (UNW-DPC). Since WMO and other organizations were partners in both initiatives, there were several complementary objectives and goals with the IDMP.

The UN-Water Initiative was initially funded by the UNW-DPC and the other organizations also provided funds to enable the organization of six regional workshops (Eastern European, July 2013, Romania; Latin America and the Caribbean, December 2013, Brazil; Asia-Pacific, May 2014, Viet

Nam; Eastern and Southern Africa, August 2014, Ethiopia; Near East and North Africa, November 2014, Egypt; and West and Central Africa, May 2015, Ghana). These workshops were organized with participants from drought-prone countries with a focus on developing countries and transition economies (UN-Water Decade Programme on Capacity Development, 2015a).

The overarching goals of the UN-Water Initiative were to enhance capacities of key government stakeholders dealing with drought issues in countries and ensure effective coordination within all levels of governments in order to develop more drought-resilient societies by reducing the risk associated with the incidence of drought in the future.

Based on these goals, the key targets of the workshops were as follows:

- Improve the awareness of drought issues and countries' needs to establish strategies for NDMP based on the principles of risk reduction.
- Equip key government stakeholders concerned with drought with tools and strategies to support decision-making and for risk assessments of vulnerable sectors, population groups and regions.
- Advance NDMP by taking into account long-term benefits of risk-based and proactive approaches that address drought and water scarcity problems at large and to move beyond short-term planning, which addresses drought as a crisis.
- Promote collaboration between the various sectors (agriculture, water resources, meteorological/climatological, ecological, urban) at country and regional levels.

Each of the partner organizations were responsible for one of the main thematic sessions across the workshops. While these workshop were not a direct IDMP activity, the 10-step process (WMO & GWP, 2014) described above was used during workshops. The UN-Water initiative has officially ended but the organizations are discussing ways to assist countries individually with developing national drought policies with the Integrated Drought Management Help Desk, described above, as a potential vehicle to deliver sustained support to the national process where needed; and there is currently discussion on how the lessons and experience of the UN-Water Initiative can be further developed by the IDMP.

During the course of these workshops, there were several challenges identified by the various participants. One of the main challenges was the availability of relevant data. Many countries concluded that the availability of data on drought characterization in countries was scarce. Data issues that needed to be resolved included developing a country-level database on past drought incidences and impacts; promoting exchange and integration of data needed for drought monitoring; developing assessment tools and approaches to quantify drought impacts; and increasing density of rain gauges and sensors or stations for drought-related parameters such as stream flow, soil moisture and reservoir levels. Many countries reported that there was no consistent methodology for assessing drought impacts or archiving this information in a database. Other significant challenges included the lack of political will, which can hinder progress on NDMP and the lack of funding which limited developing and implementing national drought policy.

The participants also listed several next steps that will be needed at a national level to successfully implement national drought policies. These steps included: to continue improving human and institutional capacities; to improve the understanding of the economics of drought; raise awareness of the ineffectiveness of the current approach to drought management; and strengthen cooperation at all levels.

Regional and national application of outputs

While it is too early to see results in implementing this approach, there has been progress on several activities: (1) in CEE, there were 20 national consultations in 10 countries over two years based on which guidance has been developed and drought management plans in the regional policy framework elaborated; (2) establishment of several regional projects in Eastern Africa, West Africa, and in South Asia and Central America; and (3) development of the two publications, *National Drought Management Policy Guidelines – A template for action* and the *Handbook on Drought Indices and Indicators*.

Central and Eastern Europe

The longest established regional programme is the IDMP CEE, managed by GWP CEE, bringing over 40 partners from the region together to implement the programme. It has, since 2013, provided practical advice on how droughts can be managed with the goal to increase the capacity and ability of countries in CEE for adaptation to climate variability and change by enhancing resilience to drought. Outputs are a compendium of good practices, support for a drought information exchange platform, demonstration projects testing innovative solutions for better resilience to drought as well as capacity building training and workshops at the national and regional level. The actions of the IDMP CEE also focused on the status and implementation of drought management plans. A first analysis (GWP CEE & Falutova, 2014) showed that the majority of countries had not produced drought management plans and the implementation of key elements of the drought management plans, namely, indicators and threshold establishing different drought stages, measures to be taken in each drought stage and the organizational framework for drought management were lacking.

Based on the realization that drought management plans in the policy framework of CEE were lacking or insufficient and the guidance with the 10-step process for drought management policies produced by WMO & GWP (2014), the IDMP CEE set out to support countries to develop drought management plans. Part of this process included two rounds of consultations in the ten participating countries, Bulgaria, Czech Republic, Hungary, Lithuania, Moldova, Poland, Romania, Slovakia, Slovenia and Ukraine. The consultations involved key actors involved in drought management in each country, including government ministries and competent authorities, hydrometeorological services and universities as well as affected stakeholders, including farmers, energy utilities and fisheries. The first round of consultation focused on analysing the current state of drought policy in the individual countries. Draft guidelines for the preparation of drought management plans within the context of the European Union Water Framework Directive river basin management plans, as the overarching policy context of the region, were developed. The second round of national consultations aimed to gather national experiences and information relevant to drought planning and develop the draft guidelines from the first round of consultations further.

This process of using the input from the global level with the 10-step process and adapting it to the regional context through the 20 national consultations led to the publication of guidelines for the preparation of drought management plans in the context of the EU Water Framework Directive (GWP CEE, 2015) and the definition of seven steps that are specifically tailored for the EU Water Framework Directive, the regional policy context:

1. Develop a drought policy and establish a drought committee
2. Define objectives of drought risk-based management policy

3. Make inventory of data for drought management plan development
4. Produce/update drought management plan
5. Publicize drought management plan for public involvement
6. Develop scientific and research programme
7. Develop educational programmes.

The steps are forming the basis for a third round of consultations currently ongoing in 2016 to develop drought management plans. The publication has also been translated to a majority of the national languages in the countries participating in their development.

It was found that that the adaptation of the step-by-step planning process proposed by the National Drought Management Policy Guidelines (WMO & GWP, 2014) for the EU and/or accession countries in the context of the EU Water Framework Directive by linking them to river basin management plans has synergistic effects in achieving environmental objectives of these regional policy frameworks.

Horn of Africa

The IDMP has established a number of regional initiatives and programmes to support the practical application of its principles at the regional and national level. In 2015, the implementation of the IDMP in the HOA and in West Africa commenced. Both programmes are aiming to close the gap and provide an impetus to existing drought management initiatives in these regions. The GWP Country Water Partnerships will play a crucial role to bring the key actors not only from the water community but also from the agriculture and energy community together.

The Greater HOA is at high risk to extreme climate events such as droughts and floods. This region has developed an innovative way to address these issues at a regional level.

In 1989, 24 countries across Eastern and Southern Africa established a Drought Monitoring Centre with its headquarters in Nairobi. In 2003, the Eastern African regional Intergovernmental Authority on Development (IGAD) adopted this centre as a specialized IGAD institution. In 2007, the name of this institution was changed to the IGAD Climate Prediction and Application Centre (ICPAC) in order to better reflect its mandates, mission and objectives within the IGAD system (ICPAC, 2016; UN-Water Decade Programme on Capacity Development, 2015b).

This centre is responsible for the 11 countries in the Great HOA: Burundi, Djibouti, Eritrea, Ethiopia, Kenya, Rwanda, Somalia, South Sudan, Sudan, Tanzania and Uganda and works closely with the NMHSs of member countries as well as regional and international centres for data and information exchange. Its main objectives are to provide timely early warning information on climate change and to support sector-specific applications for the mitigation of poverty and the management of environment and sustainable development in relation to the impact of climate variability; to improve the technical capacity of producers and users of climatic information; to develop an improved, proactive, timely, broad-based system of information/product dissemination and feedback; and to expand the climate knowledge base and applications within the sub-region in order to facilitate informed decision-making on climate risk-related issues (ICPAC, 2016).

One of the outputs of the IDMP HOA regional project was the publication of the *Assessment of Drought Resilience Frameworks on the Horn of Africa* (GWP Eastern Africa, 2015). This publication provides an overview of drought policy and institutional frameworks for Djibouti, Ethiopia, Kenya, Somalia, South Sudan, Sudan and Uganda.

This assessment noted that after the severe regional drought in 2010/2011, the Summit of Heads of State and Government of the IGAD and the East African Community (EAC) met in Nairobi in September 2011. This summit decided to address the effects of recurring droughts on vulnerable communities in the IGAD region, calling for increased commitment by affected countries and development partners to support investments in sustainable development especially in arid and semi-arid areas. According to the assessment, Ethiopia, Kenya and Uganda have national policies on disaster risk management while South Sudan is in the process of developing similar policies. Djibouti, Somalia, and Sudan either have policies that focus more on emergency responses or they do not yet have disaster risk management policies (see Table 2). The assessment concluded that even the existing policies in many countries of the HOA are not yet comprehensive enough to fully address integrated drought management.

Most countries in the region have a government institution responsible for leading and coordinating the implementation of disaster risk management but the structure of the arrangements varies among the countries. For example, Kenya has established a National Drought Management Authority (NDMA) (GoK, 2012). South Sudan has a Ministry of Environment which includes Humanitarian Affairs and Disaster Management. Ethiopia has established a Disaster Risk Management and Food Security Sector led by Minister of State under the Ministry of Agriculture. Uganda and Somalia have higher levels of coordination under their respective Office of the Prime Minister.

IGAD has developed an IGAD Drought Disaster Resilience and Sustainability Initiative (IDDRSI) framework that member states can use to prevent, mitigate and adapt to the adverse impacts of drought (IGAD, 2013b). The approach developed and recommended by IDDRSI combines relief with development interventions in dealing with drought and related emergencies in the region through the IGAD Secretariat.

The IDDRSI Framework developed the IGAD Regional Programming Paper which is the programmatic and implementation arm of the initiative in order to operationalize drought-related actions at both country and regional levels (IGAD, 2013a). With IDDRSI in place, member states have developed Country Programming Papers (CPPs) which can serve as the planning, coordination and resource mobilization tools for projects and investments required to assist in ending drought emergencies. The CPPs have been able to identify the root causes of vulnerability to drought, areas of intervention and investments and establish adequate national coordinating mechanisms to implement drought resilience programmes (IGAD, 2012). For example, the Ethiopia CCP identified the following challenges: recurrence of drought; population growth versus shrinking resources; low levels of infrastructure; low implementation capacity; violent conflicts and climate change (Government of Ethiopia, 2012).

The IDMP HOA regional project is working closely with IDDRSI to assist Eritrea, Ethiopia, Kenya, Sudan and Uganda. Djibouti, Somalia and South Sudan will be considered for some IDDRSI components of the programme.

While the region has been taking some positive measures towards building drought resilience, the assessment noted the following major gaps that still need to be addressed through promotion of integrated drought management (GWP Eastern Africa, 2015):

(1) Limitations in human and institutional capacities that are needed to coordinate and implement drought risk management and resilience building initiatives.
(2) Inadequate policy and legislative frameworks for disaster risk management and particularly on drought risk management.
(3) Lack of information on water and other natural resources in the arid and semi-arid areas of the countries.

Table 2. Policies, strategies/plans related to drought management in the HOA countries (GWP Eastern Africa, 2015).

Countries	National policies, strategies, plans related to drought management	Major priority areas identified
Djibouti	• National Programme for Food Security • The Public Investment Plan (PIP)	• Investments in food security in the country • Around 15% of the PIP in 2013 was allocated to interventions focused on the priority areas of the CPP, including in areas of environment
	• National Plan for Climate Change Adaptation	
Ethiopia	• National Policy on Disaster Risk Management • Strategic Program and Investment Framework	• Voluntary settlement • Provision of suitable social services and expansion of infrastructure
	• Pastoral Development Policy and Strategy Framework	• Emergency response and coordination
	• Early Warning and Emergency Coordination Centre (under formation)	• Early warning systems
	• Climate Resilient Green Economy • Strategy and Framework (2011)	• Climate resilience building
Kenya	• Disaster management policy and strategy • Vision 2030 has mainstreamed DRM in all its key pillars • The Second Medium Term Plan (2013–2017) • National Climate Change Response Strategy (2010) and Action Plan (2013)	• Disaster risk management • Climate change adaptation
Somalia	• Following the establishment of the Federal Government of Somalia in August 2012 – a fresh momentum and steady progress being made on the coordination for drought intervention and livelihood programmes in Somalia	• Applying integrated water resource management
		• Promoting alternative renewable sources of energy • Develop capacity for geo-information management and services for early warning of hazards, environmental impact assessments and monitoring of water resources
South Sudan	• National Disaster Management Policy (under development) • Five Years Strategic Plan (2013–2018) on Disaster Management	• Disaster risk management
Sudan	• There are a number of sectoral policies, laws, strategies and programmes related to drought but they are yet to be properly coordinated	• The understanding and coordinated implementation of IDDRSI at country and regional level is a major priority based on the national drought management policy
Uganda	• The Constitution 1995 provides for drought risk management	• Disaster risk management
	• The National Climate Change Policy, 2013; Disaster Management and Preparedness Policy, 2010 • The Rangeland Management Policy, 2001 • National Water Policy, 1999	• Climate change adaptation

(4) Weak market, communication and transport infrastructure in the drought vulnerable areas.
(5) Weak early warning systems to inform vulnerable communities on weather and disasters and alerts for effective preparedness and response.
(6) Low level of educational information coupled with a strong adherence to traditional ways of keeping large herds of livestock by pastoralist communities.
(7) Limitation of resources to finance drought risk management and resilience building initiatives.
(8) Inadequate participatory infrastructure in drought management programmes.
(9) Continued reactive crisis management approach to drought management including an over-reliance on relief aid.

Based on these gaps, the IDMP HOA regional project has aligned various interventions to address the gaps based on prioritization by the countries in the region. The IDMP HOA is implementing a capacity development programme. The support is tailored for countries to develop and revise their drought plans and develop practices that increase drought resilience.

These gaps highlight the main challenge in the region which is to reverse the growing human vulnerability against environmental hazards, such as droughts and man-made disturbances such as conflicts and economic crises (GWP Eastern Africa, 2015). Other challenges include the threat to the pastoral and agricultural production systems due to rapid population growth, migration, environmental degradation, land reallocation, fragmentation of rangelands, decreasing mobility for herds and growing competition in using scarce pasture and water resources. Since land access and water rights are not sufficiently regulated, conflicts can arise between different competing stakeholders, especially among cross-border communities. The combination of these adverse factors is accelerating environmental degradation and therefore exacerbating the vulnerability of societies. When droughts occur, the whole agro-pastoral production system can collapse with disastrous consequences for the affected populations. Large financial resources are then needed for humanitarian aid and even more to recover the production systems and livelihoods of the drought affected communities.

In the past, efforts were more concentrated in managing the drought disaster and related humanitarian emergencies. The new approach will focus on the underlying causes of the need for humanitarian aid and will approach disaster management through pro-active and preventive solutions.

The following opportunities and steps taken were identified for the promotion and implementation of an integrated approach to drought management in the HOA (GWP Eastern Africa, 2015):

- Existence of IGAD to establish regional and international mechanisms for cooperation to address regional drought issues.
- Existence of the IDDRSI framework including the adopted CPPs for drought resilience and sustainable development.
- Availability of political will and commitment to drought risk reduction by governments in the region.
- Existing national implementing and coordination structures and institutions for drought management.
- Availability of relevant national policies, strategies, and initiatives on drought management in the HOA countries.
- Availability of institutions with experience and well-developed frameworks in implementing programmes and projects which can provide examples of good practices.
- Interest of development partners, IGAD member countries and the private sector to support national and regional initiatives to enhance drought resilience.

The IDMP HOA assessment of drought resilience status shows that there is commitment by governments to combat drought for sustainable national growth and development. The IGAD initiative, IDDRSI, to strengthen food security and drought resilience in the region, has also resulted in national programmes with specific institutional frameworks for their implementation. In addition, development partners are ready to support actions that are aimed at strengthening drought resilience instead of reactive emergency and relief operations. The assessment shows that the IDMP HOA project can support the various drought resilience related initiatives and can contribute to the identified priorities in the Regional and CPPs.

Based on the assessments, the following priority areas were recommended for building drought resilience in the HOA region:

1. Demonstrating innovative drought resilience cases
2. Capacity development for drought management and resilience building
3. Promoting partnership for integrated drought management
4. Facilitating regional cooperation/collaboration for drought management in the HOA region
5. Facilitating policy development for integrated drought management
6. Mainstreaming drought mitigation and adaptation strategies in relevant government sector ministries and agencies
7. Strengthening early warning systems.

Innovative drought resilience case studies would include using lessons learned and best practices that use an approach of integrated water resources management from countries.

Kenya

Given the above summary on the HOA, we can show a brief example of Kenya in order to illustrate the issues at country level. In the IDMP HOA assessment report, there were individual country summaries. The following is a summary of the Kenya report in GWP Eastern Africa (2015) and additional information is provided from the Kenya country reports from the Eastern and Southern Africa Regional Workshop held from 5 to 8 August 2014 (UN-Water Decade Programme on Capacity Development, 2015b).

The main impacts of drought in Kenya in the various sectors are the following: agriculture: crop failure, reduced animal/fish production, death of animals; tourism: decline of wildlife; energy: reduced hydroelectric power generation and the increased cost of power generation; and water: reduced water volumes in water bodies and groundwater table (GWP Eastern Africa, 2015). Kenya has faced several droughts in the past 20 years with the main occurrences in 1983/84, 1991/92, 1995/96, 1999/2001, 2004/2005, 2009 and 2011 (McSweeney *et al.*, 2007; Bassi *et al.*, 2011).

The institutions within the Government of Kenya that are involved with drought issues are the NDMA in the Office of the President which coordinates all matters related to drought management in Kenya. There is also the Kenya Food Security Meeting and National Disaster Operations Centre (drought monitoring and early warning) and the Kenya Meteorological Department with structures at the national and local levels (GWP Eastern Africa, 2015; UN-Water Decade Programme on Capacity Development, 2015b).

The Government of Kenya has initiated a Draft National Disaster Management Policy and National Climate Change Response Strategy (GWP Eastern Africa, 2015) and has also created the Kenya Vision 2030, which is the national long-term development policy that aims to transform Kenya into a newly industrializing, middle-income country providing a high quality of life to all its citizens by 2030 in a clean and secure environment (Kenya Vision 2030, 2008). Within the Kenya Vision 2030, drought and climate change are widely addressed under risk management. Within this Vision 2030, the following main types of emergency interventions are provided: food relief for affected people, with special food formulas provided for the most affected such as children, the elderly and mothers; human disease control and treatment; animal feed and supplements; water for humans and livestock; cash transfers; food/cash for work/assets; livestock disease control; shelter; debt relief; destocking; restocking; distribution of seeds; supplementary feeding for livestock, especially breeding stock; rehabilitation of water points; and agricultural credit (UN-Water Decade Programme on Capacity Development, 2015b).

During drought emergencies, rapid response teams from the government are activated to implement pre-planned interventions. Decisions on actions to be taken are recommended by the Kenya Food Security Meeting and the Kenya Food Security Steering Group based on information gathered regularly by multidisciplinary teams and guided by internationally set common principles and universal minimum standards for the delivery of high quality humanitarian responses (UN-Water Decade Programme on Capacity Development, 2015b).

Other Government of Kenya Initiatives in drought management include: the Kenya Drylands Livestock Development Programme (2010–2013); the National Agricultural and Livestock Extension Programme; Enhancement of Food Security through Water Harvesting Project; the Kenya Agricultural Productivity Project; and the Kenya Adaptation to Climate Change in Arid and Semi-Arid Lands (GWPEA, 2015).

The IDMP HOA regional project, based on its own work and that of other institutions, jointly assessed with the main government and civil society actors of all countries of the HOA the national drought resilience frameworks (GWPEA, 2015). For Kenya, the shared recommendations for future work include the following:

- Demonstrating innovative drought resilience case studies to learn lessons and best practices.
- Strengthening institutional and community capacity in drought management.
- Partnership building for integrated drought management with already existing institutions and programmes.
- Supporting the timely delivery of climate information services to users.

This summary provides some insight into what countries have developed but it should be noted that the IDMP and other partners are available to assist countries to maintain and further develop these programmes.

Other regions

The IDMP West Africa focuses on the targeted localities of Burkina Faso, Niger and Mali and then shares lessons learned with other neighbouring countries through the Country Water Partnerships. In South Asia, the IDMP is collaborating with the International Water Management Institute (IWMI) and GWP South Asia in developing a South Asian Drought Monitoring System. In Central America, based on a shared understanding with the Central American Committee of Hydraulic Resources (in Spanish: *Comité Regional de Recursos Hidráulicos* (*CRRH*), training on the standardized precipitation index (SPI) has been carried out by GWP Central America and a study on the socioeconomic impacts of the 2014 drought in the region has been published (GWP Central America, 2016).

In Mexico, due to recurrent droughts from 2010 to 2013, the President of Mexico announced the National Programme Against Drought (Spanish acronym PRONACOSE) in January 2013, which is being coordinated by the National Water Commission (CONAGUA). The objective of PRONACOSE is to develop tools based on a proactive and preventive approach for integrated drought management at the level of the basin council. The PRONACOSE framework incorporates an evaluation mechanism to assess the effectiveness of each implemented activity/strategy and ensure sustainability by including continuous feedback and lessons learned in the various implementation phases. The PRONACOSE activities are structured with the following themes:

- Formulation and implementation of measures to prevent and mitigate drought impacts, including monitoring and early warning.
- Establishment of a legal framework to ensure continuous drinking water supplies during droughts.
- Coordination of institutional response towards drought mitigation measures.

In the beginning, CONAGUA developed 26 programmes on drought prevention and mitigation measures (PMPMS) for each basin council. These programmes addressed the drought characteristics and vulnerability of each basin. A guide was also developed and CONAGUA staff and researchers from 12 national institutions were trained to standardize activities and contents of the PMPMS. The PRONACOSE work plan calls for programme implementation during the second and third years, evaluation in the fourth and fifth, and improvements and further implementations again in the sixth year. The overall aim is to ensure ownership of the programmes by the basin councils and a continued gradual implementation beyond the sixth year. The IDMP has been providing technical support and guidance to PRONACOSE (WMO & GWP, 2014).

Conclusions

The High-Level Meeting on National Drought Policies in 2013 has put the spotlight on what needs to be done and has provided several key pillars essential for developing national drought policies. It has resulted in a concerted effort through various initiatives to move forward. However, much remains to be done in terms of developing fully operational drought management policies that fulfil the multiple objectives of being proactive, mitigating impacts, improving public awareness, employing appropriate economic instruments and linked to local and national development frameworks. To support countries in putting these recommendations into practice, the IDMP uses information from its partners to provide a status of national drought policies and activities from around the world, working in partnership to close the gaps to move from crisis to risk management and foster horizontal integration – among different sectors and actors – and vertical exchange – learning from the local through to the global and vice versa. The IDMP does not try to coordinate all drought activities around the world but rather aims to synthesize and apply existing knowledge and approaches to integrated drought management in collaboration with many international and regional organizations. IDMP and its many partners provide a framework and commensurate technical support to countries but the actual development and implementation of national drought plans and policies needs to be done by the governmental ministries and national stakeholders. The IDMP has already used these partnerships to create several early outputs in assisting countries to develop more proactive drought policies and plans with practical guidance and applications tailored to regional and national circumstances.

References

Bassi, A. M., Deenapanray, P. N. K. & Tan, Z. (2011). *Strengthening Institutional Capacity for Integrated Climate Change Adaptation and Comprehensive National Development Planning in Kenya.* Final Report. Millennium Institute, Washington, DC, USA.

Government of Ethiopia (2012). *Ethiopia Country Programming Paper to End Drought Emergencies in the Horn of Africa.* Ministry of Agriculture, Addis Ababa, Ethiopia, November 2012. Available at: http://resilience.igad.int/attachments/article/243/CPP%20Ethiopia.pdf.

Government of Kenya (GoK) (2012). *Programming Framework to End Drought Emergencies in the Horn of Africa – Ending Drought Emergencies in Kenya: Country Programme Paper.* Ministry of State for Development of Northern Kenya and other Arid Lands, Government of Kenya, Nairobi, Kenya.

GWP CEE (2015). *Guidelines for the Preparation of Drought Management Plans.* Development and implementation in the context of the EU Water Framework Directive, Global Water Partnership Central and Eastern Europe, p. 48.

GWP CEE & Falutova, E. (2014). *Report on Review of the Current Status of Implementation of the Drought Management Plans and Measures.* Global Water Partnership Central and Eastern Europe.

GWP Central America (2016). Socio-economic Analysis of the Sectoral Impacts of the 2014 Drought in Central America.

GWP Eastern Africa (2015). *Assessment of Drought Resilience Frameworks in the Horn of Africa.* Integrated Drought Management Program in the Horn of Africa (IDMP HOA), Global Water Partnership Eastern Africa, Entebbe, Uganda.

ICPAC (2016). *Website of the IGAD Climate Prediction & Application Centre (ICPAC).* Available at: http://www.icpac.net/?page_id=4.

IGAD (2012). *IDDRSI Country Programming Papers.* Available at: http://resilience.igad.int/index.php/programs-projects/national.

IGAD (2013a). *The IGAD Drought Disaster Resilience and Sustainability Initiative (IDDRSI). Regional Programming Paper. Final Draft.* Intergovernmental Authority on Development (IGAD), Djibouti.

IGAD (2013b). *Drought Disaster Resilience and Sustainability Initiative (IDDRSI): The IDDRSI Strategy.* Intergovernmental Authority on Development (IGAD), Djibouti.

Kenya Vision 2030 (2008). *Website of Kenya Vision 2030.* Available at: http://www.vision2030.go.ke/.

McSweeney, C., New, M. & Lizcano, G. (2007). *UNDP Climate Change Country Profiles: Kenya.* University of Oxford, Oxford, UK.

UNCCD (2016). *Windhoek Declaration For Enhancing Resilience to Drought in Africa.* Boon, Germany. Available at: http://www.unccd.int/Documents/Windhoek%20Declaration%20Final%20Adopted%20by%20the%20ADC%20of%2015-19%20August%202016.pdf.

UNCDD, FAO, WMO (2013). *High Level Meeting on National Drought Policy (HMNDP) Policy Document: National Drought Management Policy.* Available at: http://www.droughtmanagement.info/literature/WMO_HMNDP_policy_document_2012.pdf.

UN/ISDR (2007). *Drought Risk Reduction Framework and Practices: Contributing to the Implementation of the Hyogo Framework for Action.* United Nations Secretariat of the International Strategy for Disaster Reduction, Geneva, Switzerland.

UN-Water Decade Programme on Capacity Development (2015a). Capacity Development to Support National Drought Management Policies, 11 pp. Available at: http://www.ais.unwater.org/ais/pluginfile.php/516/course/section/168/NDMP-Synthesis.pdf.

UN-Water Decade Programme on Capacity Development (2015b). Proceedings of the Regional Workshops on Capacity Development to Support National Drought Management Policies for Eastern and Southern Africa and the Near East and North Africa Regions, 155 pp. Available at: http://www.ais.unwater.org/ais/pluginfile.php/516/course/section/168/proceedings-no-14_WEB.pdf.

Wilhite, D. A. (1991). Drought planning: a process for state government. *Water Resources Bulletin 27*, 29–38.

Wilhite, D. A., Hayes, M. J., Knutson, C. & Smith, K. H. (2000). Planning for drought: moving from crisis to risk management. *Journal of the American Water Resources Association 36*, 697–710.

Wilhite, D. A., Hayes, M. J. & Knutson, C. L. (2005). Drought preparedness planning: building institutional capacity. In: *Drought and Water Crises: Science, Technology, and Management Issues.* Wilhite, D. A. (ed.). CRC Press, Boca Raton, Florida, USA, pp. 93–136.

WMO (2013). *High Level Meeting on National Drought Policy (HMNDP) Final Declaration.* World Meteorological Organization, Geneva, Switzerland.

WMO & GWP (2014) *National Drought Management Policy Guidelines; A Template for Action* (D. Wilhite). Integrated Drought Management Programme (IDMP) Tools and Guidelines Series 1. World Meteorological Organization, Geneva, Switzerland and Global Water Partnership, Stockholm, Sweden.

WMO & GWP (2016). *Handbook of Drought Indicators and Indices. Integrated Drought Management Programme* (IDMP) (M. Svoboda and B. A. Fuchs). Integrated Drought Management Tools and Guidelines Series 2. World Meteorological Organization, Geneva, Switzerland and Global Water Partnership, Stockholm, Sweden.

Strengthening drought risk management and policy: UNESCO International Hydrological Programme's case studies from Africa and Latin America and the Caribbean

Koen Verbist, Abou Amani*, Anil Mishra and Blanca Jiménez Cisneros

International Hydrological Programme, UNESCO, 7 place de Fontenoy, F-75352 Paris 07 SP, France

**Corresponding author. E-mail: a.amani@unesco.org*

Abstract

Droughts have resulted in significant socio-economic impacts in the regions of Africa and Latin America and the Caribbean (LAC), especially in developing countries. The main gaps to mitigate its effects, identified in both Africa and LAC regions, include a lack of human and institutional capacity, a lack of access to relevant early warning information for decision-making, the identification of vulnerable communities within the countries and the integration of these two components into drought management policies. UNESCO International Hydrological Programme (UNESCO-IHP) has been providing support to enhance human capacity, policy guidance and tools to the countries to address drought-related challenges and this paper presents some examples. Through capacity building at regional institutions in Western, Eastern and Southern Africa, drought monitoring and early warning tools have been transferred and validated for inclusion into national climate risk management plans. In LAC, a drought atlas was produced to identify the frequency of meteorological droughts and the exposure of population to droughts. Also in LAC, national drought observatories were developed in two pilot countries, providing locally relevant and actionable drought monitoring and early warning information, socio-economic vulnerabilities and appropriate drought indicators for decision-making to strengthen current drought policies for these countries.

Keywords: Africa; Capacity building; Decision-making; Disaster; Drought; Early warning; Latin America; Risk management; Tools; UNESCO

Introduction

Climate change is likely to increase the frequency and magnitude of meteorological droughts (less precipitation) and agricultural droughts (less soil moisture) in presently dry regions by the end of the 21st century under the IPCC RCP 8.5 scenario (IPCC, 2014). This is also likely to increase the occurrences

doi: 10.2166/wp.2016.223

of short hydrological droughts (less surface and groundwater) (Lyon & DeWitt, 2012; IPCC, 2014; Masih *et al.*, 2014). Furthermore, the IPCC Fifth Assessment Report (IPCC, 2014) also emphasizes that impacts from recent climate-related extremes, such as heat waves, and droughts, reveal significant vulnerability and exposure of some ecosystems and many human systems to current climate variability.

Droughts are even more damaging in developing countries, where drought mortality is concentrated (UN, 2008). Absolute economic losses are largest in developed regions but in relative terms, developing countries are those suffering the biggest economic losses (UNISDR & CRED, 2015). In a number of countries, drought wiped out significantly more than 5% of the previous year's GDP (UN, 2008). In countries that rely heavily on agriculture, the impact of droughts on crop production can be disastrous for the economy as well as to low-income populations, often inducing famines in vulnerable communities (Guha-Sapir *et al.*, 2004).

Recent advancement in hydrological modelling, observation technologies, including from remote sensing, allows countries to develop or strengthen appropriate drought policies. Monitoring drought development and providing timely seasonal forecasts are essential for drought risk reduction and to address climate variability (Amissah-Arthur, 2003; Tarhule & Lamb, 2003; Hayes *et al.*, 2004; Hansen *et al.*, 2011; Sheffield *et al.*, 2014). Currently, there are a number of different global and regional drought monitoring systems developed by different institutions, such as the Global Drought Observatory (Vogt *et al.*, 2016), the African Drought Observatory (Sheffield *et al.*, 2014), the European Drought Observatory (Vogt *et al.*, 2011), the South Asian Drought Monitoring System (IWMI, 2015), the North America Drought Monitor (Svoboda *et al.*, 2002) and the SPEI Global Drought Monitor (Vicente-Serrano *et al.*, 2010), among others. However, integration of these tools into drought management policies, especially in countries from the developing world, still lags behind (e.g., Patt *et al.*, 2007). Additionally, these monitoring systems often focus on one aspect of drought, while drought management needs to address all affected sectors. To differentiate the types of drought impacts, they are often classified into meteorological, agricultural, hydrological and socio-economic droughts (UNISDR, 2007; Wilhite *et al.*, 2014). The challenge remains to address all these drought types simultaneously in a comprehensive drought management strategy.

Since 2013, and especially after the High-level Meeting on National Drought Policy (Sivakumar *et al.*, 2014), a greater awareness has been internationally raised to identify the gaps and needs at the national level regarding integrated drought risk management (IDRM). IDRM has been promoted as a response to move from crisis to risk management (Wilhite *et al.*, 2000, 2014; UNISDR, 2007), and aims at integrating the different components that are essential for an effective drought management. Baethgen (2010) identified four pillars to manage drought risks, these being: (a) identification of vulnerabilities and potential opportunities; (b) quantification of uncertainties in climate information; (c) identification of technologies and practices; and finally (d) identification of interventions, institutional arrangements and best practices. This IDRM approach is also comprehensively presented in the three pillars of IDRM (Figure 1). The first component of the IDRM focuses on monitoring and early warning capacities as the foundation of the drought plan. The second is linked with vulnerability/resilience and impact assessment in order to identify those areas and livelihoods which are most prone to the impact of droughts. Finally, the third component, mitigation and response planning, provides preparation and mitigation actions triggered by monitoring and early warning, prioritizing the most vulnerable areas.

Shifting from a crisis management approach to a proactive and more integrated risk-based approach is crucial in lowering the loss and damage resulting from droughts (Grobicki *et al.*, 2015). This requires the development of human and institutional capacities in the countries and design and implementation of

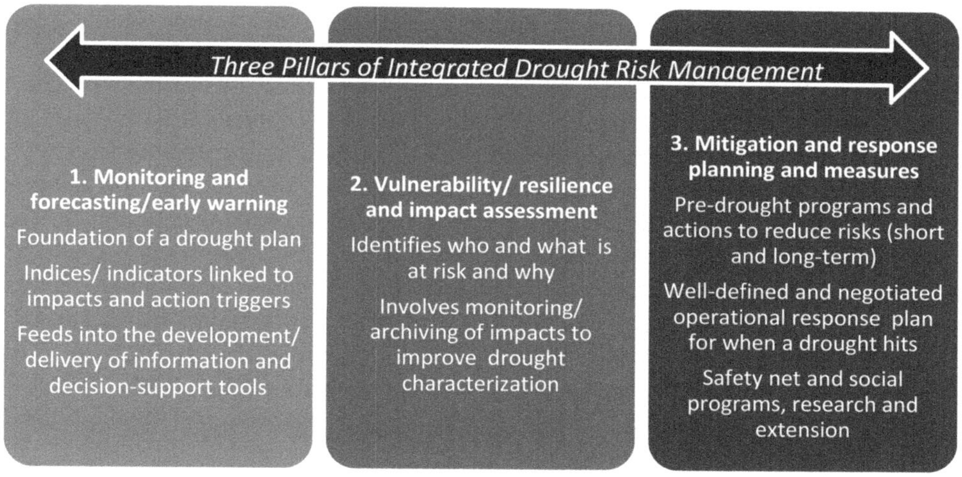

Fig. 1. The three pillars of IDRM (adapted from The World Bank Group, based on Gutierrez *et al.* (2014)).

adequate national drought policies. Moreover, because droughts do not respect boundaries and as they are a cause for migration (European Commission, 2015), international scientific cooperation is essential to achieve implementation of effective drought management policies.

In the case of Africa, historical evidence and general circulation models clearly show that the continent is experiencing increased droughts and aridity (Masih *et al.*, 2014). The projected combined impacts of climate change and population growth also suggest an alarming increase in water scarcity for many countries, with 22 of the 28 countries considered likely to face water scarcity or water stress by 2025 (Cooper *et al.*, 2008). The slow progress of drought risk management and the lack of adequate drought policies, along with increasing population and land degradation pose challenges to address the vulnerability of the region to drought impacts. Therefore, there is an urgent need in Africa to address human and institutional capacities to equip countries with means to shift towards a proactive drought risk management. In the Latin America and the Caribbean (LAC) region, there is a comparatively larger human capacity to develop effective drought monitoring and management solutions, but adequate tools and institutional capacity are still lacking, hindering the implementation of effective drought management and response systems.

This paper summarizes the current situation of drought management and policies in Africa and LAC, with emphasis on specific needs and gaps in both regions. It also focuses on UNESCO-IHP's contribution in strengthening the countries' capacities, and the specific drought management tools developed to support them, considering the different types of drought.

The UNESCO-IHP approach for drought management

The UNESCO International Hydrological Programme (UNESCO-IHP) is the only intergovernmental programme of the United Nations system devoted to the scientific, educational and capacity building aspects of hydrology. The UNESCO-IHP, through its Global Network on Water and Development Information in Arid Lands (G-WADI) and International Drought Initiative (IDI), supports the countries in identifying and addressing the drought management gaps and needs by strengthening global, regional and local capacities to manage water resources and by providing access to data and policy recommendations for a more integrated drought management. UNESCO also promotes international scientific cooperation to facilitate knowledge sharing and transfer for the optimal use of the available resources.

UNESCO-IHP collaborated with Princeton University in the development of an experimental drought monitoring and forecast system for Africa (Sheffield *et al.*, 2014) and more recently for LAC, accessible at http://stream.princeton.edu/. The system tracks drought conditions in near real-time using remote sensing precipitation and atmospheric analysis data. The historic and real-time data are calculated using the Variable Infiltration Capacity (VIC) land surface hydrological model (Liang *et al.*, 1994). This system allows monitoring of meteorological, hydrological and agricultural droughts in developing regions where institutional capacity is generally lacking and access to information and technology prevents the development of systems locally. It has the advantage of providing a standardized format for any of the components of the water balance, providing a comprehensive analysis for any point location within the Monitor's domain (currently covering Africa, LAC and the United States), while providing an overview of the regional, transboundary extent of drought hazards.

In a similar vein, UNESCO-IHP has collaborated with the Center for Hydrometeorology and Remote Sensing (CHRS), University of California, Irvine, on the development of tools to provide near real-time global satellite precipitation estimates at high spatial and temporal resolutions, including the Precipitation Estimation from Remotely Sensed Information using Artificial Neural Networks-Cloud Classification System (PERSIANN-CCS) (Hsu *et al.*, 2010). These tools include the G-WADI GeoServer (http://hydis.eng.uci.edu/gwadi/), that provides real-time precipitation estimates for water resources managers. By providing updated precipitation observations at the global level, the GeoServer is a direct support to meteorological drought monitoring and early warning worldwide, and a contribution to the Global Framework for Climate Services (GFCS) hosted by WMO (Hewitt *et al.*, 2012).

Supporting drought risk management in LAC

Gaps and needs identified

The occurrences of drought hazards have been increasing significantly in LAC, affecting over 60 million people over the last three decades (EM-DATA, 2015). In order to move away from reactive crisis management and work towards proactive climate risk management, preventing and mitigating the impacts of drought is a top priority of many governments in the LAC region (Wieriks & Vlaanderen, 2015). This requires insight into the causes and characteristics of drought events and their processes, as well as the identification of the vulnerability of livelihoods to drought hazards. UNESCO-IHP has engaged with the countries to identify the most urgent needs and existing gaps of drought risk management, identifying several countries requiring the development of national drought monitoring and early warning systems (MEWS).

A common problem in LAC when developing these MEWS, is the decentralized data collection, scattered over multiple agencies that are dependent on different ministries. This requires collaboration across ministries through a multi-sectorial approach, which often cannot be effectively implemented without direct support from high-level policymakers. Although most countries have foreseen the development of such monitoring systems in their legislation, it often remains underdeveloped and inappropriate for decision-making.

A second problem in LAC that was identified during the development of these national monitoring systems and that is connected to the first, is the range of different data formats that need to be handled to allow integration of different data sources. MEWS require combining data sources from national weather and hydrological services, agricultural extension services and public databases, as well as data streams from international partners providing remote sensing datasets or global/regional weather and climate model outputs. This requires technological solutions that allow the integration of multiple data sources with different temporal and spatial resolutions. An additional challenge is that they often have a complex data structure and data exchange formats.

To develop the MEWS for this region, an important requirement is to provide actionable information for decision-making in order to become an effective tool in drought management and to be useful for drought policy. Data and information available in those MEWS should fulfil four conditions (adapted from Baethgen (2010)) as follows:

a) Updated: the drought hazard needs to be monitored in (near) real time.
b) Relevant: conditions of drought need to be detectable at the local (municipality) level.
c) Understandable: the drought information needs to be communicated in a format which is in line with the decision-making context.
d) Reliable: data sources need to be available continuously and need to be validated by the data source provider.

As an intermediate strategy, regional MEWS have been under development to support national governments with drought information over the last decade, but this approach poses its own problems. In most cases, regional systems have a strong bias in data provision, focusing on a particular aspect of drought (meteorological, hydrological or agricultural), and they are often too coarse to be actionable at the local scale. Therefore, the development of national MEWS remains a key challenge to connect the information available in those MEWS with effective drought policies.

We also identified drought vulnerability assessment, the second pillar of drought management (Figure 1), as a need for the region as it provides an essential link between MEWS and drought management and policy. If well-developed, the vulnerability maps of a country serve two purposes. First, they provide the means to quantify the differences in vulnerability between local communities, allowing the identification of those communities that need to be prioritized for the implementation of mitigation actions when droughts occur. Second, they can also provide a metric to validate the impact of public policy to reduce vulnerability when maps are updated regularly. By comparing successive vulnerability maps, the vulnerability index of individual communities should progressively become lower in those localities where governmental policies have been implemented. Used in such a way, these vulnerability maps have the potential to become a strong instrument to steer and adjust public spending to effectively improve drought resilience of vulnerable communities. They are also helpful to inform society on the impact of national drought management policies and public drought mitigation programmes. Nevertheless, few countries have currently developed such vulnerability analyses at the national level. In Mexico, the National Drought Programme PRONACOSE (Federman *et al.*, 2014) was launched as one of the first integrated drought management plans at the national scale in the LAC region. This programme has advanced significantly to identify drought vulnerabilities in a standardized, comparable manner, incorporating drought policy explicitly in its framework. Other recent examples include Peru (MINAGRI, 2012), Northeastern Brazil (De Nys & Engle, 2015) and Ecuador (Ycaza *et al.*, 2012).

Drought management policies exist in most LAC countries, but are often loosely tied to objective indicators, making it difficult to generate a standardized response to reach the most vulnerable populations proactively and, as such, have no feedback mechanism to adjust the policies. As an example, the 2016 Peruvian Drought Management Plan (MINAGRI, 2016) that was developed as a response to the 2015–2016 El Niño event, indicates a set of actions to be taken proactively to mitigate the impact of drought, suggesting that the Standardized Precipitation Index (SPI) (McKee *et al.*, 1993) should be used to trigger those responses, without indicating what threshold should be used. Additionally, SPI only considers meteorological aspects of drought, making the indicator less useful to monitor potential impacts of agricultural and hydrological aspects of drought. A similar gap has been identified in the drought declarations of the Chilean Ministry of Agriculture, that uses a range of indicators to identify hotspots of drought at the communal level, but the final area that will be declared for receiving financial support is defined by local and regional policymakers. This has resulted in significant over-allocation of resources to regions that have only suffered mild impacts of (meteorological) droughts, urging a revision of the current declaration procedure based on objective indicators.

The above description indicates that different aspects of drought management have been developed in the different countries, but the challenge to integrate the different components, as defined in Figure 1, to work seamlessly in a national integrated drought management strategy is still needed.

In order to identify the needs and gaps in drought management and the means to address them, UNESCO-IHP and its partners organized the International Symposium on Drought Management Tools in 2014 in Santiago de Chile to identify advances of the different countries and to identify remaining gaps and needs. As an outcome of the meeting, the Santiago Declaration on Drought Management Tools was established. This defines a roadmap for action in the region, identifying a set of remaining gaps to be addressed:

(a) the lack of effective agricultural drought monitoring indicators in current drought monitoring tools;
(b) that drought-related information should be made available with sufficient detail, equally considering spatially and temporally well-distributed hydro-meteorological data and the local water demand/abstractions as an integral component of hydrological drought management;
(c) that in areas with steep topography and/or snow, groundwater and reservoir storage, these indicators need to be considered in hydrological drought assessment and management;
(d) a knowledge gap in currently available socio-economic indicators for drought monitoring, which are essential to establish a clear evaluation of long-term drought impacts and to allow assessing the cost of inaction;
(e) that drought impacts communities depending on their vulnerability, and that there is a need to identify the vulnerability of communities as an integral part of national socio-economic drought management;
(f) that there is a need to shift from reactive drought crisis management towards proactive drought risk management and to identify a central role of drought MEWS to trigger proactive responses;
(g) that regional and global drought monitoring and early warning tools are useful, but that there is a need for local calibration and validation of these products to support drought management at the watershed level.

Tools and solutions provided in LAC

Based upon the gaps identified in the previous section, the UNESCO-IHP strategy consisted in supporting the countries to reduce these gaps through its G-WADI and IDI activities focused on institutional capacity development and the development of adapted tools and solutions to provide drought information and to support IDRM. Notably, effective drought MEWS have been developed to increase drought management capacities at the regional, national and watershed level in LAC.

Development of regional drought tools for LAC. At the regional level, a drought monitoring and flood forecast system for LAC was developed by Princeton University with the support of UNESCO-IHP, with similar characteristics as the African version (Sheffield *et al.*, 2014). It provides information on current and expected drought (and flood) conditions. Based on micro-scale hydrological modelling (Liang *et al.* 1996), the system analyses available data to provide real-time assessment of the water cycle and drought and flood conditions, and puts this in the context of the long-term record dating back to 1950. The Flood and Drought Monitor can be accessed from http://stream.princeton.edu/. The portal holds information on precipitation, temperature, radiation and wind speed, drought indicators (SPI, soil moisture, NDVI, evapotranspiration) and flood indicators (surface runoff and streamflow). The information can be obtained either spatially or for point locations (Figure 2), for specific dates, months or annual timescales, and is compared with the normally expected conditions or percentiles. The Flood and Drought Monitor system also produces short and longer seasonal forecasts.

Complementary to the Drought Monitor, UNESCO collaborated with the countries in the LAC region to develop a regional drought atlas. The atlas identifies the drought frequency using a regional frequency analysis with L-moments (Hosking & Wallis, 2005), which is the probability of drought occurrence with a certain magnitude and duration at any given place in the area of interest. This methodology was first used in the US Drought Atlas (Willeke *et al.*, 1994), and further elaborated to construct the Chilean Drought Atlas as a pilot version (Nuñez *et al.*, 2011). This methodology was further expanded to cover the whole LAC region (Nuñez *et al.*, 2016). The Drought Atlas for LAC contributes to identify the livelihood vulnerability providing an effective tool to raise awareness on the exposure to drought. It also identifies the variability of rainfall deficits in countries of LAC and allows visualizing how climatic

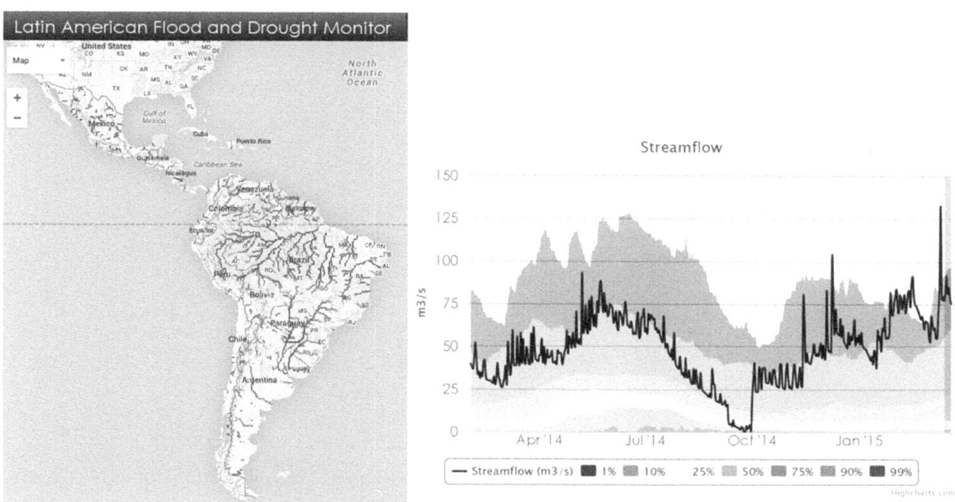

Fig. 2. The Latin American Flood and Drought Monitor system, showing the average daily streamflow and its corresponding historic values expressed as percentiles.

variability differs spatially within the countries, even at short distances. Specific deliverables focus on hazard maps for precipitation deficit (drought) and excess (floods). Additionally, the atlas can be used to support decision-making for drought risk management by answering the following questions: 'How rare is the current drought?' 'How large a drought should we plan for?' and 'How rare is the drought of record?' The atlas can be accessed at www.climatedatalibrary.cl/CAZALAC/maproom/.

Establishment of national drought observatories in pilot countries. While data and information are often available from national and international data sources on drought hazards, these are not always provided in a relevant, timely and actionable manner. With the development of national drought observatories, informed decision-making related to droughts, as well as other climatic hazards such as heat waves, frost and extreme rainfall events, is made possible. In this manner, vulnerability to droughts and other agroclimatic events can be reduced by providing relevant and rapid information to allow timely responses.

In close collaboration with the Chilean Ministry of Agriculture, the Food and Agriculture Organization (FAO) and the International Research Institute for Climate and Society (IRI), the Chilean Agroclimatic Observatory (www.climatedatalibrary.cl/UNEA/maproom/) was launched in June 2013. The observatory is a collection of maps and other figures that monitors present drought conditions. It provides near-future seasonal forecasts and allows putting the current droughts into a historical context (Figure 3). The system allows the creation of compound indices, taking into account a number of

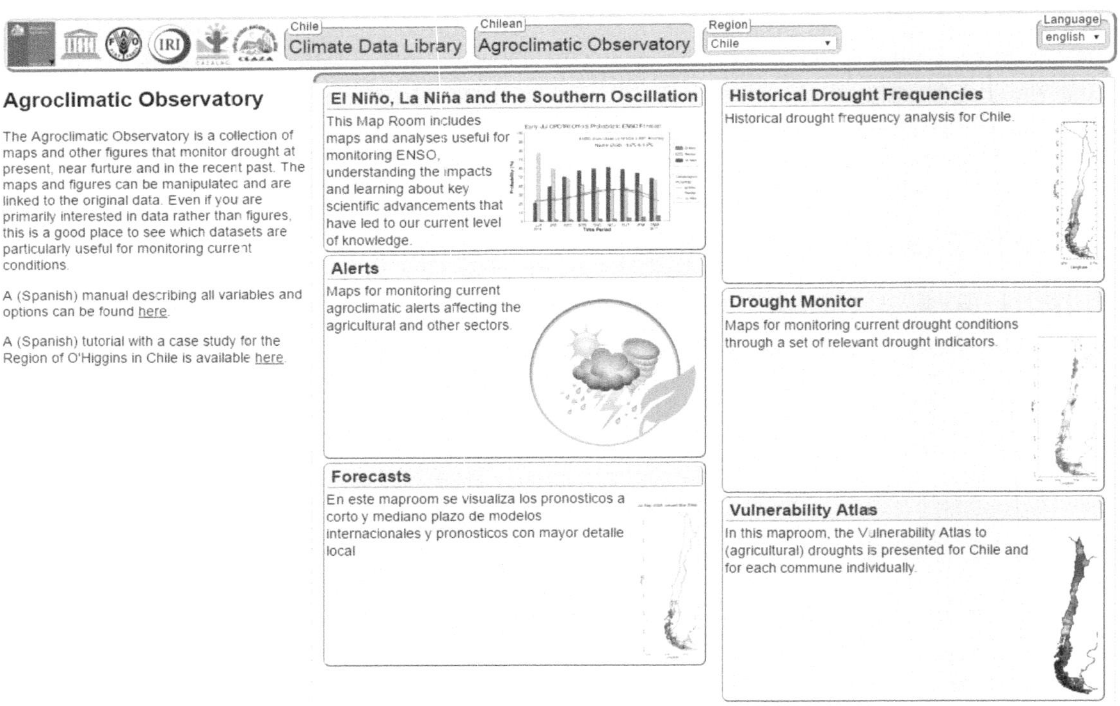

Fig. 3. Interface of the Chilean Agroclimatic Observatory, showing the different components: present drought conditions and alerts, near-future seasonal forecasts, historical drought analyses, ENSO observations and impact data, as well as drought vulnerability maps.

different drought indicators at the same time; information that helps to avoid false alarms for drought events arising as a result of external causes such as crop diseases, wildfires, etc.

The Chilean Agroclimatic Observatory builds upon the Climate Data Library (CDL), a tool that allows the collection of all raw databases of national and international institutions relevant to drought monitoring (Del Corral *et al.*, 2012). Data of numerous formats can be added, with which additional indicators can be calculated using advanced arithmetic or geo-statistical functions. In order to provide effective decision support tools, a user-friendly interface was built on top of the CDL, called the 'map-room', which holds relevant drought indices on meteorological, hydrological and agricultural drought, and combines information from national and international datasets.

A similar process was followed in Peru, where a consortium of partners developed the National Drought Observatory (ONS in Spanish) using the same CDL technology, accessible at http://ons.snirh.gob.pe/Peru/maproom/. The ONS has been fully integrated in the National Water Resources System (SNIRH), becoming a cornerstone of its drought management strategy. The system collects daily levels of rivers, reservoirs, streamflow, hydropower production, precipitation, temperature, vegetation conditions, as well as information on past droughts (the Peruvian Drought Atlas) and near-future seasonal forecasts. It also has a particular focus on El Niño Southern Oscillation (ENSO), due to the high correlation between droughts in the country and the ENSO phenomenon.

Remote sensing data are a key component of these observatories in both Chile and Peru. To provide tailored satellite precipitation estimates for these pilot countries, UNESCO-IHP collaborated with its partners to develop calibration methodologies for the G-WADI GeoServer developed at CHRS, University of California. Using a quantile mapping and Gaussian weighting interpolation approach, the biases could effectively be removed, providing a near real-time daily precipitation estimate for Chile coherent with observed values from more than 400 rain gauges (Yang *et al.*, 2016).

A more recent component of UNESCO-IHP efforts focused on the vulnerability and drought impact assessment (Figure 1), supporting the development of an (agricultural) drought vulnerability atlas for Chile. The focus was put on the vulnerability of rural communities using indicators related to agricultural production and rural poverty. Sixteen indicators were used to calculate the overall agricultural drought vulnerability map (Figure 4), grouped into three categories (Sensitivity, Exposure and Adaptation Capacity) and covering socio-economic, biophysical and institutional aspects. The individual indicators, as well as the final vulnerability maps are all accessible through the Chilean Agroclimatic Observatory.

In order to provide an objective and actionable drought index, information from meteorological, hydrological and agricultural drought indicators were integrated into a combined drought index (CDI) for Chile (Figure 5), which was based upon the CDI developed for the European Drought Observatory (Sepulcre-Canto *et al.*, 2012). For the Chilean CDI, the underlying indicators and their thresholds were calibrated using drought impact information obtained over the 2008–2015 period and validated with claims of crop losses from agricultural insurance policies over the same period (Agroseguros, 2016). As such, the resulting CDI provides three standardized drought alert phases that are defined by their drought intensities (Figure 4). By standardizing these alert phases for the whole country, the implementation of a more objective drought management response is made possible.

Impact on drought policy in pilot countries of LAC

The different activities implemented along the three pillars of IDRM (Figure 1) have impacted drought policy in LAC along different pathways. The regional monitor has been transferred and

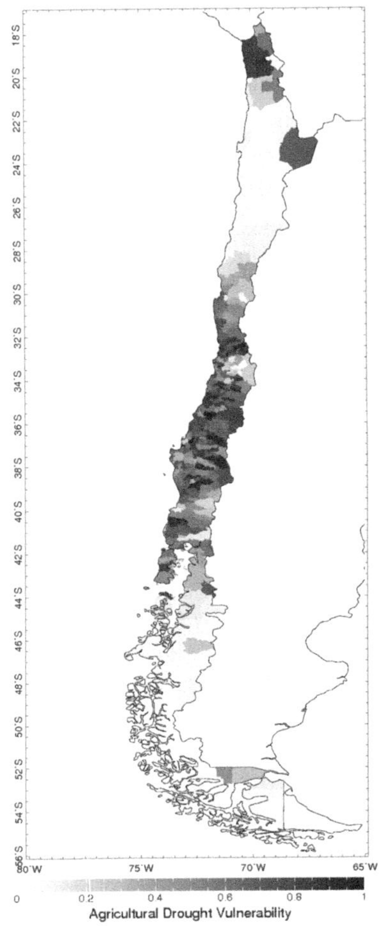

Fig. 4. The Chilean Agricultural Drought Vulnerability map.

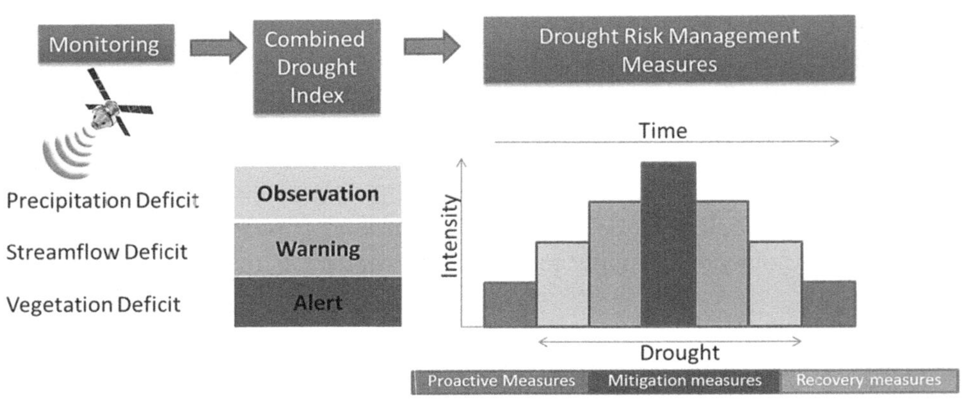

Fig. 5. The CDI for decision-making.

tested by 22 hydrometeorological agencies from countries of LAC for operational use. Currently, the flood and drought monitor system is being integrated in the national meteorological and water resources agencies in Bolivia, Chile, Ecuador, Peru and Uruguay. In addition, both in Chile and Peru, the national observatories have been adopted in the most recent national drought policy plans (e.g., MINAGRI, 2016) as tools to steer agricultural drought mitigation protocols. In the case of Peru, this has led in 2016 to the integration of the drought response measures into the ONS, showing a feedback mechanism of how the third pillar on drought measures and policy can be coupled to drought MEWS.

The Drought Atlas is also incorporated in the monthly 'Agroclimatic Outlook' provided by the Chilean Ministry of Agriculture. This allows the current droughts to be put into context by expressing the rainfall deficit in terms of return periods (for example, 'the reduced rainfall amount this last month occurs every 5 years on average'). This way, drought indicators become more actionable, as the data are translated into a product which is easily understandable by policy and decision-makers. Drought indicators also highlight the events which are unusual or extreme, raising awareness on natural variability by correcting the false perception that droughts are more recurrent and more severe than normally expected.

The more recent development of vulnerability maps for agricultural drought in Chile is providing a more socio-economic approach to drought management, and a very promising link between the biophysical work on objective drought indicators and the policy aspects of drought management. This allows identification of the areas in which to prioritize proactive drought measures, as the communities that are hardest hit by drought are identified. In this respect, it is expected that the Chilean Vulnerability Atlas will eventually become a tool to validate the impact of drought policies aiming at increasing the resilience of vulnerable communities and to tackle intermediate aspects that are relevant in a drought context (poverty, education, access to water, health coverage, investments in R&D, etc.). In this way, the atlas also contributes to the overall 2030 agenda aiming to reduce poverty.

Finally, the CDI was developed in 2016 in collaboration with the Chilean Ministry of Agriculture as a key component to streamline the (agricultural) drought declarations and to provide a baseline to identify a minimum number of affected communities, based on objective drought indicators. This shows a direct pathway of how MEWS can become an integral part of drought policy and management.

Supporting drought risk management in Africa

Gaps and needs identified

Drought is one of the leading impediments to development in Africa (African Risk Capacity, 2015). Much of the continent is dependent on rain-fed agriculture, which makes it particularly susceptible to climate variability (Cooper *et al.*, 2008). Prolonged and frequent occurrences of droughts present significant challenges to agriculture, forestry, water resources management, urban planning and food security. Southern Africa has been experiencing since 2015 a drought caused by the El Niño phenomenon, and which led to the declaration of a state of emergency by many countries in the sub-region (FEWS-NET, 2016). In 2010–2011, a drought occurred in Eastern Africa, considered to be the worst drought in 60 years, affecting more than 13 million persons within the sub-region, and causing the loss of lives and livelihoods (IGAD, 2013).

Monitoring drought and providing near real-time and seasonal forecasts are essential for an IDRM and risk reduction in Africa (Tadesse *et al.*, 2008). Drought monitoring over Africa is challenging because of the sparseness of observational data, either historically or in real time. Currently available approaches in Africa are very limited, in part because of unreliable monitoring networks. Operational seasonal climate forecasts are often reliant on statistical regressions, which are unable to provide detailed information relevant for drought assessment (Patt *et al.*, 2007; Sheffield *et al.*, 2014). However, the wealth of data from satellites and recent advancements in large scale hydrological modelling and seasonal climate model predictions have enabled the development of state-of-the-art monitoring and prediction systems that can help address many of the problems inherent to drought-prone regions in Africa. Satellite remote sensing combined with ground truth data and modelling is capable of overcoming differences in data gaps across political boundaries that have historically hindered drought monitoring in Africa (Grimes & Diop, 2003; Grimes *et al.*, 2003).

Mitigating the impacts of droughts and famine is one of the primary aims of many governments and humanitarian organizations in the African region, which requires considering the resilience of communities. However, it cannot be effective if the causes and characteristics of drought events and processes are not well understood. To this end, a drought monitoring system is essential to assist water users in making climate-informed decisions (Diouf *et al.*, 2002; Hellmuth *et al.*, 2007; Cooper *et al.*, 2008).

Many areas in Western, Eastern and Southern Africa, particularly the arid and semi-arid areas are drought-prone areas. Regional institutions have been put in place in Western, Eastern and Southern Africa to address the challenges caused by recurrent droughts. In West Africa, the Inter-State Permanent Committee to Combat Drought in the Sahel was established in 1974, following the devastating droughts of 1973–1974, with its two technical specialized agencies, namely, AGRHYMET for agro-hydro-meteorological monitoring and capacity building and the Institute of Sahel (INSAH) for socio-economic related dimension. Member Countries of the Permanent Interstates Committee for Drought Control in the Sahel (CILSS) include Benin, Burkina Faso, Cape Verde, Chad, Cote d'Ivoire, Gambia, Guinea, Guinea Bissau, Mali, Mauritania, Niger, Senegal and Togo. In response to the devastating weather-related disasters, countries in Eastern and Southern Africa established in 1989 a Drought Monitoring Centre with its headquarters in Nairobi (DMCN) and a sub-centre in Harare (Drought Monitoring Centre Harare, DMCH). The DMCN was adopted later as a specialized institution of Intergovernmental Authority on Development (IGAD) and became the IGAD Climate Prediction and Applications Centre (ICPAC). DMCH was replaced by the Southern Africa Development Community (SADC) climate service centre covering 15 countries (Angola, Botswana, Democratic Republic of Congo, Lesotho, Madagascar, Malawi, Mauritius, Mozambique, Namibia, Seychelles, South Africa, Swaziland, United Republic of Tanzania, Zambia and Zimbabwe). The IGAD region comprises eight countries within the Greater Horn of Africa (GHA) including Djibouti, Eritrea, Ethiopia, Kenya, Somalia, Sudan, South Sudan and Uganda. The region has about 60% of the land classified as arid. Following the 2010–2011 droughts in Eastern Africa, the countries of the sub-region adopted the IGAD Drought Disaster Resilience and Sustainability Initiative (IDRISSI) to end drought emergencies within the sub-region, with emphasis on building resilience of communities and promoting equitable and significant disaster risk management, preparedness and effective response in the IGAD region.

One of the objectives of these regional climate-related institutions is to provide appropriate and timely climate early warning information and support specific sector applications for the mitigation of the impacts of climate variability and change for poverty alleviation, management of environment and sustainable development. These centres are producing and disseminating climate products to various users, including decadal, monthly climate bulletins and already for more than a decade the climate outlook, which includes climate seasonal forecasts with a focus on the expected precipitation for the rainy seasons (Patt *et al.*, 2007). Despite the data and information platforms put in place by these regional centres, there is still a need for more data and analysis tools at the regional level to complement existing systems. Particularly, near real-time and seasonal forecasts of climate extremes, including droughts and floods, are still lacking. The main gaps facing African countries are related to data and human capacity. Hydrometeorological networks have been deteriorating over the last decades, impacting the quality of the climate information for decision-making. Also, the different regional centres and national meteorological and hydrological services in many African countries are generally lacking skilled and sufficient human resources for addressing drought-related challenges.

Tools and solutions provided in Africa

The African Flood and Drought Monitor, developed by Princeton University with the support of UNESCO-IHP, was deployed in 2010 in combination with extensive training in Western Africa in 2011 and 2013 and in Eastern Africa in 2012 (Sheffield *et al.*, 2014). An extension of this work to include the Southern Africa sub-region is planned for the end of 2016. In West and Eastern Africa, the deployed drought monitor has been providing near real-time information on drought conditions (soil moisture indexes and discharges) and is used as complementary information to the existing monitoring systems at AGRHYMET (Diouf *et al.*, 2002) and ICPAC, thus improving their monitoring capabilities.

In addition to the drought monitor system, the G-WADI GeoServer near real-time rainfall estimation developed by the CHRS of the University of California, Irvine has been used by AGRHYMET in West Africa and Namibia as complementary information for drought and flood monitoring. UNESCO-IHP has also deployed in Eastern Africa the hydrological seasonal forecast approach already used in West Africa which is now being considered within the climate outlook forums in the region.

Impact on drought policy in pilot countries of Africa

Most of the important perennial rivers and streams of Namibia have their origin outside of the country: the Kavango, Kwando and Zambezi rivers in the northeast originate in Angola and Zambia, while the Orange River in the south originates in Lesotho and South Africa. Water resources planning and emergency preparedness for floods and droughts requires information on upstream conditions of these rivers. As one important tool in such information gathering, the Namibia Hydrological Services (NHS) of the Ministry of Agriculture, Water and Forestry incorporates precipitation estimates from the G-WADI GeoServer. In conjunction with limited stream gauging data, GeoServer information assists the NHS to make more accurate judgements as to future water supply for multiple uses, and provides early warning for floods and droughts in areas located in these and other national and transboundary river basins. The flood bulletins and other flood information are made available on the Namibia Flood Dashboard (http://matsu.opencloudconsortium.org/namibiaflood) and disaster-related information is made available by the Namibian Directorate of Disaster Risk Management (http://www.ddrm.gov.na/).

The current African drought monitoring system has been installed and used at regional level by the different regional centres to complement their existing monitoring systems. An essential component to ensure actual uptake of this climate information for decision-making is capacity building and outreach to decision-makers and farming communities (Patt *et al.*, 2007; Cooper *et al.*, 2008). Therefore, UNESCO-IHP has focused on training of the national hydrological and meteorological services, as well as the regional centres, to empower them with the tools to further develop and adapt the drought monitor by incorporating national datasets available within the countries and the centres. This, in turn, should contribute to improve the quality of the information provided at national and local levels and strengthen the human capacities in those countries directly to disseminate the improved information on climate-related disaster risks to the regional, national and local levels.

Conclusions

The severe impact of droughts on vulnerable communities in the Latin American, Caribbean and African regions urge the adoption of an integrated climate risk management approach (Figure 1) that focuses on proactive drought management and policy. This approach consists in the development of monitoring and early warning tools, combined with the assessment of the vulnerabilities of local communities to droughts, and by integrating proactive planning and mitigation measures into national drought policies.

UNESCO-IHP, through its G-WADI and IDI activities, supports the countries in these regions by identifying the needs and gaps for an integrated drought management and aims in filling those gaps by providing the countries with tools, data and methodologies and by providing platforms for international scientific and political cooperation.

A set of regional tools was developed for this purpose, including the high spatial and temporal resolution global precipitation estimates from satellite data (the G-WADI GeoServer), a regional drought and flood monitoring monitor for both the African and LAC region, and a regional drought atlas for LAC. At the national level, examples of national drought observatories were developed for pilot countries (Chile and Peru) as case studies covering all three components of IDRM. In these platforms, both globally available and local datasets can be combined and integrated providing a more tailored environment focused on the needs of local decision-makers. A pilot vulnerability assessment was also developed for one of the countries, providing relevant information regarding actual drought policy and mitigation instruments, as well as allowing identification of those communities that still are vulnerable to current and future climatic variability and change. Finally, drought policies are under development in these two countries that integrate the objective drought indicators provided through monitoring and early warning, in order to streamline the national response to drought hazards and to allow proactive drought measures to be implemented.

Nevertheless, significant challenges still remain in the Latin American, Caribbean and African regions that need to be addressed, many of which were summarized as part of the Santiago Declaration. Technology transfer and capacity building of local human resources in those regions remain essential to empower further development and uptake of the integrated drought management approach, and are part of the mandate of UNESCO-IHP. An additional focus area includes local calibration and validation of currently available (regional) drought information, to ensure reliable early warning and forecast at the local level, which is currently under development with support from the countries from these regions.

Acknowledgements

The authors wish to recognize Princeton University, the Centre for Hydrometeorology and Remote Sensing (CHRS), University of California at Irvine and the International Research Institute for Climate and Society (IRI) at Columbia University for their collaboration with UNESCO-IHP. The authors wish also to thank their colleague Yo Nishimura for his revision and valuable suggestions in reviewing and formatting this paper.

References

African Risk Capacity (2015) The cost of drought in Africa, 4 pp. Available at: http://www.africanriskcapacity.org/documents/350251/371107/ARC_Cost_of_Drought_EN.pdf (accessed 20 June 2016).

Agroseguros (2016). National AgroInsurance Information System (SISA). Available at: http://sisa.agroseguros.gob.cl/ (accessed 15 May 2016).

Amissah-Arthur, A. (2003). Targeting climate forecasts for agricultural applications in sub-Saharan Africa: situating farmers in user-space. *Climatic Change 58*, 73–92.

Baethgen, W. E. (2010). Climate risk management for adaptation to climate variability and change. *Crop Science 50*(2), 70–76.

Cooper, P. J. M., Dimes, J., Rao, K. P. C., Shapiro, B., Shiferaw, B. & Twomlow, S. (2008). Coping better with current climatic variability in the rain-fed farming systems of sub-Saharan Africa: an essential first step in adapting to future climate change? *Agriculture, Ecosystems & Environment 126*, 24–35.

De Nys, E. & Engle, N. L. (2015). *Living with the Semi-Arid and Proactive Drought Management in Northeast Brazil: A New Perspective.* World Bank, Washington, DC, USA.

Del Corral, J., Blumenthal, M., Mantilla, G., Ceccato, P., Connor, S. & Thomson, M. (2012). Climate information for public health: the role of the IRI climate data library in an integrated knowledge system. *Geospatial Health 6*(3), S15–S24.

Diouf, M., Nonguierma, A. & Amani, A. (2002). Experience of AGRHYMET Regional Centre for the monitoring of environment in the Sahel region. In: *Gestion intégrée des ressources naturelles en zones inondables tropicales (Integrated Management of Natural Resources in Flood-Prone Tropical Areas).* Orange, D., Arfi, R., Kuper, M., Morand, P. & Poncet, Y. (eds). Actes du Colloque Girnzit, Bamako, 20–23 June 2000, pp. 917–928. IRD, collection colloques et séminaires, Paris.

EM-DATA (2015). Guha-Sapir, D., Below, R. & Hoyois, Ph. EM-DAT, the CRED/OFDA International Disaster Database – www.emdat.be. Université Catholique de Louvain, Brussels, Belgium, accessed December 2015.

European Commission (2015). *Science for Environment Policy Migration in Response to Environmental Change.* Thematic Issue 51. Issue produced for the European Commission DG Environment by the Science Communication Unit, UWE, Bristol, UK. Available at: http://ec.europa.eu/science-environment-policy (accessed 12 June 2016).

Federman, D. K., Cortés, F. & Pérez, M. (2014). Constructing a framework for national drought policy: the way forward in Mexico. *Weather and Climate Extremes 3*, 90–94.

FEWS-NET (2016). Illustrating the extend and severity of 2015-2016 drought in Southern Africa. Southern Africa Special Report. 8 pp. Available at: http://www.fews.net/sites/default/files/documents/reports/FEWS%20NET_Southern%20Africa%202015_16%20Drought%20Map%20Book_20160318_0.pdf (accessed 15 June 2016).

Grimes, M. & Diop, M. (2003). Satellite-based rainfall estimation for river flow forecasting in Africa. I: rainfall estimates and hydrological forecasts. *Hydrological Sciences Journal 48*(4), 567–584.

Grimes, D. I. F., Coppola, E., Verdecchia, M. & Visconti, G. (2003). A neural network approach to real-time rainfall estimation for Africa using satellite data. *American Meteorological Society*, 1119–1133.

Grobicki, A., MacLeod, F. & Pischke, F. (2015). Integrated policies and practices for flood and drought risk management. *Water Policy 17*(S1), 180–194.

Guha-Sapir, D., Hargitt, D. & Hoyois, P. (2004). *Thirty Years of Natural Disasters 1974–2003: The Numbers.* Presses Universitaires de Louvain, Louvain-la-Neuve, Belgium.

Gutiérrez, A. P. A., Engle, N. L., De Nys, E., Molejón, C. & Martins, E. S. (2014). Drought preparedness in Brazil. *Weather and Climate Extremes 3*, 95–106.

Hansen, J. W., Mason, S. J., Sun, L. & Tall, A. (2011). Review of seasonal climate forecasting for agriculture in sub-Saharan Africa. *Experimental Agriculture 47*, 205–240.

Hayes, M. J., Wilhelmi, O. V. & Knutson, C. L. (2004). Reducing drought risk: bridging theory and practice. *Natural Hazards Review 5*, 106–113.

Hellmuth, M. E., Moorhead, A., Thomson, M. C. & Williams, J. (2007). *Climate Risk Management in Africa: Learning From Practice.* International Research Institute for Climate and Society (IRI), Columbia University, New York, USA.

Hewitt, C., Mason, S. & Walland, D. (2012). The global framework for climate services. *Nature Climate Change 2*, 831–832.

Hosking, J. R. M. & Wallis, J. R. (2005). *Regional Frequency Analysis: An Approach Based on L-Moments.* Cambridge University Press, Cambridge, UK, 244 pp.

Hsu, K., Behrangi, A., Iman, B. & Sorooshian, S. (2010). Extreme precipitation estimation using satellite-based PERSIANN-CCS algorithm. In: *Satellite Rainfall Applications for Surface Hydrology.* Gebremichael, M. & Hossain, F. (eds). Springer, New York, USA, pp. 49–67.

IGAD (2013). IGAD Drought Disaster Resilience and Sustainability Initiative (IDRISI), the IDDRISI Strategy, 40 pp.

IPCC (2014). Climate change 2014: synthesis report. In: *Contribution of Working Groups I, II and III to the Fifth Assessment Report of the Intergovernmental Panel on Climate Change.* Core Writing Team, Pachauri, R. K., Meyer, L. A. (eds). IPCC, Geneva, Switzerland, 151 pp.

IWMI (2015). Development of South Asia Drought Monitoring System Colombo, Sri Lanka, 22 pp. Available at: http://www.droughtmanagement.info/wp-content/uploads/2015/07/SADMS-TechnicalDevelopment-Phase1.pdf (accessed 12 June 2016).

Liang, X., Lettenmaier, D. P., Wood, E. F. & Burges, S. J. (1994). A simple hydrologically based model of land surface water and energy fluxes for general circulation models. *Journal of Geophysical Research 99*(D17), 14415–14428.

Liang, X., Wood, E. F. & Lettenmaier, D. P. (1996). Surface soil moisture parameterization of the VIC-2L evaluation and modification. *Global Planet. Change 13*, 195–206.

Lyon, B. & DeWitt, D. G. (2012). A recent and abrupt decline in the East African long rains. *Geophysical Resolution Letters 39*, L02702.

Masih, I., Maskey, S., Mussá, F. & Trambauer, P. (2014). A review of droughts on the African continent: a geospatial and long-term perspective. *Hydrology and Earth System Sciences 18*(9), 3635–3649.

McKee, T. B., Doesken, N. J. & Kliest, J. (1993). The relationship of drought frequency and duration to time scales. In: *Proceedings of the 8th Conference on Applied Climatology, Anaheim, CA, USA*, 17–22 January, Boston, MA, USA, pp. 179–184.

MINAGRI (2012). Plan de Gestión de Riesgos y Adaptación al Cambio Climático en el Sector Agrario Periodo 2012-2021 - PLANGRACC-A (Risk management plan and climate change adaptation plan for the agricultural sector, Period 2012-2021 - PLANGRACC-A). Ministry of Agriculture, Peru, 98 pp. Available at: http://minagri.gob.pe/portal/plangracc (accessed 16 June 2016).

MINAGRI (2016). *Plan MultiSectorial de Prevención y Reducción de Riesgos ante Sequías 2016 (MultiSectorial Drought Risk Prevention and Reduction Plan 2016).* Gobierno de Perú, Lima, 63 pp. Available at: http://www.redhum.org/documento_detail/17685 (accessed 22 June 2016).

Nuñez, J. H., Verbist, K., Wallis, J., Schaeffer, M., Morales, L. & Cornelis, W. M. (2011). Regional frequency analysis for mapping drought events in north-central Chile. *Journal of Hydrology 405*, 352–366.

Nuñez, J. H., Hallack-Alegría, M. & Cadena, M. (2016). Resolving regional frequency analysis of precipitation at large and complex scales using a bottom-up approach: the Latin America and the Caribbean drought Atlas. *Journal of Hydrology 538*, 515–538.

Patt, A. G., Ogallo, L. & Hellmuth, M. (2007). SUSTAINABILITY: Learning from 10 years of climate outlook forums in Africa. *Science 318*, 49–50.

Sepulcre-Canto, G., Horion, S., Singleton, A., Carrao, H. & Vogt, J. (2012). Development of a combined drought indicator to detect agricultural drought in Europe. *Natural Hazards and Earth System Sciences 12*, 3519–3531.

Sheffield, J., Wood, E., Chaney, N., Guan, K., Sadri, S., Yuan, X., Olang, L., Amani, A., Ali, A., Demuth, S. & Ogallo, L. (2014). A drought monitoring and forecasting system for Sub-Sahara African water resources and food security. *Bulletin of the American Meteorological Society 95*(6), 861–882.

Sivakumar, M. V. K., Stefanski, R., Bazza, M., Zelaya, S., Wilhite, D. & Magalhaes, A. R. (2014). High level meeting on national drought policy: summary and major outcomes. *Weather and Climate Extremes 3*, 126–132.

Svoboda, M., LeComte, D., Hayes, M., Heim, R., Gleason, K., Angel, J., Rippey, B., Tinker, R., Palecki, M., Stooksbury, D., Miskus, D. & Stephens, S. (2002). The drought monitor. *Bulletin of the American Meteorological Society 83*(8), 1181–1190.

Tadesse, T., Haile, M., Senay, G., Wardlow, B. & Knutson, C. (2008). The need for integration of drought monitoring tools for proactive food security management in sub-Saharan Africa. *Natural Resources Forum 32*(4), 265–279.

Tarhule, A. & Lamb, P. (2003). Climate research and seasonal forecasting for West Africans: perceptions, dissemination, and use? *Bulletin of the American Meteorological Society 84*(12), 1741–1759.

UN Department of Economic and Social Affairs (2008). *Trends in Sustainable Development 2008–2009*. United Nations, New York, USA. Available at: http://www.un.org/esa/sustdev/publications/trends2008/drought.pdf (accessed 15 June 2016).

UNISDR (2007). *Drought Risk Reduction Framework and Practices: Contributing to the Implementation of the Hyogo Framework for Action*. United Nations Secretariat of the International Strategy for Disaster Reduction (UNISDR), Geneva, Switzerland, 98 + vi pp.

UNISDR & CRED (2015). The human cost of weather related disasters 1995-2015. Available at: http://www.unisdr.org/2015/docs/climatechange/COP21_WeatherDisastersReport_2015_FINAL.pdf (accessed 16 June 2016).

Vicente-Serrano, S. M., Beguería, S., López-Moreno, J. I., Angulo, M. & El Kenawy, A. (2010). A new global 0.5° gridded dataset (1901–2006) of a multiscalar drought index: comparison with current drought index datasets based on the Palmer Drought Severity Index. *Journal of Hydrometeorology 11*, 1033–1043.

Vogt, J., Barbosa, P., Hofer, B. & Singleton, A. (2011). Developing a European drought Observatory for Monitoring, Assessing and Forecasting Droughts across the European Continent. *Geophysical Research Abstracts 13*, 6350.

Vogt, J., de Jager, A., Carrão, H., Magni, D., Mazzeschi, M. & Barbosa, P. (2016). A Global Drought Observatory for Emergency Response. *EGU General Assembly Conference Abstracts* Vol. *18*, p. 2582.

Wieriks, K. & Vlaanderen, N. (2015). Water-related disaster risk reduction: time for preventive action! Position paper of the High Level Experts and Leaders Panel (HELP) on water and disasters. *Water Policy 17*(S1), 212–219.

Wilhite, D. A., Hayes, M. J., Knutson, C. & Smith, K. H. (2000). Planning for drought: moving from crisis to risk management. *Journal of the American Water Resources Association 36*, 697–710.

Wilhite, D. A., Sivakumar, M. V. K. & Pulwarty, R. (2014). Managing drought risk in a changing climate: the role of national drought policy. *Weather and Climate Extremes 3*, 4–13.

Willeke, G., Hosking, J. R. M., Wallis, J. R. & Guttman, N. B. (1994). *The National Drought Atlas. Institute for Water Resources Report* 94-NDS-4, U.S. Army Corps of Engineers.

Yang, Z., Hsu, K., Sorooshian, S., Xu, X., Braithwaite, D. & Verbist, K. M. J. (2016). Bias adjustment of satellite-based precipitation estimation using gauge observations: a case study in Chile. *Journal of Geophysical Research: Atmospheres 121*, 3790–3806.

Ycaza, P., Friend, F. & Zambrano, C. (2012). *Implementation of A Sectorial Vulnerability Information System for the Guayas Province to Address Climate Change and Climate Variability*. CIIFEN, Guayaquil, Ecuador, 20 pp.

IWA Publishing's authorised EU representative for General Product Safety Regulations is Diane D'Arras, 15 rue Duret 75116 Paris, France, e-mail: safety@iwap.co.uk.

Printed and bound by CPI Group (UK) Ltd, Croydon, CR0 4YY
12/05/2026
02108866-0020